COMPREHENSIVE ANALYTICAL CHEMISTRY

ELSEVIER SCIENCE B.V.
Sara Burgerhartstraat 25
P.O. Box 211, 1000 AE Amsterdam, The Netherlands

First edition 2002

Library of Congress Cataloging in Publication Data
A catalog record from the Library of Congress has been applied for.

British Library Cataloguing in Publication Data
A catalogue record from the British Library has been applied for.

ISBN: 0-444-50737-X
ISSN: 0166-526X

∞ The paper used in this publication meets the requirements of ANSI/NISO Z39.48-1992 (Permanence of Paper).
Printed in The Netherlands.

Wilson & Wilson's

COMPREHENSIVE ANALYTICAL CHEMISTRY

Edited by

D. BARCELÓ

Research Professor
Department of Environmental Chemistry
IIUQAB-CSIC
Jordi Girona 18-26
08034 Barcelona
Spain

Wilson & Wilson's

COMPREHENSIVE ANALYTICAL CHEMISTRY

VOLUME XXXVIII

COUNTERCURRENT CHROMATOGRAPHY
The Support-Free Liquid Stationary Phase

Edited by

ALAIN BERTHOD

2002
ELSEVIER
Amsterdam – Boston – London – New York – Oxford – Paris
San Diego – San Francisco – Singapore – Sydney – Tokyo

CONTRIBUTORS TO VOLUME XXXVIII

Alain Berthod
Laboratoire des Sciences Analytiques, CNRS, Université Claude Bernard-Lyon 1, Bat. CPE 308, 69622 Villeurbanne, France
Béatrice Billardello
Laboratoire des Sciences Analytiques, CNRS, Université Claude Bernard-Lyon 1, Bat. CPE 308, 69622 Villeurbanne, France
Les Brown
MicroSolv Technology Corporation, 101 Brighton Avenue, Long Branch, NJ 07740, USA
Yu Chuan Chen
Department of Applied Chemistry, National Chiao Tung University, Hsinchu 30050, Taiwan, ROC
Alain P. Foucault
GEPEA, UMR CNRS 6144, Centre de Recherche et de Transfert de Technologies, BP 406, 44602 Saint Nazaire Cedex, France
Pilar Franco
Laboratorio de Química Farmaceútica, Facultad de Farmacía, Universidad de Barcelona, Avenida Juan XXIII, 08028 Barcelona, Spain
Shi-Jun Han
NorthWest Institute of Nuclear Technology, P.O. Box 69-14, 710 024 Xi'an, China
Guy H. Harris
MERCK Research Laboratories, Mail Stop R80Y-360, P.O. Box 2000, Rahway, NJ 07065, USA
Olivier Intes
Laboratoire de Pharmacognosie, Université de Reims, Moulin de la Housse, 51097 Reims Cedex, France
Yoishiro Ito
Laboratory of Biophysical Chemistry, National Institutes of Health, Bldg. 50, Room 3334, Bethesda, MD 20892, USA
Yu-Ren Jin
NorthWest Institute of Nuclear Technology, P.O. Box 69-14, 710 024 Xi'an, China
Jack Legrand
GEPEA, UMR CNRS 6144, Centre de Recherche et de Transfert de Technologies, BP 406, 44602 Saint Nazaire Cedex, France

Gilda Guimarães Leitão
Núcleo de Pesquisas de Produtos Naturais, Universidade Federal do Rio de Janeiro, CCS, Bloco H, Ilha do Fundão, 21.941-590, Rio de Janeiro, RJ, Brasil
Suzana Guimarães Leitão
Departamento de Produtos Naturais e Alimentos, Faculdade de Farmácia, Universidade Federal do Rio de Janeiro, CCS, Bloco A, Ilha do Fundão, 21.941-590, Rio de Janeiro, RJ, Brasil
Alexandre Maciuk
Laboratoire de Pharmacognosie, Université de Reims, Moulin de la Housse, 51097 Reims Cedex, France
Luc Marchal
GEPEA, UMR CNRS, Centre de Recherche et de Transfert de Technologies, BP 406, 44602 Saint Nazaire Cedex, France
Cristina Minguillón
Laboratorio de Química Farmaceútica, Facultad de Farmacía, Universidad de Barcelona, Avenida Juan XXIII, 08028 Barcelona, Spain
Jean-Marc Nuzillard
Laboratoire de Pharmacognosie, Université de Reims, Moulin de la Housse, 51097 Reims Cedex, France
Laureano Oliveiros
C.N.A.M., 292 rue Saint Martin, 75141 Paris Cedex 03, France
Gerard Patissier
Laboratoire de Mécanique, IUT de Nantes, 44000 Nantes, France
Jean-Hugues Renault
Laboratoire de Pharmacognosie, Université de Reims, Moulin de la Housse, 51097 Reims Cedex, France
Jean-Michel Rosant
Laboratoire de Hydrodynamique, UMR CNRS 6598, Ecole Centrale, 44000 Nantes, France
Ian A. Sutherland
Brunel Institute of Bioengineering, Brunel University, Uxbridge, Middlesex UB8 3PH, UK
Tiing Yu
Department of Applied Chemistry, National Chiao Tung University, Hsinchu 30050, Taiwan, ROC
Li-Xing Zhang
NorthWest Institute of Nuclear Technology, P.O. Box 69-17, 710 024 Xi'an, China

Lian-Zhong Zhang
NorthWest Institute of Nuclear Technology, P.O. Box 69-17, 710 024 Xi'an, China
Tianyou Zhang
Beijing Institute of New Technologies and Applications, Xizhimen South Street No 16, 100 035 Beijing, China

WILSON AND WILSON'S

COMPREHENSIVE ANALYTICAL CHEMISTRY

VOLUMES IN THE SERIES

Vol. IA	Analytical Processes
	Gas Analysis
	Inorganic Qualitative Analysis
	Organic Qualitative Analysis
	Inorganic Gravimetric Analysis
Vol. IB	Inorganic Titrimetric Analysis
	Organic Quantitative Analysis
Vol. IC	Analytical Chemistry of the Elements
Vol. IIA	Electrochemical Analysis
	Electrodeposition
	Potentiometric Titrations
	Conductometric Titrations
	High-Frequency Titrations
Vol. IIB	Liquid Chromatography in Columns
	Gas Chromatography
	Ion Exchangers
	Distillation
Vol. IIC	Paper and Thin Layer Chromatography
	Radiochemical Methods
	Nuclear Magnetic Resonance and Electron Spin Resonance Methods
	X-Ray Spectrometry
Vol. IID	Coulometric Analysis
Vol. III	Elemental Analysis with Minute Sample
	Standards and Standardization
	Separation by Liquid Amalgams
	Vacuum Fusion Analysis of Gases in Metals
	Electroanalysis in Molten Salts
Vol. IV	Instrumentation for Spectroscopy
	Atomic Absorption and Fluorescence Spectroscopy
	Diffuse Reflectance Spectroscopy
Vol. V	Emission Spectroscopy
	Analytical Microwave Spectroscopy
	Analytical Applications of Electron Microscopy
Vol. VI	Analytical Infrared Spectroscopy
Vol. VII	Thermal Methods in Analytical Chemistry
	Substoichiometric Analytical Methods
Vol. VIII	Enzyme Electrodes in Analytical Chemistry
	Molecular Fluorescence Spectroscopy
	Photometric Titrations
	Analytical Applications of Interferometry
Vol. IX	Ultraviolet Photoelectron and Photoion Spectroscopy
	Auger Electron Spectroscopy

	Plasma Excitation in Spectrochemical Analysis
Vol. X	Organic Spot Tests Analysis
	The History of Analytical Chemistry
Vol. XI	The Application of Mathematical Statistics in Analytical Chemistry
	Mass Spectrometry
	Ion Selective Electrodes
Vol. XII	Thermal Analysis
	Part A. Simultaneous Thermoanalytical Examination by Means of the Derivatograph
	Part B. Biochemical and Clinical Applications of Thermometric and Thermal Analysis
	Part C. Emanation Thermal Analysis and other Radiometric Emanation Methods
	Part D. Thermophysical Properties of Solids
	Part E. Pulse Method of Measuring Thermophysical Parameters
Vol. XIII	Analysis of Complex Hydrocarbons
	Part A. Separation Methods
	Part B. Group Analysis and Detailed Analysis
Vol. XIV	Ion-Exchangers in Analytical Chemistry
Vol. XV	Methods of Organic Analysis
Vol. XVI	Chemical Microscopy
	Thermomicroscopy of Organic Compounds
Vol. XVII	Gas and Liquid Analysers
Vol. XVIII	Kinetic Methods in Chemical Analysis
	Application of Computers in Analytical Chemistry
Vol. XIX	Analytical Visible and Ultraviolet Spectrometry
Vol. XX	Photometric Methods in Inorganic Trace Analysis
Vol. XXI	New Developments in Conductometric and Oscillometric Analysis
Vol. XXII	Titrimetric Analysis in Organic Solvents
Vol. XXIII	Analytical and Biomedical Applications of Ion-Selective Field-Effect Transistors
Vol. XXIV	Energy Dispersive A-Ray Fluorescence Analysis
Vol. XXV	Preconcentration of Trace Elements
Vol. XXVI	Radionucleide X-Ray Fluorescence Analysis
Vol. XXVII	Voltammetry
Vol. XXVIII	Analysis of Substances in the Gaseous Phase
Vol. XXIX	Chemiluminescence Immunoassay
Vol. XXX	Spectrochemical Trace Analysis for Metals and Metalloids
Vol. XXXI	Surfactants in Analytical Chemistry
Vol. XXXII	Environmental Analytical Chemistry
Vol. XXXIII	Elemental Speciation – New Approaches for Trace Element Analysis
Vol. XXXIV	Discrete Sample Introduction Techniques for Inductively Coupled Plasma Mass Spectrometry
Vol. XXXV	Modern Fourier Transform Infrared Spectroscopy
Vol. XXXVI	Chemical Test Methods of Analysis
Vol. XXXVII	Sampling and Sample Preparation for Field and Laboratory
Vol. XXXVIII	Countercurrent Chromatography: The Support-Free Liquid Stationary Phase

Contents

Foreword

Countercurrent chromatography (CCC) was named after the countercurrent distribution method (CCD) that operates essentially by the same principle of partitioning solutes between two liquid phases. CCD was developed in the early 1940s mainly by the effort of Dr. L.C. Craig, and during the 1950s the Craig CCD apparatus was widely used for separation and purification of a variety of natural products. While his apparatus was able to process a large amount of pure compound, it required a long separation time and the size of the apparatus was such that a large unit occupied all the space in the whole room. My intention in developing countercurrent chromatography was to overcome these problems by utilizing a centrifugal force field. Shortly after we built the first prototype helical CCC centrifuge (later called the toroidal coil centrifuge) at the Laboratory of Technical Development of the National Heart Institute (presently called National Heart, Lung, and Blood Institute), our director, Dr. Robert L. Bowman, invited Dr. Craig to see our prototype. Dr. Craig was amazed to observe his 20 ml-capacity CCD partition unit being reduced to a few microliter capacity in a segment of coiled Teflon tubing. After this, Dr. Craig strongly supported our continued research on CCC, inviting me to the Gordon Conference in New Hampshire and the Anniversary Symposium on Chromatography in New York, both held in 1972. I still remember that during one session of the Gordon Conference some in the audience strongly argued that the name "CCC" was inappropriate, because the two liquid phases do not literally flow countercurrent to each other. Dr. Craig stood up and defended me by stating that countercurrent chromatography did indeed use the same principle as that of CCD. My research steadily advanced, finally achieving an efficient preparative separation and it has almost entirely replaced the countercurrent distribution method. The CCC technology, which eliminates the solid support of the conventional chromatographic column, has now this "solid support" from a number of outstanding scientists worldwide; many of these are authors of this monograph. In 2000, the First International Conference on CCC was held in London and the Second International CCC Conference was held in Beijing in April 2002. This monograph will be published shortly after this second CCC conference to celebrate the growing CCC technology that originated from the countercurrent distribution method developed by the early workers.

Yoichiro Ito
Laboratory of Biophysical Chemistry
National Institutes of Health
Bethesda, Maryland, USA

Series Editor's Preface

The idea for including this book by Alain Berthod and co-authors in the *Comprehensive Analytical Chemistry* series occurred to me three or four years ago, when I was evaluating several research projects. One of the new projects was based on Countercurrent Chromatography (CCC). Today, I am sure that together with the reviewers I made the right choice since the book covers a technology that is growing in importance in analytical chemistry. It contains twelve chapters that cover the fundamentals of CCC, operating the machine, solvent systems and special uses of the support-free liquid stationary phase. Attention is also given to real-world applications such as the analysis of natural products and inorganic compounds. The growing potential of CCC in chiral separations is shown and the final chapter deals with the use of supercritical fluid as the mobile phase for this technique.

Overall, this is a timely book, appearing as it does just after the Second International CCC Conference. I am sure that it will be useful to a broad audience, including experts in the field, postgraduate students and newcomers who want to know more about this technology. Just to make things easier for the reader, I like the easy-to-use definition given by Berthod in the introduction: CCC is just an LC technique with a support-free liquid stationary phase.

Finally, I would like to thank all the authors who participated in this reference work on CCC. Their time and efforts in preparing their contributions is highly appreciated.

D. Barceló

Preface

Countercurrent chromatography (CCC) was invented in the late 1960s by Dr. Yoichiro Ito in his laboratory at the National Institutes of Health in Bethesda, Maryland (USA). The originality of the technique is that the stationary phase is a liquid. Since it is a liquid chromatography (LC) technique, the mobile phase is also a liquid. Biphasic liquid systems are used, but there is not necessarily a countercurrent motion of one phase versus the other. Dr. Ito dedicated his professional life to his technique. He published several hundreds of articles on CCC, establishing the triple C acronym so well that it is accepted worldwide and will no doubt remain.

Unfortunately, the very name of the technique has led people who do not know it to keep asking when visiting my CCC group: "I just see one pump on each of your CCC set-ups, where is the second pump needed to establish the countercurrent flow?" And I start to explain that CCC is a classical, one pump, LC technique with just a support-free liquid stationary phase. The subtitle of this book points out the beauty of the CCC technique. Let me say it again: CCC is just an LC technique with a support-free liquid stationary phase.

As can be guessed, the problem is to obtain a stable support-free liquid stationary phase. Here is the real difficulty of the CCC technique. The CCC column cannot be a simple tube with two frits at each end. In all modern CCC "columns", centrifugal fields are used to hold the liquid stationary phase when the liquid mobile phase is pushed through it. Centrifugal fields imply rotating parts, rotors, gears, spools, rotating seals, motors, speed regulators, noise and wearing problems—all things that hinder the development of CCC. It should be clearly understood that the whole CCC machine is just a chromatographic column.

The first reason for this new book on CCC is to explain the hardware in a simple and, it is hoped, straightforward way. It will be the goal of the first two chapters to describe the fundamentals of CCC and the ways of operating a CCC machine. Why is it of interest to have a support-free liquid stationary phase? The first advantage is that the solutes have access to the whole stationary phase volume, rather than the interface between the mobile phase and the surface of the solid stationary phase as in many other chromatographic techniques. The second advantage is that any one of the two phases of the biphasic system can be the mobile phase. The phase role can even be changed during a run.

Some knowledge of solvent systems is necessary; this is the topic of the third chapter. Again, there is no countercurrent circulation of fluids in CCC. There are, however, two types of CCC machines, both of which were initiated by Yoichiro Ito who later specialized in the second type. The first type of CCC machine uses a constant centrifugal field, rotary seals and geometric and periodic forms to retain the liquid stationary phase. These machines are called Centrifugal Partition Chromatographs (CPC). They are described and their properties elucidated in Chapter 4 by one of the experts of the CPC field. The flow pattern, mass transfer and solute residence time distribution in CPC differ from that in the second type of CCC machine, and this is discussed for CPC and HSCCC in Chapters 5 and 6, respectively.

High speed CCC (HSCCC) is one of the acronyms used to refer to the hydrodynamic machines which make up the second type of CCC machine. In these machines, an open Teflon tube is coiled on one or several spools that are rotated in different ways (angle, speed, axles) around a central axle in such a way that no rotary seals are used.

Some advantages of a support-free liquid stationary phase have already been shown. More advantages and original possibilities opened by the liquid nature of the stationary phase are covered in Chapter 7. These possibilities lead to real-world applications of CCC especially in the rich field of separation and purification of natural products (Chapter 8) and inorganic compounds (Chapter 9). CCC is moving out of the research laboratory; the pharmaceutical industry was the first to realize the powerful capabilities of CCC as shown in Chapter 10. The last two chapters describe advanced and original uses of the CCC techniques: Chapter 11 shows that CCC has enough resolution power to be able to separate enantiomers, sometimes on a preparative scale; Chapter 12 describes the simple modifications to make to a CCC machine so that it can work with the pressures necessary to use a supercritical CO_2 mobile phase.

It is my pleasure to finish by dedicating the book to Dr. Yoichiro Ito and thanking him for his kind words in support of this collective work. All its authors have tried to make the book useful for the analytical chemist who wants to know more about CCC. I thank them all for their hard and timely work which is reflected in the quality of the book.

Alain Berthod

Chapter 1

Fundamentals of countercurrent chromatography

Alain Berthod

1.1 INTRODUCTION

The usual way of classifying chromatographic techniques is to use the nature of the mobile phase. Gas chromatography encompasses all techniques in which the mobile phase has a gaseous state; supercritical fluids are used as mobile phases in supercritical fluid chromatography; and liquid chromatography (LC) includes all techniques using a liquid mobile phase, i.e., HPLC, size exclusion chromatography, thin layer chromatography and ionic and ion-exchange chromatography. In countercurrent chromatography (CCC), the mobile phase is a liquid, so CCC is a liquid chromatography technique. The unique feature of CCC is that, unlike all other techniques, the stationary phase is also a liquid without any solid support.

1.2 A LIQUID STATIONARY PHASE

Countercurrent chromatography uses a *support-free* liquid stationary phase. Most separations are done in gas chromatography (GC) with a liquid stationary phase, but the GC liquid stationary phase is maintained by a solid support that can either be particles (filled columns) or a long tubing (capillary and macrobore GC columns). Furthermore, by definition, the mobile phase is a gas with a very low friction coefficient and high diffusion coefficient compared to a liquid. In CCC, there is no solid support; centrifugal fields are used to keep the liquid stationary phase steady while the liquid mobile phase is pushed through it.

Comprehensive Analytical Chemistry XXXVIII
A. Berthod (Ed.)

A typical CCC set-up comprises: a mobile phase reservoir, a pump, an injection valve, a column, a detector, and a recorder or data processor. This is the exact description of any HPLC unit. The difference is the column: in CCC all "columns" contain rotating parts needed to create and maintain a centrifugal field that will work with a support-free liquid stationary phase.

Yoichiro Ito, the originator of CCC, spent years looking for the best possible way of obtaining a very stable liquid stationary phase [1–3]. He described more than 20 different CCC machines [3]. Today, the Ito Scheme IV, also called *coil planet centrifuge*, is the basis of the most reliable and easy to construct CCC apparatuses. The quest for the best CCC machine will not be described here; the interested reader should refer to the excellent and authoritative 1989 book written by Walter Conway who exhaustively describes the numerous Ito CCC designs existing at the time [4]. Due to the lack of reliable rotating seals, Ito developed machines using gears and a continuous open Teflon® tube. This type of CCC "column" is called a *hydrodynamic* machine. A Japanese company, Sanki Engineering Ltd, developed another type of CCC "column", based on channels interconnected by ducts and placed in a centrifuge rotor with two rotating seals. This type of CCC "column" is called a *hydrostatic* apparatus. These two types of "column" are describe extensively in Chapters 4, 5 and 6 and the Appendix.

1.3 CHROMATOGRAPHIC THEORY

CCC is an LC technique with a special column that enables the stationary phase to be a liquid phase. It is worth pointing out that there is no countercurrent liquid circulation in CCC. The countercurrent chromatography term and the CCC acronym were coined by Y. Ito, the founder of the technique, to refer to the countercurrent distribution method which also used two liquid phases. CCC obeys the classical chromatographic equations and the basic plate theory. These will be briefly recalled here.

1.3.1 Fundamental retention equation

Liquid–liquid partition is the only phenomenon responsible for solute retention in CCC. However, CCC is not an automated liquid–liquid extraction process in the same way that HPLC is not an automated solid phase extractor. In the chromatographic process, the solutes undergo a number of back-and-forth liquid phase exchanges. The plate count is a theoretical measure of this number. The solutes separate because of their differing affinity for the stationary phase. The solutes enter and exit the column *in the mobile phase*. The solute retention volume, V_R, is given by:

$$V_R = V_M + K_D V_S \tag{1.1}$$

V_M and V_S are the mobile and stationary phase volumes, respectively, inside the CCC apparatus; V_M corresponds to the "dead" volume in HPLC; K_D is the distribution constant or partition ratio of the solute; and K_D is the ratio of the solute concentration in the stationary phase over the solute concentration in the mobile phase. For example, if the stationary phase is octanol and the mobile phase is water, the K_D distribution constant is the octanol/water partition coefficient, $P_{o/w}$, widely used in hydrophobic measurements [5–6].

A unique property of CCC is that the internal machine volume, the "CCC column", contains the mobile phase and the liquid stationary phase only. No silica support is present. The "column" volume, V_C is:

$$V_C = V_M + V_S \tag{1.2}$$

Since the CCC apparatus internal volume is always known (either given by the manufacturer, or measured by filling the machine with a single phase), the mobile and stationary phase volumes are not two independent parameters: Eq. (1.2) relates them to each other. The stationary phase volume retention ratio, *Sf*, is defined as:

$$Sf = V_S/V_C \tag{1.3}$$

Using Eqs. (1.2) and (1.3), Eq. (1.1) can be rewritten as:

$$V_R = V_C + (K_D - 1)V_S = V_C\,[1 + (K_D - 1)\,Sf] \tag{1.4}$$

Since the phase volumes inside the CCC machine, and/or the *Sf* parameter, are easily measured, these equations show that the position of the peak corresponding to a given solute can be fully predicted if its distribution coefficient, K_D, is known, or the distribution coefficient, K_D, of a solute can be calculated using its retention volume, V_R [4–6].

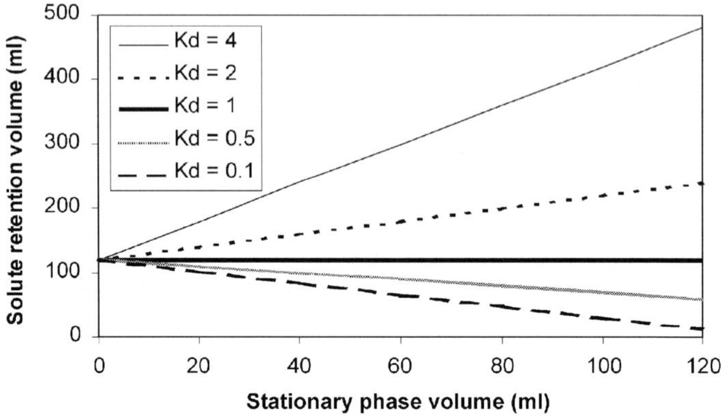

Fig. 1.1. Retention volume of a solute with a distribution coefficient K_d versus the stationary phase volume contained in a 120 ml CCC column.

Figure 1.1 shows the retention volumes calculated for solutes with distribution constant, K_D, ranging from 0.1 to 4 and phase volumes, V_M and consequently (Eq. 1.2) V_S, varying from 0 ml to 120 ml, the machine volume. Three points are worthy of comment:

- The K_D = 1 value is an interesting limit (thick line). An equal partitioning of the solute between the mobile and the stationary phase (K_D = 1) produces a constant retention volume equal to the apparatus volume, V_C (Eq. 1.4) for any stationary phase volume.
- If a solute has a high affinity for the mobile phase, its K_D value is lower than 1 and its retention volume increases when part of the liquid stationary phase is lost (= V_M increases). On the other hand, the solutes with a high affinity for the stationary phase ($K_D > 1$) are less well retained when the stationary phase volume decreases.
- The solute retention volume increases linearly with its distribution constant, K_D. It will be shown that liquid composition changes may produce dramatic changes in the K_D values. A K_D value of 100,000 (log K_D = 5) is possible. The solute retention volume would be prohibitive (in the litres range) even with a low stationary phase volume.

1.3.2 Retention factor and selectivity

The retention factor in liquid chromatography is denoted k and is expressed by:

$$k = (V_R - V_M) / V_M \tag{1.5}$$

Using Eqs. (1.1) and (1.4), we get:

$$k = K_D V_S/V_M = K_D [(1 - Sf)/Sf] = K_D \Phi \tag{1.6}$$

with $\Phi = V_S/V_M$, the phase ratio.

The retention factor is directly proportional to the distribution constant. However, k is not a typical reference retention parameter in CCC as it is in HPLC. Indeed, the Φ parameter is constant in HPLC, but not in CCC. This is the reason why the distribution ratio or partition coefficient, K_D, is always preferred in CCC.

The selectivity factor, α, is the ratio of the retention factor of two successive peaks:

$$\alpha = k_2 / k_1 \tag{1.7}$$

By convention, the selectivity ratio is always equal or higher than one, i.e. k_2 is the retention factor of the more retained peak.

1.3.3 Resolution

The resolution factor

The quality of a separation is quantified by the resolution factor, *Rs*. A 1.5 resolution value means that two adjacent peaks are separated with a baseline return. A resolution factor lower than 1.5 means there is some peak overlap. Resolution factors higher than 1.5 mean that there is space between the two peaks. *Rs* is expressed as the ratio of the distance between the two peak maxima to the mean value of the peak width, *W*, at the baseline.

$$Rs = \frac{V_2 - V_1}{(W_2 + W_1)/2} \tag{1.8}$$

The chromatographic efficiency, *N*, is defined as the ability of a "column" to keep the concentration bands of the solutes as narrow as possible. As already seen, it can also be considered as a measure of the number of times the solute was exchanged between the mobile and stationary phase. *N* is related to the peak width. The efficiency unit is the theoretical plate number. The classical equation for a Gaussian shaped peak is:

$$N = 4(V_R / W_{0.6h})^2 \quad (1.9)$$

in which $W_{0.6h}$ is the peak width at 60% of the peak height. $W_{0.6h}$ corresponds exactly to 2σ, the variance, if the peak is perfectly Gaussian. The Gaussian peak base width, W, is equal to 4σ or 2 $W_{0.6h}$.

Using Eqs (1.1) and (1.7), and assuming the efficiency, N, is constant, Rs can be rewritten as:

$$Rs = \frac{1}{4}\sqrt{N}\,\frac{K_{D2} - K_{D1}}{\frac{V_M}{V_S} + \frac{K_{D2} + K_{D1}}{2}} \quad (1.10)$$

The resolution factor can be expressed with retention factors instead of partition coefficients. Substituting Eq. (1.8), one obtains:

$$Rs = \frac{1}{4}\sqrt{N}\,\frac{k_2 - k_1}{1 + \frac{k_2 + k_1}{2}} \quad (1.11)$$

Notably, the phase ratio has disappeared. Using the selectivity ratio, the Snyder resolution equation, popular in HPLC, can be obtained:

$$Rs = \frac{1}{4}\sqrt{N}(\alpha - 1)\frac{k_1}{1 + k_1} \quad (1.12)$$

Equations (1.10) and (1.12) represent the same resolution factor. However Eq. (1.10), containing the V_S / V_M parameter, clearly shows the critically important role of the liquid stationary phase volume in CCC. This point allows us to delineate the difference between HPLC and CCC.

Phase ratio and resolution factor

The phase ratio, V_S/V_M, is very different in HPLC and in CCC. In CCC, a minimum of 10% stationary phase retention is required to do a separation, the V_S/V_M ratio is 0.10 or log $V_S/V_M = -1$. With a well-retained liquid system, the stationary phase retention can reach 80%, producing a V_S / V_M ratio of 4 (log V_S / V_M = 0.6). In reversed phase HPLC, a common value of the bonding density is 3 μmol/m^2; with an average 200 m^2/g surface area, the estimated bonding material amount is about 6×10^{-4} mol/g [7]. A heavy, commonly bonded moiety is octadecane, C18, with a

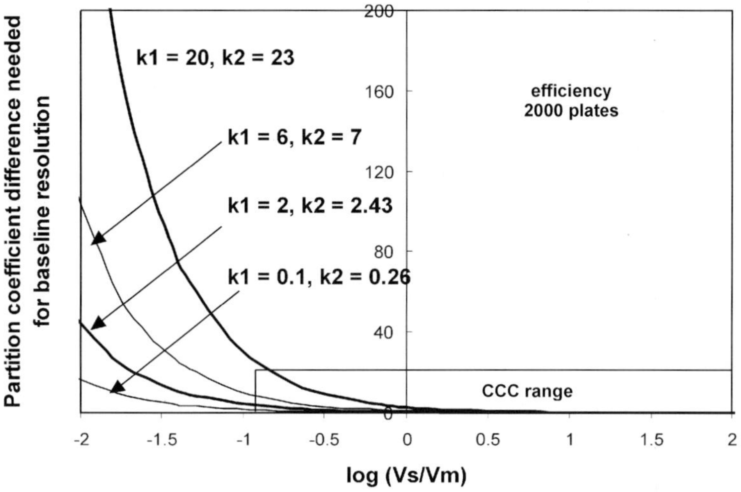

Fig. 1.2. Difference in partition coefficient ($K_{D2} - K_{D1}$) needed to obtain a baseline separation (Rs = 1.5) between Solute 2 and Solute 1. The left part of the figure (log (V_s/V_m) < –1) corresponds to the HPLC work range; the right part (square) corresponds to the CCC work range.

molar weight of 253; it leads to 0.15 g of bonded phase per g of packing, usually expressed as 13% carbon content [8]. This is a volume of about 0.19 cm^3 of organic bonded layer per gram of packing, which can be taken as the solute accessible stationary phase volume. The mobile phase volume per gram of LC packing is usually in the cm^3 range [8]. The V_S/V_M ratio is 0.19 (log V_S/V_M = –0.7). With a HPLC packing phase less densely grafted and/or with a lower surface area and/or using a lighter moiety (C10, C4, C1), the V_S/V_M ratio can be as low as 0.01 (log V_S/V_M = –2). In normal-phase LC, V_S corresponds to the superficial polar layer of unbounded porous silica. An unknown amount of water and/or polar molecules is tightly bound to the surface silanol groups, making it impossible to know the exact estimate of the V_S volume, but it is, no doubt, very low.

Figure 1.2 shows the distribution ratio difference, $K_{D2} - K_{D1}$, needed to obtain a baseline resolution, Rs = 1.5, between the peak of Solute 1 and that of Solute 2, plotted versus the logarithm of the phase ratio. Four sets of k values were examined. It is recalled that the retention factor, k, depends on the phase ratio and the distribution ratio, K_D (Eq. 1.6). Figure 1.2 clearly shows that the distribution ratio difference needed to obtain a baseline resolution is much higher in HPLC than in CCC, due to the low phase ratio V_S/V_M in HPLC. This becomes more critical as the retention factor, k, increases. For example, with a 2000 plate efficiency, a baseline separation between two peaks around $k = 20$ is obtained when the distribution ratio difference $K_{D2} - K_{D1}$ is equal to 30 units on a densely grafted LC phase with $V_S/V_M = 0.1$. The $K_{D2} - K_{D1}$ difference can be as high as 200 when a moderately grafted HPLC short chain ($V_S/V_M = 0.016$) is used [8]. Since in HPLC only the retention factors k are used, this point will be ignored. With the densely grafted HPLC column ($V_S/V_M = 0.1$), a baseline resolution is obtained between Solute 1 with $k_1 = 20$ and Solute 2 with $k_2 = 23$. The fact that the respective partition coefficients are 200 and 230 is ignored. With the moderately grafted column ($V_S/V_M = 0.016$), two solutes with k values of 20 and 23 are also baseline separated. But their respective partition coefficients are now 1250 and 1450. These solutes are certainly not the same on the two columns since their partition coefficient are more than an order of magnitude apart. In CCC, the partition coefficient difference is lower than 2 if the stationary phase retention factor, *Sf*, is higher than 50% (i.e., the V_S/V_M is higher than 1).

1.4 CCC AND HPLC

Figure 1.2 shows the operating range of the CCC technique. It does not overlap much with the HPLC operating range which is with V_S/V_M ratios lower than 0.1 ($\log V_S/V_M < -0.1$) and high distribution coefficient differences (left part of Fig. 1.2). CCC operates with a significant stationary phase volume ($\Phi > 0.1$). HPLC operates with a very low stationary phase volume ($\Phi < 0.1$). This critical difference should be pointed out to understand that the two techniques are complementary.

1.4.1 Plate count working range

The difference in stationary phase volume between HPLC and CCC explains why the two techniques do not need the same number of plates to operate efficiently. Figure 1.2 was prepared theoretically using the same efficiency (2000 plates) for the two techniques. It should be pointed out that a 2000 plate efficiency is very poor for an LC column. Compared to CCC, HPLC compensates for the low V_S/V_M ratio with high efficiency. Figure 1.3 shows the number of theoretical plates needed to obtain a

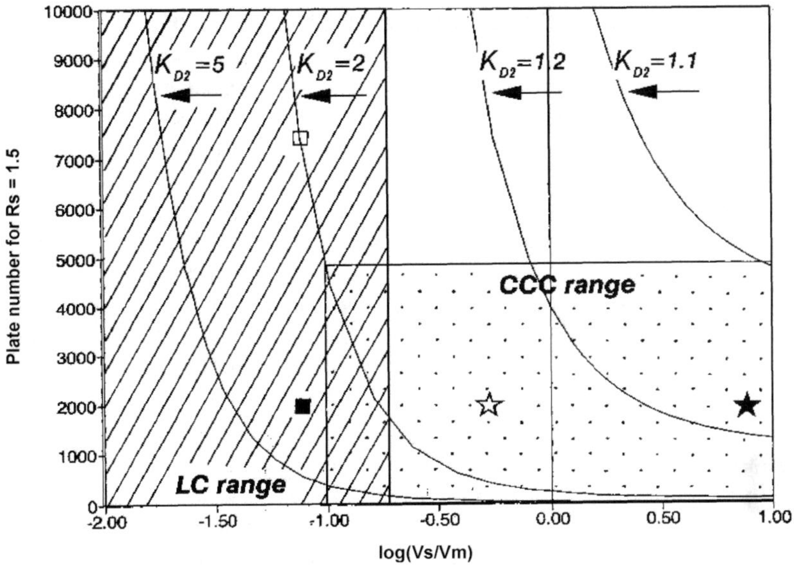

Fig. 1.3. The chromatographic efficiency, in number of plates, needed to obtain a baseline separation (Rs = 1.5) between a solute of distribution constant K_{D1} = 1 and a solute of distribution constant K_{D2} = 1.1, 1.2, 2 or 5. Dotted area: CCC working range; hatched area: HPLC working range. The open and closed squares and stars correspond to the chromatograms of Fig. 1.4.

baseline resolution between Solutes 1 and 2. The K_{D1} value is 1; the K_{D2} value is shown in Fig. 1.3. The maximum efficiency obtained so far with a CCC column is below 5000 plates. This limits the range of use of the CCC technique as showed by the dotted area of Fig. 1.3. The phase ratio limits the LC range, as indicated by the hatched area of Fig. 1.3.

Figure 1.4 illustrates the significance of Figs. 1.2 and 1.3, and shows four chromatograms. The complete sets of experimental values used to obtain the four chromatograms are listed in Table 1.1. The closed square LC chromatogram (Fig. 1.4) shows a poor separation of the two solutes. The separation is easily enhanced by an efficiency increase from 2000 to 7600 plates (open square). Figure 1.4 (bottom) shows the CCC chromatograms corresponding to Table 1.1 data. The two solutes are completely separated even with a 33% stationary phase retention (open star) which is a poor *Sf* value in CCC. A baseline separation would still be observed with only 200 plates (chromatograms not shown). The CCC technique can work efficiently with a surprisingly low plate count.

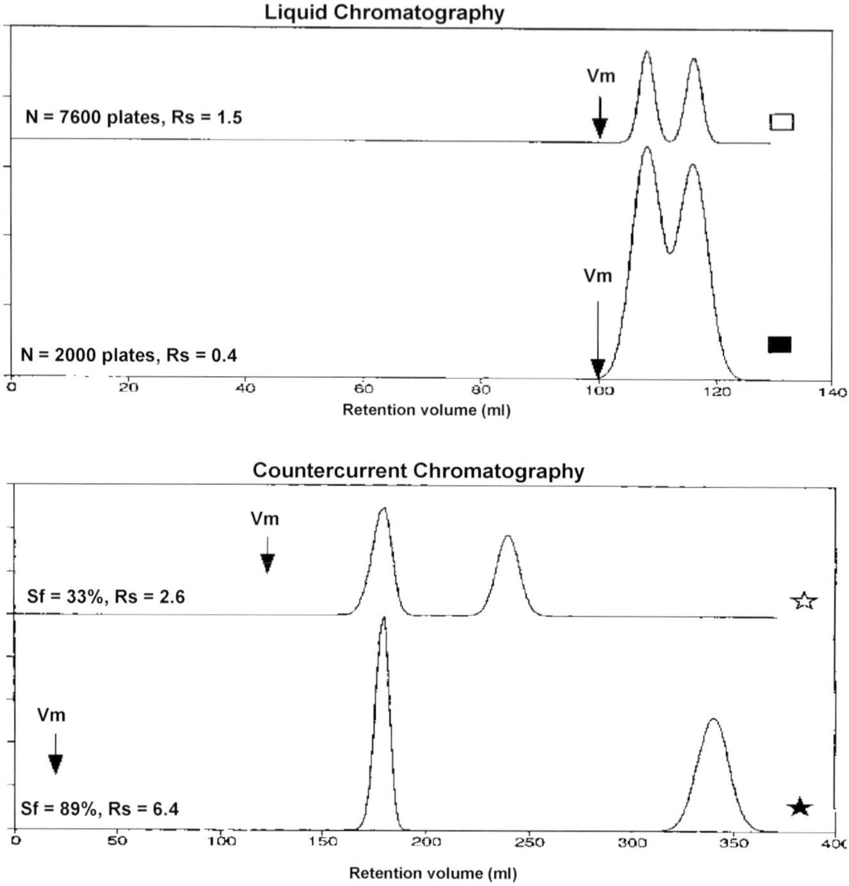

Fig. 1.4. Chromatograms corresponding to the data listed in Table 1.1. Top: HPLC (squares) improves resolution by increases in efficiency. Bottom: CCC (stars) improves resolution by increasing the stationary phase volume in the column.

1.4.2 CCC is fundamentally a preparative technique

Table 1.2 compares the general features of the two chromatographic separation techniques. The organic layer bonded on the silica particles used as HPLC packing is considered as a solid. This layer is more or less impregnated by mobile phase molecules and has a somewhat liquid state nature [8]. The solute–stationary phase interaction and maximum load

TABLE 1.1

Chromatographic parameters used to prepare Fig. 1.4

Parameter	Symbol (Unit)	CCC		LC	
Symbol in Fig. 1.4		★	☆	■	❑
Column volume	V_T (ml)	180	180	250	250
Stationary phase volume	V_S (ml)	160	60	8	8
Mobile phase volume	V_M (ml)	20	120	100	100
Phase ratio	$\log V_S/V_M$	0.9	–0.3	–1.1	–1.1
Partition coefficient: Solute 1	K_{D1}	1	1	1	1
Partition coefficient: Solute 2	K_{D2}	2	2	2	2
Retention factor: Solute 1	k'_1	8	0.5	0.08	0.08
Retention factor: Solute 2	k'_2	16	1	0.16	0.16
Retention volume: Solute 1	V_{R1} (ml)	180	180	108	108
Retention volume: Solute 1	V_{R2} (ml)	340	240	116	116
Efficiency	N (plates)	2000	2000	2000	7600
Resolution	Rs	6.4	2.6	0.4	1.5

The LC column is a 20 cm semi-prep column with 4 cm internal diameter.

TABLE 1.2

CCC and HPLC features

Parameter	CCC	LC
Stationary phase	liquid simple	solid complex
Mechanism	liquid–liquid partition	partition, adsorption, silanol interaction, pore size exclusion, ion-exchange
Solute interaction	volume	interface
Solute loadability	high	low–medium
Efficiency	low	high
Operation	difficult	easy
Cost	cheap	expensive

are linked. In CCC the solute can access the whole volume of the liquid stationary phase. In HPLC, the solute cannot penetrate into the solid silica matrix, it accesses the solid–liquid interface covered by the organic bonded layer. This explains why overloading occurs much more rapidly in HPLC than in CCC. It is possible to inject up to 20% of the CCC column volume. CCC will be used to isolate and or to purify large amounts of material [9–10]. Practical examples of the preparative capability of the CCC technique are given in Chapters 7, 8 and 10.

1.4.3 Two complementary techniques

Countercurrent chromatography and preparative LC are two complementary techniques. Figures 1.2 and 1.3 show that their respective operating ranges do not overlap much. To be fair to HPLC, it should be pointed out that the HPLC–CCC comparison was made assuming that only partitioning was responsible for solute separation. However, this is only true for CCC. The HPLC solute separation process involves partitioning along with mobile phase solvophobic effects, adsorption, silanol interactions, ion-exchanges and pore size exclusion [8]. The complexity of the LC separation process renders the separation predictions difficult but it enhances the selectivity. However, the HPLC column-to-column reproducibility is somewhat weak, whereas it is not a problem in CCC.

As has been pointed out, CCC is a preparative technique: it is not adapted to the separation of microlitre complex samples; it cannot perform the efficient analyses done by analytical HPLC with 5–25 cm columns and 5–20 μl sample injection. CCC should be compared to preparative LC only [10].

The main problem with CCC is that it is difficult to keep a support-free liquid stationary phase motionless. However, the advantages of a liquid phase compensate well for the technological problems. The three main advantages are: (1) a high loadability in a low-cost, economical liquid phase; (2) phase reversal capability, i.e., the light phase or the denser liquid phase can be used as the stationary phase, no irreversible solute adsorption is possible; (3) unique selectivity due to the infinite number of possible biphasic liquid systems offering a wide polarity range.

1.5 HYDRODYNAMIC AND HYDROSTATIC CCC COLUMNS

The early droplet CCC apparatuses, in which the liquid stationary phase was maintained only by gravity, are no longer used today. Extremely low rates for these devices were measured in ml/hours. Solute retention times were measured in days. Only the few biphasic liquid systems that could form droplets which were stable in the earth's gravitational field could be used [3–4]. All modern CCC apparatuses use a centrifugal field to maintain the liquid stationary phase. Flow rates in the ml/min range are commonly used producing retention times in the order of a few hours or less. Commercially available CCC columns can be sorted in two classes: hydrodynamic and hydrostatic machines.

1.5.1 Hydrodynamic CCC columns

The two characteristics of hydrodynamic CCC devices are (1) spools containing coiled Teflon tube, and (2) no rotary seals and a gear assembly with two rotation axes generating a variable centrifugal field. The hydrodynamic devices were almost exclusively developed by Yoichiro Ito [1–4]. They will be referred to as high speed countercurrent chromatographs or HSCCC, the term "high speed" referring to the time scale of the droplet CCC machines with days as the retention time unit. Compared to a modern HPLC analysis which is performed in a few minutes, a preparative CCC separation may still need several hours.

Liquid motion in a hydrodynamic spool
Figure 1.5 illustrates how the two liquid phases contact in an equilibrated hydrodynamic spool. Due to the planetary motion of the apparatus spools, the centrifugal field changes in intensity and direction [3,4].

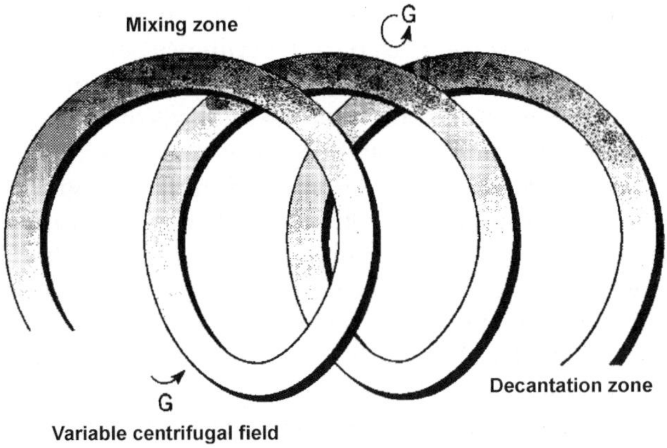

Fig. 1.5. Liquid behaviour in a hydrodynamic machine. The two liquid phases separate when the centrifugal field is high (decantation zones) and are mixed when the centrifugal field inverts. The two liquid phases are in contact throughout the tubing.

When the centrifugal field is high, phase decantation occurs. When the centrifugal field direction reverses, the separated liquid phases commingle in an emulsion-like state. The mixing–decantation process occurs twice for each rotor revolution. Alternating decantation and mixing zones appear in the spool [4]. The two liquid phases remain in contact throughout the tube length. There is little pressure build-up.

As explained in detail in Chapter 6, the two liquid phases are pushed toward one side of the coil called the head. The other side is called the tail. The upper phase always tends to gather on the head side, the lower phase accumulates behind. So, when the upper phase is used as the mobile phase, it should enter the CCC column through the tail. This is called the tail-to-head direction (T → H) [3–5]. When doing so, it will go through the denser liquid stationary phase to exit the head of the column. If the denser liquid phase is the mobile phase, the head-to-tail direction should be used. The lower liquid phase should enter the CCC column through the head to cross through the lighter stationary phase and exit the CCC column through the tail side.

Hydrodynamic CCC machines

The numerous different designs of hydrodynamic machines by Y. Ito have been extensively described in several books and review articles [3,4,11–13]. The commercially available machines are briefly described in the technical appendix at the end of this book. Most of these CCC machines are two-axle coil planet centrifuges derived from the Scheme IV design by Ito. A rotor holds at least one spool containing a coiled Teflon tube. The spools are geared to a stationary central toothed wheel (first axis of rotation) so that the rotor rotation produces a planetary motion of the spools around their axis (second axis of rotation). Different tubing arrangements allow the avoidance of any rotary seal.

Hydrodynamic machines can be as small as 5 ml and as large as 2 l. Very often two Teflon tubes are coiled together on the same spool which results in a machine with a variable volume. The problem of hydrodynamic machines was the noise generated by the gear assembly. Modern machines are enclosed in a thick box that reduces the noise and allow the temperature to be regulated. Figure 1.6 shows the SFCC2000 machine built by the Société Francaise de Chromatographie et Colonnes from 1989 to 1994 and discontinued when the company was bought by Shandon. The author still uses two models of this sturdy hydrodynamic three-spool CCC column (V_C = 165 ml) [14]. Anti-twisting connectors were used to avoid tube rupture during the rotor rotation. It was

Fig. 1.6. The SFCC2000 hydrodynamic machine built from 1986 to 1992. The top electronic unit controls the entrance pressure, the rotor rotation speed and the temperature; the intermediate unit contains a heat exchanger and a fan. The rotor bears three spools of 55 ml each with three anti-twisting connectors. The left valve allows selection of head-to-tail mode or tail-to-head mode. The right valve is the injection valve.

possible to work with either one spool (V_C = 55 ml) or three spools serially connected (V_C = 165 ml).

1.5.2 Hydrostatic CCC columns

The two characteristics of CCC hydrostatic columns are: (1) a geometrical volume, tube, channel, locule, in which the two liquid phases are contacted; the volume repeats itself through connecting tubing or channels, forming a pattern; and (2) a single axle of rotation generating a constant gravity field. For almost twenty years commercial hydrostatic machines were exclusively made by the Japanese company Sanki Engineering Ltd. Their CCC columns were called "Centrifugal Partition Chromatographs", with the acronym CPC (which confused people working in the CCC field since CPC is also used for Coil Planet Centrifuge).

CCC experiments performed with these machines were called Centrifugal Partition Chromatography [15]. Chapter 4 gives information on the Sanki CCC machines.

Liquid motion in a hydrostatic machine

Figure 1.7 shows a schematic view of the liquid motion in the channels and ducts of a hydrostatic machine. As is seen from Chapter 5, this crude view does not represent the liquid motion in a hydrostatic machine in all cases. However, it does show that the connecting ducts only contain the mobile phase without contact with the stationary phase. This is always true and is unfavourable for solute exchange between phases, i.e., the connecting ducts increase band broadening during the chromatographic process. Also, due to the presence of only one phase in the connecting ducts, a small hydrostatic pressure builds up from channel to channel explaining the denomination of this type of CCC machine.

In a similar way to hydrodynamic machines, the lighter liquid phase is pushed toward the top of the channels or tubes contained in the hydrostatic machines. The denser liquid phase gathers at the bottom of the geometrical volumes. If the upper phase is used as the mobile phase, it should enter through the bottom of the last channel to move against the constant gravitational field through the lower liquid stationary phase. This is not called tail-to-head but ascending mode [4,15]. On the other hand, when the denser phase is the mobile phase, it should enter through the top of the first channel to move with the gravitational field through the liquid upper stationary phase (descending mode).

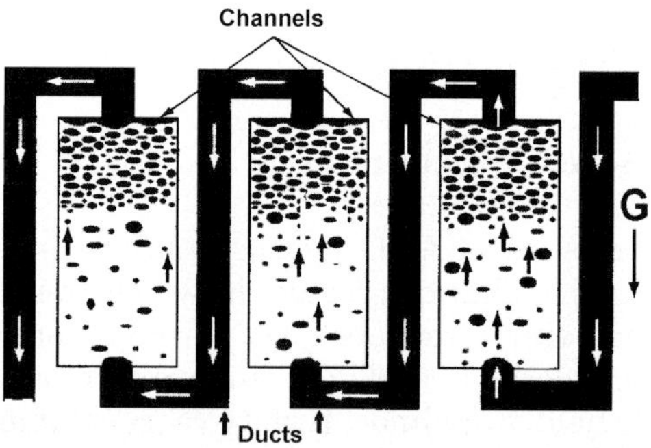

Fig. 1.7. An oversimplified view of the liquid motion in the channels of a hydrostatic CCC machine. The stationary phase is the denser phase. The mobile phase is the lighter phase moving in the ascending mode. The scheme shows that the mobile phase does not contact the stationary phase in the ducts.

Commercial hydrostatic machines

The early CCC columns were hydrostatic machines. The droplet CCC machine, the rotation locular CCC machine and the early toroidal CCC machine all have a single axis of rotation with geometrical tubes, locules and spires, respectively [3,4]. The modern hydrostatic machines contain geometrical channels interconnected by small ducts that minimize the dead volumes. Two rotating seals are necessary at the top and bottom of the centrifuge.

Figure 1.8 shows the rotor of the Kromaton hydrostatic machine with the two rotating seals. The rotor contains disks of a polymeric material in which channels and connecting ducts are engraved.

Fig. 1.8. The Kromaton FCPC has a rotor containing ten disks of PTFE in which channels and ducts are engraved (see Fig. 1.7). The valve allows ascending or descending mode to be selected. The top rotary seal is apparent on the photograph. A similar rotary seal is at the bottom of the apparatus. Apparatus volume: 200 ml.

Chapters 4 and 5 give all information on the hydrostatic machines or centrifugal partition chromatogaphs.

1.5.3 Pressure drop in CCC machines

It was demonstrated that the observed pressure drop, P_d, in operating a CCC machine was the sum of two terms [16]:

$$P_d = A\,\Delta d\,\omega^2\,Sf + B\,\eta\,F \tag{1.14}$$

in which *A* and *B* are constants depending on the machine and liquid system used, Δd is the liquid phase density difference, ω is the rotor spin rate, *Sf* is the stationary phase retention fraction, η is the mobile phase viscosity and *F* is the mobile phase flow rate. The first term of Eq. (1.14) is the hydrostatic term, the second term is the hydrodynamic or viscous term.

With Ito coiled tube CCC machines, the *A* term is very small ($A < 6\times10^{-4}$ cm^2 with densities in kg/cm^3, spin rate in rd/s and pressure in bar) because both liquid phases contact throughout the machine, minimizing the hydrostatic pressure build-up. It will make less than one-fourth of the observed pressure drop that should stay below 5–6 bars (70–90 p.s.i.) to avoid Teflon tube ruptures. The hydrodynamic term is the main responsible of the pressure drop observed with hydrodynamic machines. It should be noted that the *A* term may become negative when the upper phase is the mobile phase used in the tail-to-head direction. This is due to the Archimedean screw effect (see Chapter 6). With low viscosity liquid systems, in the head-to-tail direction, the CCC column can act like a pump by itself. Adding a restrictor (a small length of narrow bore tubing) at the detector outlet will solve this problem.

With hydrostatic machines, the *A* term is significant ($A > 0.01$ cm^2). If the density difference between the two liquid phases is low (e.g., $\Delta d \sim 0.1$ g/cm^3, heptane-methanol), there is a balance between the weight of the two terms of Eq. (1.14). At low flow rates and high spin rates, most of the pressure drop is due to the first hydrostatic term, and *vice versa*. With a significant density difference (e.g., hexane–water, $\Delta d = 0.34$ g/cm^3), the pressure drop increases dramatically with the rotation speed of hydrostatic machines. The hydrodynamic viscous term is minor [16].

Equation (1.14) shows that the pressure drop depends on *Sf*, the stationary phase retention ratio. This is only true if the equilibrium is reached in the machine. At the beginning of the equilibration of the CCC apparatus, only the stationary phase is present in the machine. The pressure drop is low, whatever liquid system is used, due only to the hydrodynamic viscous term (but with the stationary phase viscosity in Eq. (1.14) since the stationary phase liquid is displaced and moving in the machine). Then, as more and more spires (hydrodynamic machines) or channels (hydrostatic machines) are equilibrated, the pressure drop increases linearly with time [16]. This effect is limited with hydrodynamic machines that have a low A factor. It is important to watch the pressure during the equilibration of a CCC hydrostatic column. If it is noted that the pressure will pass the maximum pressure that the rotary seals can withstand, then the rotor spin rate, and not the flow rate, should be decreased. For hydrostatic machines, the effect of the flow rate on pressure drop is less dramatic than the spin rate except when the mobile phase viscosity is high. Once the CCC column is equilibrated, the pressure drop is a good indicator of column stability. If it decreases, it means that part of the stationary phase leaches, and the *Sf* factor decreases (Eq. 1.14).

1.6 CONCLUSION

Countercurrent chromatography is a liquid chromatography technique that uses the very same equipment used for preparative HPLC: pumps, injection valves, detectors and fraction collectors. The difference is the column. Since the stationary phase is a liquid in CCC, special columns should be used. There are two kinds of CCC machine: hydrodynamic machines, mostly made with rotating spools containing coiled Teflon tubing, and hydrostatic machines, with rotary seals and interconnected channels. The main point to keep in mind is that a stationary phase is needed to perform any separation. It is necessary to know and maximize the *Sf* factor in order to obtain satisfactory results in CCC. Also, CCC is fundamentally a preparative technique.

REFERENCES

1 Y. Ito, M.A. Weinstein, I. Aoki, R. Harada, E. Kimura and K. Nunogaki, *Nature*, 212 (1966) 985.

2 Y. Ito, J. Suaudeau and R.L. Bowman, *Science*, 189 (1975) 999.

3 N.B. Mandava and Y. Ito, *Countercurrent Chromatography, Theory and Practice*, Chromatographic Science Series, Vol. 44. Marcel Dekker, New York, 1985.

4 W.D. Conway, *Countercurrent Chromatography, Apparatus, Theory & Applications*. VCH Publishers, Weinhein, 1990.
5 A. Berthod, In: A.P. Foucault (Ed.), *Centrifugal Partition Chromatography*, Chromatographic Science Series, Vol. 68. Marcel Dekker, New York, 1995, p. 167.
6 A. Berthod, Y.I. Han and D.W. Armstrong, *J. Liq. Chromatogr.*, 11 (1988) 1441.
7 A. Berthod, *J. Chromatogr.*, 549 (1991) 1.
8 L.R. Snyder and J.J. Kirkland, *Introduction to Modern Liquid Chromatography*. Wiley, New York, 1979.
9 A. Berthod, B. Billardello and S. Geoffroy, *Analusis*, 27 (1999) 750.
10 A. Berthod and B. Billardello, *Adv. Chromatogr.*, 40 (2000) 503.
11 Y. Ito, In: J.C. Giddings, E. Grushka and J. Cazes (Eds.), *Advances in Chromatography,* Vol. 24. Ch. 6. Marcel Dekker, New York, 1984, p. 181.
12 Y. Ito, *Countercurrent Chromatography*, Journal of Chromatography Library Series, Vol. 51A. Elsevier, Amsterdam, 1992, p. 69.
13 Y. Ito, In: Y. Ito and W.D. Conway (Eds.), *High Speed Countercurrent Chromatography*, Chemical Analysis, Vol. 132. Wiley, New York, 1996, p. 3.
14 A. Berthod and M. Bully, *Anal. Chem.*, 63 (1991) 2508.
15 A. Foucault, *Centrifugal Partition Chromatography*, Chromatographic Science Series, Vol. 68. Marcel Dekker, New York, 1995, p. iii.
16 A. Berthod and D.W. Armstrong, *J. Liq. Chromatogr.*, 11 (1988) 547.

Chapter 2

Operating a countercurrent chromatography machine

Alain Berthod, Les Brown, Gilda G. Leitão and Suzana G. Leitão

2.1 INTRODUCTION

In HPLC, the choice of a particular column, whether a normal-phase polar column or a reversed-phase apolar column, is linked to the mobile phase selection. Most often, a silica-based stationary phase is selected. Then the mobile phase composition is chosen, taking into account the properties of the stationary phase: dissolution of silica precludes a mobile-phase pH higher than 10, and cleavage of the bonding precludes a mobile-phase pH below 2. When these requirements are fulfilled, many mobile-phase compositions can be prepared. The stationary phase should not be affected. The stationary–mobile phase interaction in HPLC can be well defined.

The situation is completely different in countercurrent chromatography (CCC). A biphasic liquid system is used and the more polar liquid phase *or* the less polar liquid phase can be used as the stationary phase. The phases' roles can even be interchanged during the chromatographic separation—this is called dual-mode CCC. The physical chemistry of biphasic liquid systems should always be kept in mind. The mobile phase and the stationary phase are two liquid phases in equilibrium with one another, and changing the mobile-phase composition may change the stationary-phase composition even more.

The purpose of this chapter is to focus on the particular ways in which a CCC machine should be used. Practical consideration is given to the liquid system and the preparation of the apparatus. Next, the

Comprehensive Analytical Chemistry XXXVIII
A. Berthod (Ed.)

dramatic effect of the *Sf* factor (volume of stationary phase) on the resolution factor will be illustrated using a particular separation. We also show, using a practical example, the results obtained when a separation obtained on one machine is transposed to another machine. The dual-mode use of a CCC machine and its important capabilities will also be exemplified under different experimental conditions. The chapter will conclude with details of the special care that should be taken when working with gradients in CCC.

2.2 LIQUID SYSTEM PREPARATION

The liquid system is the heart of any CCC separation since it forms the stationary phase and the mobile phase of the column. Before preparing the liquid system, one should know which solvent association, providing a biphasic liquid system, will be best to separate the sample. The different systematic methods for evaluating the sample polarity and optimizing the solvent selection are discussed in Chapter 3. Most often, more than two solvents are used to prepare the biphasic liquid system—two immiscible solvents form the two liquid phases, and a third solvent that partitions between the two phases allowing one to adjust their relative polarities.

2.2.1 Basic properties of ternary phase diagrams

Ternary phase diagrams are used to represent all possible mixtures of three solvents [1]; they are described in Chapter 3. Here, we shall indicate how they should be used to minimize the solvent consumption. Figure 2.1 (top) shows the methanol–chloroform–water ternary phase diagram with the tie-lines in the biphasic domain. Five particular compositions are shown in the diagram: Table 2.1 lists their respective compositions, and they all belong on the same tie-line. So, by definition, they separate into two liquid phases that have exactly the same compositions. The only difference is the phase volume ratio. Whatever liquid composition belonging to the tie-line is prepared, the aqueous phase is Composition 1 and the organic phase is Composition 5, both on the binodal curve on both ends of the tie-line. It is very important to realize that a CCC separation performed with the 5:9:7, v/v/v, chloroform–methanol–water system [2] will give exactly the same results as one done with the 13:7:4, v/v/v, system [3]. The two systems belonging to the same

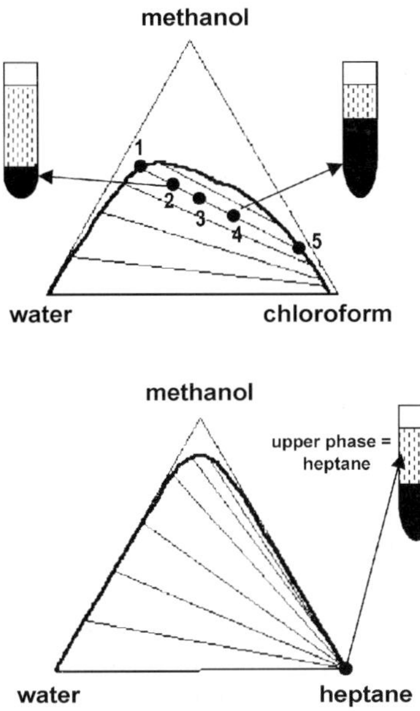

Fig. 2.1. Ternary phase diagrams. Top: system $CHCl_3$–MeOH–H_2O; the exact solvent proportions of the five compositions indicated are listed in Table 2.1. Bottom: system heptane–MeOH–H_2O; the upper phase of any of the mixtures in this system is pure heptane.

tie-line separate into two phases that have exactly the same composition. The difference is that Composition 2 (rich in methanol and water) separates in 77% of the light aqueous phase and only 23% of the heavy chloroform-rich organic phase (left tube in Fig. 2.1). The chloroform-rich 13:7:4 system (Composition 3) separates in the 35% aqueous phase and 65% heavy organic phase (Table 2.1, and right tube in Fig. 2.1). The amount of phases is different, but their compositions are identical. Since the volume of mobile phase used for a separation is larger than the volume of stationary phase, Composition 2 will be prepared if the selected mobile phase is the aqueous phase. In contrast, Composition 4 is appropriate if the mobile phase is the organic lower phase. Compositions 1 and 5 will be prepared, in both cases and respectively, if the separation needs more mobile phase than the amount initially prepared.

The heptane–methanol–water ternary system should be mentioned because it is extremely convenient for use in CCC [4,5]. As shown in Fig. 2.1 (bottom), all its tie-lines converge toward the heptane apex. This means that whatever proportions of heptane–methanol–water are mix-

TABLE 2.1

Composition of the chloroform–methanol–water liquid systems belonging to the tie-line shown in Fig. 2.1.

Mixture	Composition ($CHCl_3$–MeOH–H_2O v/v/v)				Upper phase volume (mixture 1)	Lower phase volume (mixture 5)
	Integer ratio	v%	v%	v%		
1	Aqueous phase	7.5	50.1	42.4	100%	0%
2	5 : 9 : 7	23.8	43	33.2	77.4%	22.6%
3	4 : 4 : 3	36.4	36.4	27.2	60.1%	39.9%
4	13 : 7 : 4	54.2	29.2	16.6	35.4%	64.6%
5	Organic phase	79.8	17.6	2.6	0%	100%

ed, the upper phase is always pure heptane (bottom right tube in Fig. 2.1). Furthermore, if the water-to-methanol volume percentage ratio is higher than 20%, it can be considered that there is practically no heptane in the lower phase (Fig. 2.1). In practice, if the separation can be carried out with the heptane–methanol–water 1:1:1, v/v/v, system, it is possible to prepare one phase by just mixing equal amounts of methanol and water; the other phase is pure heptane. With this system, when heptane is the mobile phase, pure heptane can be used without pre-equilibrating it with the methanol–water stationary phase. If heptane is the stationary phase, a gradient with increasing amounts of methanol, up to 80% v/v methanol, can be run without affecting the stationary phase [6].

2.2.2 Liquid system equilibrium

Once the proportion of the solvents is chosen, a graduated cylinder can be used to prepare the biphasic liquid system. It is good practice to prepare the liquid system in advance, because the two liquid phases may need some time to equilibrate fully. Also, the mixing of solvents very often produces temperature changes. For example, the mixing of methanol and water is exothermic, and the temperature increases. Heptane and ethyl acetate mix with absorption of heat (endothermic). Also, the solubility of gases is lower in methanol–water mixtures than in pure methanol and pure water, so the mixture should be degassed.

Once the biphasic system is prepared, it is recommended not to separate the phases. Solvents have different vapour pressures, and evaporation can change the phase compositions. Closed containers should be used. It is good practice to keep the two phases together or to maintain in each phase a small amount of the other phase. Solvent stability should be considered; for example, ethyl acetate is a very popular solvent in CCC (see the Arizona systems in Chapter 3). It is too often forgotten that ethyl acetate is slowly hydrolyzed (producing acetic acid and ethanol) when water is present. Hydrolysis reduces the pH of the phase, and may explain the low reproducibility of separations made with aged mixtures.

2.2.3 Effect of temperature

With most biphasic liquid systems, the mutual solubility of the solvents increases with temperature. This means that the ternary phase diagrams change with temperature. This issue is critical in water-free biphasic systems (e.g., dimethylformamide–heptane–toluene or dimethyl sulfoxide–hexane–THF). The liquid-phase viscosity decreases when the temperature increases. This may change the stationary phase retention ratio and the observed peak efficiencies.

The solute partitioning is also affected by temperature changes. Heat may be generated by rotating seals in hydrostatic machines, and gear ball-bearings and rotating connecting tubing in hydrodynamic machines. Temperature regulation units are being included in recent commercial machines.

In our experience of working with machines that do not have temperature control, reproducible results are obtained with biphasic liquid systems containing significant proportions of water as one of the solvents, and with isocratic conditions [4,6]. However, when linear and step gradients are used, a temperature regulation system greatly improves the reproducibility of the results.

2.3 APPARATUS PREPARATION

2.3.1 Classical "column" preparation

The apparatus is first filled with the stationary phase in the correct mode, i.e., tail-to-head (ascending) if the stationary phase is the denser phase or head-to-tail (descending) if the stationary phase is the upper liquid phase. A high flow rate can be used without rotation. Once the CCC machine is full of stationary phase, the pump is stopped. An empty graduated cylinder is placed at the column exit. The pump is rinsed and primed by the mobile phase (or a second pump is used). Then the rotor rotation is started at the desired value and the mobile phase is pumped into it again in the correct mode. Only the stationary phase exits the machine, as long as equilibrium is not reached. In a hydrostatic machine, the pressure increases linearly as more channels equilibrate, building up hydrostatic pressure. For a hydrodynamic machine, the first term of Eq. (1.14) is small, so the pressure increase is also limited. When the first drops of mobile phase exit the apparatus, equilibrium is reached and the pressure stabilizes. The volume of stationary phase displaced is measured; it corresponds to the volume of mobile phase, V_M, inside the CCC column. When everything is stable, including the detector baseline, the column is ready and the sample can be injected.

2.3.2 Direct injection

The first step of the direct injection method is identical to the classical method; the column is filled with the liquid stationary phase (high flow rate, no rotation). Next, the pump is rinsed and primed with the liquid mobile phase. The rotor is started. The sample is loaded in the injection valve and injected as soon as the pump, with the mobile phase, is turned on. When the mobile phase solvent-front passes in the detector, the volume is noted and V_M is calculated. The advantage of this method is that it is faster than the classical method. The drawback is that the volume, V_M, is less accurately determined.

2.3.3 Sample injection

Although injecting the sample in an HPLC system is a simple action, it may pose problems in CCC. Since CCC is a preparative technique, large amounts of material very often need to be loaded in the CCC column. This can disturb the stationary phase–mobile phase equilibrium so much that the "column" is destroyed: the whole stationary-phase volume is washed out of the machine by the concentrated sample plug [7]. If this happens, there is no other remedy but to dilute the injected sample. It is always better to inject a larger volume of diluted sample solution than a smaller volume of concentrated solution. The injection of a large volume does produce some band-broadening, but this is much less detrimental for solute separation than is the loss of stationary phase, as will be shown later. Since the high concentration of the sample may change the biphasic liquid system properties, it is also wise to add some stationary phase to the injected sample. This will minimize possible stationary-phase solubilization at the column entrance [7]. In a particular case with very large amounts of material loaded (18 g in a 1-l machine), the injection was successful when done at a reduced flow rate. Also, when stationary phase loss was observed, the mobile phase motion was inverted for a minute (switched from head-to-tail to tail-to-head and then back to normal). This produced a reflux of mobile phase that diluted the sample and allowed restoration of the mobile phase– stationary phase equilibrium. It induced band broadening which, again, is better than decreases in stationary phase volume [7].

2.3.4 Column stability

When the same liquid system is to be used for several analyses, several injections can be made on the same "column". It is even possible to use the same "column" for several days provided the liquid system does not contain ethyl acetate and water. To keep a "column", it is important to avoid stationary-phase leaks. The mobile-phase pump is first stopped, then the rotor is stopped. Gravity drainage or siphoning can occur. One trick that can be used to stop any liquid motion is to turn the injection valve and/or the mode valve to an intermediate position. If this is done, one should not forget to return the valves to their correct working positions before turning the pump on again. Otherwise, the liquids cannot circulate and the increasing pressure will cause leaks where the various Teflon tubes connect, or the tubes may burst. To resume a separation with an existing "column", the rotor is first started and, after an equilibration time of a few minutes, the mobile-phase pump is started at a low flow rate (0.1 ml/min). A graduated cylinder is placed at the column exit to detect stationary-phase leaks. The flow rate is progressively increased to the desired flow rate, and the pressure measured. If the pressure reading is comparable to that of previous experiments, and if there are no stationary-phase leaks, the "column" is again ready for use [8].

2.4 EFFECT OF MACHINE SPECIFICATIONS ON SEPARATIONS

The most important parameter of a commercial CCC machine is its volume (or volumes). This is given by the manufacturer and should be controlled experimentally. For example, the machine can be filled with water (no rotation needed) using an HPLC pump giving constant and known flow rates. Acetone can be injected; its retention time will give the machine volume. The other technical specifications of a CCC machine depend on the machine's nature. Beside the weight, size and power requirements, the technical specifications for hydrodynamic machines are the spool and rotor diameters, beta ratio, spool number, tubing diameter, tubing material, number of turns, winding direction, maximum rotation speed, rotation capability (clockwise and/or counterclockwise), noise level, and temperature control. These parameters will be discussed in Chapter 6. For hydrostatic machines, the technical parameters are: number of channels, channel volume, duct volume, total duct volume, material used, nature of the rotary seals, maximum pressure, maximum rotation speed, and the temperature control. Chapters 4 and 5 focus on hydrostatic machines.

2.4.1 Effect of the stationary-phase volume

It was shown in Chapter 1 that all the separation power of the CCC technique is due to the stationary phase volume, V_S, retained in the CCC machine. The V_S volume is part of the retention and resolution equations (Eqs. 1.1 and 1.10). Figure 2.2 illustrates the critical role of the stationary-phase volume. Four chromatograms were calculated for the compounds listed in Table 2.2 and four different *Sf* ratios. The ten compounds are baseline-separated when the stationary phase volume is 108 ml (*Sf* = 90%). When the stationary-phase volume decreases to 84 ml (*Sf* = 70%), Compounds 4 and 5, 6 and 7, and 8 and 9, are no longer baseline-separated and their resolution factors decrease significantly. If the stationary-phase volume becomes half of the machine volume (*Sf* = 50%), the resolution worsens so much that Compounds 4 and 6 disappear from the chromatogram (Fig. 2.2). Compound 8 is buried in Compounds 7 and 9. The lowest resolution factor is 0.6, between Compounds 6 and 7 (Table 2.2). With an *Sf* ratio of 30% (V_S = 36 ml), the loss of resolution is so severe that the mixture of ten compounds injected produces a chromatogram showing only five peaks partially separated (Fig. 2.2, bottom). Table 2.2 lists the resolution factors, which are all below 1.0.

Figure 2.2 and Table 2.2 show that the compounds with a K_d coefficient lower than 1.0 (Compounds 1 to 6) have increasing retention volumes when the stationary-phase volume decreases. Conversely, the retention volumes of Compounds 8 to 19, with $K_d > 1$, decrease when there is less stationary phase to retain them. The vertical dotted line

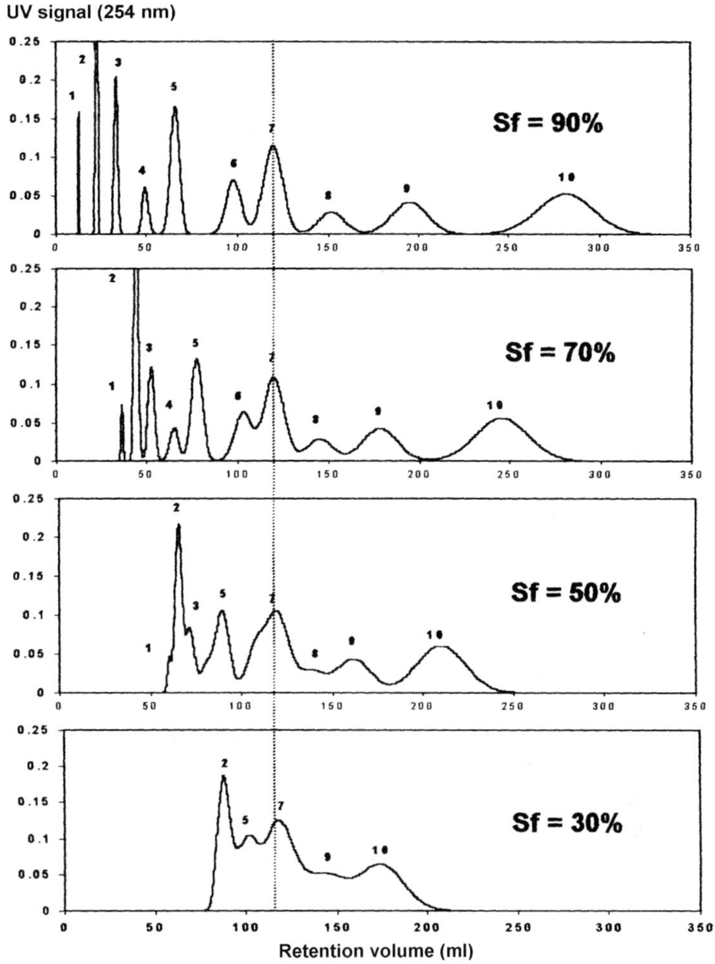

Fig. 2.2. Comparison of chromatograms obtained with the same machine, the same biphasic liquid system, and the same sample. The volume of stationary phase retained reduced from 108 ml (*Sf* = 90%) to 36 ml (*Sf* = 30%). Machine volume = 120 ml (vertical dotted line), average efficiency 500 plates. See Table 2.2 for more information.

(Fig. 2.2) corresponds to the machine volume (V_T = 120 ml) and also to the retention volume of Compound 7, which partitions equally between the two phases (K_d = 1). If the stationary-phase retention volume decreased further, all compounds would merge with Compound 7 into only one peak of 120 ml retention time.

TABLE 2.2

Hypothetical compounds used to compute the chromatograms of Fig. 2.2. Retention volumes and resolution factors.

Compounds	K_d[a]	$Sf = 0.9$ $V_S = 108$ ml		$Sf = 0.7$ $V_S = 84$ ml		$Sf = 0.5$ $V_S = 60$ ml		Sf = 0.3 $V_S = 36$ ml	
		V_r	Rs[b]	V_r	Rs[b]	V_r	Rs[b]	V_r	Rs[b]
1[c]	**0.01**	13.1		36.8		60.6		84.4	
2	**0.1**	22.8	6.6	44.4	2.4	66.0	1.1	87.6	0.5
3	**0.2**	33.6	3.5	52.8	1.6	72.0	0.8	91.2	0.4
4	**0.35**	49.8	3.1	65.4	1.7	81.0	0.9	96.6	0.5
5	**0.5**	66.0	2.0	78.0	1.3	90.0	0.8	102.0	0.4
6	**0.8**	98.4	2.5	103.2	1.8	108.0	1.2	112.8	0.6
7[d]	**1**	120.0	1.3	120.0	0.9	120.0	0.6	120.0	0.4
8	**1.3**	152.4	1.5	145.2	1.1	138.0	0.8	130.8	0.5
9	**1.7**	195.6	1.5	178.8	1.1	162.0	0.8	145.2	0.5
10	**2.5**	282.0	1.7	246.0	1.5	210.0	1.2	174.0	0.9

(a) The solute distribution coefficient, K_d, used to calculate the retention volume, V_r, using $V_r = V_T + (K_d - 1) V_S$. The machine volume, V_T, is 120 ml.
(b) R_s is the resolution factor between Compound N and Compound $(N - 1)$.
(c) Solute 1 is a dead-volume marker.
(d) Solute 7 partitions equally between the mobile and the stationary phase; its retention volume is equal to the machine volume, whatever is the volume V_S.

It is easy to express Fig 2.2 in mathematical terms by deriving the retention equation (Eq. 1.1):

$$dV_r/dK_D = Sf \tag{2.1}$$

Equation (2.1) shows that the difference in retention volume for two compounds with the same difference in distribution coefficient, K_D, is directly proportional to the volume of stationary phase retained in the machine.

2.4.2 Machine transposition

A big advantage of the CCC technique is its capability of reproducing a separation on any machine. This is due to the fact that, if the same biphasic liquid system is used, the same mobile phase is used with the same stationary phase. There is no such problem as batch-to-batch stationary phase reproducibility as seen in HPLC. However, this does not mean that the same chromatograms will be obtained using different machines. As explained previously, the machine characteristics, internal volume (V_T), efficiency (N), and ability to retain the stationary phase (Sf), are very important and affect the solute separation and resolution. In a machine with a volume V_C, the peak position, V_r, is fully predictable once the solute partition coefficient, K_D, and the stationary phase retention ratio, Sf, are known:

$$V_r = V_C\,[1 + (K_D - 1)\,Sf] \tag{2.2}$$

Equation (2.2) is a rearranged form of Eq. (1.1). The peak-shape depends on the plate-number, N, which is machine-dependent. The machine efficiency (plate-number) depends on the machine itself (tube diameter, number of tubing loops in the spool(s), beta factor), the flow rate, and the biphasic liquid system used (viscosity, interfacial tension and temperature) [1,6].

Figure 2.3 illustrates machine transposition. The same ten-solute mixture used to prepare Fig. 2.2 (Table 2.2) was again used in four different machines. The first chromatogram of Fig. 2.3 was calculated for a 120 ml and 500-plate machine with a stationary-phase retention ratio of 80% (compare with Fig. 2.2). Transposing this separation on a small-volume machine (30 ml) produced the second chromatogram. Since this machine has a good stationary-phase-retention capability (Sf = 70%) and a high efficiency, the ten solutes are baseline-separated in a very short time (one hour with a 1 ml/min flow rate). Most often, separations obtained on a small-volume machine (analytical) are transposed on a large-volume machine (preparative scale). The third chromatogram of Fig. 2.3 was calculated for a 1-l machine with a good Sf factor (85%) but a limited efficiency (200 plates). Compounds 1 to 5 are baseline-separated. The resolution of Compounds 6 to 9 is only partial. The retention volumes are large. The retention time of Compound 9 is longer than 5 hours with a 5 ml/min flow rate. It is pointed out that Compound 10 did not elute after 6 hours. The amount of material that can be injected in the 1-l machine is ten times higher than can be injected on a 100 ml machine. The last chromatogram in Fig. 2.3 was calculated for a 100 ml machine used at a high flow rate that produced a good efficiency (1000 plates) but a limited stationary-phase retention ratio (Sf = 40%). Compound 10 eluted in 32 min and was baseline separated from the other compounds. However, the nine other compounds were poorly resolved. Compounds 1, 4, 6 and 8 are buried in the more concentrated compound peaks.

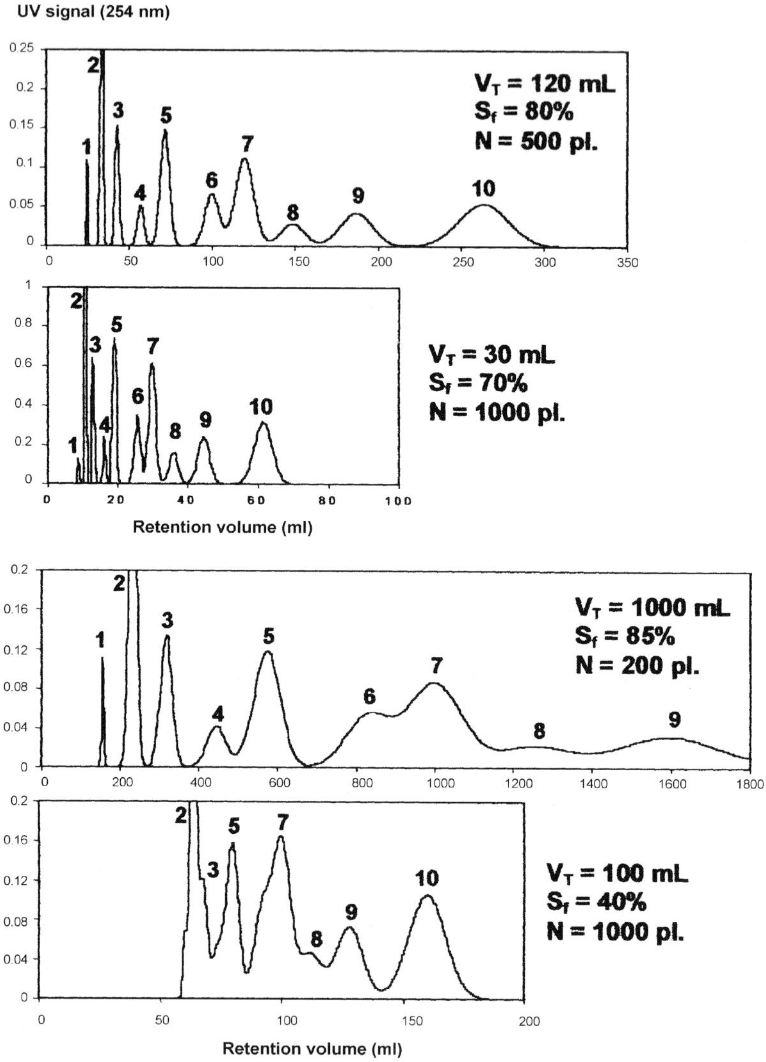

Fig. 2.3. Chromatograms of the same mixture of ten solutes obtained with the same biphasic liquid system on four different CCC machines.

Comparing Figs. 2.2 and 2.3 clearly shows that the most important parameter for maximizing the resolution power of a CCC machine is the amount of stationary phase that can be retained (*Sf* ratio). A machine capable of retaining a large amount of stationary phase has a resolution power superior to that of a more efficient machine (higher number of plates) but which retains a lower volume of stationary phase. As explained in detail in Chapter 6, the *Sf* ratio decreases when the mobile-phase flow rate increases. This means that there is always a trade-off between throughput (high flow rates but lower resolution) and purity (well separated peaks at lower flow rates).

2.5 WORKING WITH A WIDE POLARITY RANGE

2.5.1 Introduction

In CCC, as in HPLC, isocratic analysis gives a limited polarity window. Figure 2.4 shows the chromatograms obtained with a real sample and different isocratic mobile-phase compositions. Clearly, increasing the

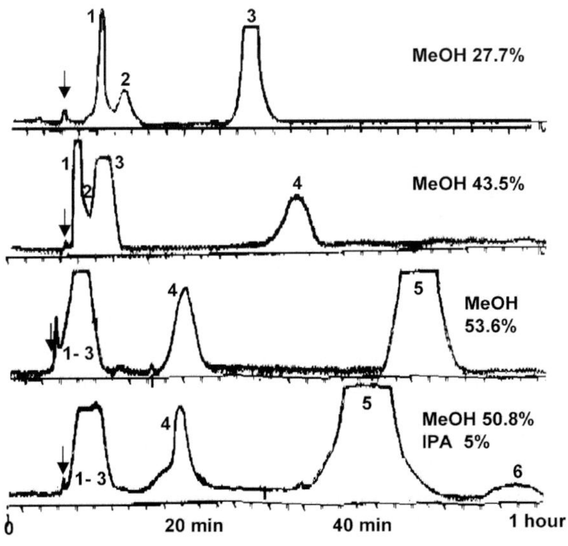

Fig. 2.4. Separation of a vegetal extract (80 mg injected) by a 120 ml hydrodynamic machine. Liquid system: hexane–EtOAc–MeOH–H_2O. Top chromatogram: 1.4:0.2:1.0:1.0 v/v. Second chromatogram: 1.4:0.2:2.0:1.0. Third chromatogram: 1.4:0.2:3.0:1.0. Bottom chromatogram: 1.4:0.2:3.0:1.0, with 0.3 v/v isopropanol added. Mobile phase: aqueous lower phase, head-to-tail direction, 2 ml/min flow rate. Numbers refer to peaks that are not necessarily pure compounds. The arrows show the solvent front (V_M volume).

eluent strength to elute highly retained compounds causes early-eluting compounds to co-elute close to the solvent front. The dual-mode and gradient-elution mode are alternative ways of using a CCC machine to separate a mixture of compounds with a wide range of polarities.

2.5.2 The dual-mode CCC chromatogram

First step: the V_{DM} volume

The liquid nature of the stationary phase allows one to fractionate a sample in a unique way: in a first step, the reversed-phase mode can be used and in a second step the normal-phase mode terminates the separation. There is no objection to doing the steps in the opposite way (normal-phase mode first). Figure 2.5 illustrates the procedure. During

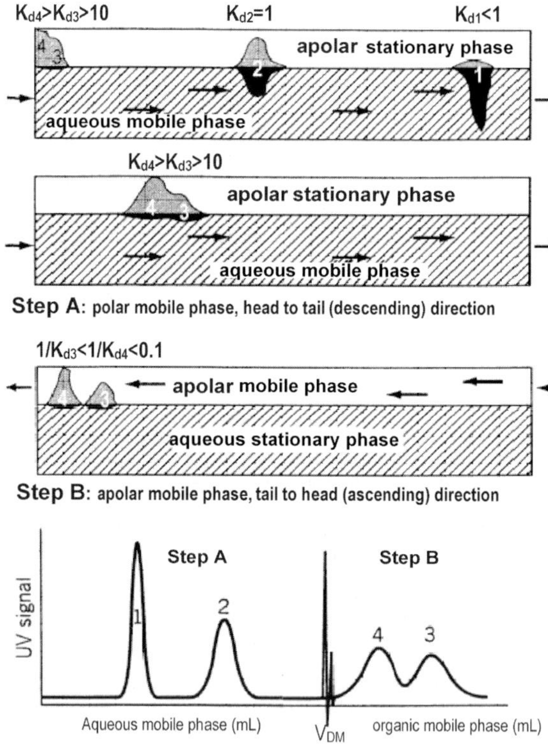

Fig. 2.5. Schematic view of a dual mode separation of four compounds with widely differing polarities: V_{DM} is the aqueous phase volume after which the phase roles were switched.

Step A, the CCC chromatograph is used in the reversed-phase mode. The mobile phase is the aqueous phase (polar), and the stationary phase is the organic apolar phase. The solutes move inside the column with a speed depending on the inverse of their partition coefficients—that is, very slow for solutes with high partition coefficients (Solutes 3 and 4). It would take too long a time to elute them. Using the advantage of a liquid stationary phase, the phase-role is switched after a volume of polar mobile phase, V_{DM}, is passed through the apparatus. The subscript DM is used for dual-mode. All the solutes with partition coefficients lower than:

$$K_{DM} = (V_{DM} - V_M)/V_S \tag{2.3}$$

left the column and were seen on the chromatogram (Solutes 1 and 2). The solutes with $K_D > K_{DM}$ are still in the column, sorted by decreasing K_D values, the lower K_D being closer to the column exit (Fig. 2.5).

Switching the phase role

During Step B, the new mobile phase is the liquid that was previously the stationary phase and it is pushed in the opposite direction. Solute 4, which moved slowest in step A, is the closest to what is now the column outlet. Solutes 3 and 4 have a high affinity for what is now the mobile phase ($K'_D = 1/K_D$). Solute 4 is the first to leave the column. All the solutes move at speeds proportional to their partition coefficients. It was demonstrated [9,10] that their retention volume, $V_{R(DM)}$, in Step B was:

$$V_{R(DM)} = V_{DM}/K_D \tag{2.4}$$

If the volume

$$V_{max} = V_{DM}/K_{DM} = V_{DM}\,V_S/(V_{DM} - V_M) \tag{2.5}$$

is eluted during Step B, all the solutes injected are eluted from the CCC column (Fig. 2.5). There is no need to pump more liquid phase—the CCC column does not contain any more injected solutes.

Selectivity

In Step A, the CCC column is used in the conventional way. The chromatographic selectivity is proportional to the stationary phase volume, V_S (Eq. 2.1). The selectivity in Step B is expressed by:

$$dV_{Step\,B}/dK_D = -V_{DM}/K_D^{\,2} \tag{2.6}$$

Equation (2.6) shows that if the V_{DM} volume increases, i.e., more compounds are eluted in the direct mode, Step A, the selectivity (distance between peaks) increases proportionately. However, the selectivity decreases quadratically with the magnitude of the distribution coefficient. In another way, Eq. (2.5) shows that if V_{DM} increases, the $V_{DM}/(V_{DM} - V_M)$ ratio tends toward unity being always higher than 1. The V_{max} volume tends to be slightly greater than the V_S volume. Pushing in Step B (Fig. 2.5) a volume V_{max} will elute all compounds, but the selectivity (and consequently the resolution) between peaks corresponding to hydrophobic compounds (very high K_D coefficient) may be low [11].

The selectivity obtained with the dual-mode use of CCC is not perfect. However, the absence of irreversible adsorption obtained using this mode, the total recovery, and the possibility of separating solutes covering a large spectrum of polarity are further big advantages obtained with the dual-mode use of the CCC technique [11].

2.5.3 Emptying a machine

Another way of extending the polarity range of compounds separated by CCC is again to exploit the liquid nature of the stationary phase to empty the machine by driving out all compounds having too strong an affinity for the stationary phase. The procedure is very simple and somewhat resembles the dual-mode method. Step A is identical to the dual-mode method (Fig. 2.5, top). When the desired mobile phase volume has been passed through the CCC machine, the rotor is stopped. The liquid stationary phase is no longer retained in the machine when more mobile phase is used to push it out of the machine. It was observed that there is little mixing of the bands of compounds formed inside the machine [6]. As in the dual-mode method, emptying all the stationary phase contained in the CCC machine ensures that no part of the injected sample remains trapped inside the machine. However, as with the dual-mode method, the selectivity is very low (Eq. 2.6). The compounds with very high K_D coefficients elute almost together.

Comparing the washing-off method and the dual-mode method, the former is easier to perform and the compounds elute in the order of increasing polarity (normal-phase mode) or of decreasing polarity (reversed-phase mode). However, continuous detection is not possible since the two phases elute together, and the machine should be re-loaded and equilibrated after the wash-off.

2.5.4 Gradient elution

Introduction

Progressively increasing the elution strength of the mobile phase allows a much broader range of polarities to be examined in one chromatogram. The choice of the gradient solvent in HPLC is easy because the stationary phase is not affected by the mobile-phase-composition changes. The situation is much more complex in CCC for several reasons, the principal reason being that any change in the composition of one phase of a biphasic liquid system may induce a change in the other phase (Fig. 2.1 and Chapter 3). Other reasons are that in most biphasic liquid systems used in CCC, more than two solvents are used. Also, the phase-density difference is an important factor in CCC, and may change during the gradient.

Modifying the composition of a biphasic liquid system

Figure 2.1 showed the heptane–methanol–water ternary phase diagram that has the property of producing a pure heptane phase for any mixture of the three solvents. Changes in the methanol content of the aqueous phase do not affect the heptane phase. This system can be used to perform gradient elution. Unfortunately, this system is not very useful in CCC, owing to the extreme difference in the heptane phase's low polarity and aqueous phase's high polarity.

The Arizona system (hexane or heptane–ethyl acetate–methanol–water systems, see Chapter 3) is much more useful. It gives a methanol–water lower phase and an ethyl acetate–heptane organic upper phase. However, when the methanol content increases, ethyl acetate tends to solubilize more and more in the aqueous phase. Figure 2.6 (top) shows the change in the phase volumes obtained when methanol is added to an initial hexane–ethyl acetate–water, 12.5:37.5:25 v/v/v, mixture. The first 10 ml of methanol go in the lower aqueous phase, and the upper organic phase is not affected. But when 20 and 30 ml of methanol are added, most of the ethyl acetate is extracted in the lower polar phase. The upper-phase volume decreases dramatically. Next, when more methanol is added, hexane itself become soluble in the lower phase. The binodal line is crossed when 120 ml of methanol (61.5% v/v) is added to the initial mixture. A clear monophasic mixture is obtained.

Figure 2.6 (bottom) shows the same experiment done with hexane–chloroform–methanol–water. In this case, the addition of methanol to the aqueous phase makes chloroform more soluble in it. Since chloro-

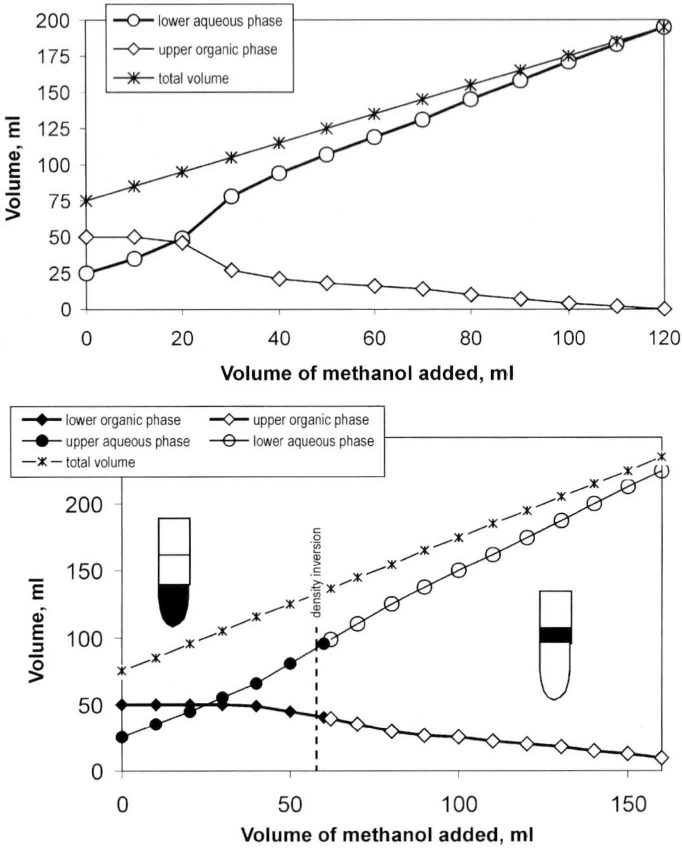

Fig. 2.6. Change in the upper and lower phase volumes upon methanol additions. Top: system hexane–EtOAc–MeOH–H_2O; initial mixture, hexane 12.5 ml, EtOAc 37.5 ml, H_2O 25 ml. Bottom: system hexane–$CHCl_3$–MeOH–H_2O; initial mixture, hexane 25 ml, $CHCl_3$ 25 ml, H_2O 25 ml.

form is denser than water, as it changes phase a density inversion occurs. Initially the hexane–chloroform organic phase is denser than the aqueous phase. When 61 ml methanol is added to the initial mixture (44.9% v/v), enough chloroform is extracted in the aqueous phase so that its density becomes higher than that of the organic phase. The upper aqueous phase then becomes the lower aqueous phase. Obviously, if such a density change occurs during a gradient elution in CCC, the "column" will be ruined.

A method for optimizing a gradient in CCC

Literature search: If one is looking for known components, a literature search for published CCC methods should first be carried out. Often, the solvent system can be found this way. If the literature search is negative, or the components are unknown, experience has shown that there is a correlation between the elution characteristics of a 0–100% acetonitrile HPLC gradient and the polarity of components. Hence, it is possible to propose a rapid method for optimizing the CCC solvent system most suitable for separating a particular sample [12].

A, B and C solvent classes: Three classes of biphasic liquid systems have been defined. Solvent systems A are highly polar biphasic systems such as the aqueous/aqueous systems: water + polymer (PEG or dextran)–water + salts (potassium or sodium phosphates) or the water–butanol–methanol systems. Solvent systems B, such as the Arizona systems (heptane– or hexane–ethyl acetate–methanol–water) have medium polarity. Solvent systems C have a non-polar phase and a polar phase (heptane– or hexane–methanol, or acetonitrile–water systems).

The HPLC screening test: Next, the sample is injected into an HPLC system with a 15 cm C18 column set to run a gradient going from 100% water to 100% acetonitrile in one hour at 1 ml/min flow rate. The sample's peak position is examined. If most of the peaks show retention times lower than 15 min, it means they are eluted by mobile phases containing less than 25% acetonitrile. CCC using the very polar Solvent system A will fractionate them. If the peak retention times are within 15 and 50 min, it means the compounds have intermediate polarities. Solvent system B will be selected for the CCC separation. If most of the peaks elutes after 50 min (with more than 80% acetonitrile in the mobile phase) solvent system C will be selected for the CCC separation. The chromatogram obtained also gives an idea of the sample's complexity.

Optimizing partition coefficients: Once the polarity of the sample is bracketed, thin layer chromatography or a test-tube-shake partition test should be performed to evaluate the distribution of the sample components in the various solvent mixtures belonging to the selected system. Distribution coefficients between 0.5 and 2 are desirable. It should be kept in mind that the ionization state of the compounds can dramatically change the solute's K_D value [13]. The mobile phase's pH should be controlled for ionizable solutes.

The gradient solvent: The last step is to choose the solvent whose composition will be varied during the gradient. This is always a solvent that partitions well between the two immiscible phases. Thus, it is methanol in the butanol–methanol–water, hexane– or heptane–methanol–water and hexane– or heptane–ethyl acetate–methanol–water systems. In the Arizona systems, it could be also ethyl acetate. Also, the gradient solvent will be the one that produces a reduction in the stationary phase volume (Fig. 2.6). Otherwise, if the stationary phase volume increases during the gradient elution, wash-off of stationary phase will be observed. This may be detrimental for the separation, and may render the continuous detection difficult or impossible.

Linear gradient versus step gradient

The mobile phase's composition can be changed linearly using two pumps. The two mobile phases are prepared and saturated with their respective stationary phases. Linear gradients have been used successfully in many applications [14], including one-step isolation of a novel bio-active compound from an untreated plant extract [12]. When CCC is used to purify a compound, it is difficult to extrapolate from the gradient separation to the isocratic composition that would best separate the compound from its impurity. This is a consequence of the continuous change in stationary phase composition during the gradient separation.

A step-gradient is performed using a single pump. The different mobile phases are all prepared in advance and saturated with their respective stationary phases. After the desired volume of the first mobile phase has been used, the pump's inlet filter is plunged into the second mobile phase for the desired time, and so on. This method limits stationary-phase wash-off and adds some versatility, since an extra solvent can be added into the final mobile phase to optimize a separation.

2.6 TWO PRACTICAL EXAMPLES

Chapter 8 is dedicated to the CCC separation of natural products and Chapter 10 presents the use of CCC in the pharmaceutical industry. The two examples given now are also isolations and purifications of vegetal bioactive compounds by CCC, and are presented to focus on the practical use of the CCC technique.

2.6.1 Purification of active alkaloids from a Brazilian plant extract

In the search for antioxidant agents from Brazilian natural sources, 71 Brazilian plant extracts belonging to sixteen species of five different families were screened [15]. The ethanol extract from the leaves of *Lantana trifolia* was found to be very active. It was partitioned between water and various organic solvents. The solvent that extracted all active principles was dichloromethane. Fractionation of the CH_2Cl_2 extract by traditional chromatographic methods, such as silica gel open-column chromatography combined with preparative reversed-phase HPLC, failed. About 40% of the crude extract binds irreversibly to the silica, which means that a significant amount of important bioactive substances may be lost. Accordingly, the use of CCC was evaluated.

The CCC gradient separation

Search for the liquid system: A small amount of the crude CH_2Cl_2 extract was dissolved in each of five different test tubes containing the biphasic solvent system composed of hexane–ethyl acetate–methanol–water (1:1:X:1, v/v/v/v), where X = 0.1 (System I); 0.5 (System II); 1.0 (System III); 1.75 (System IV), and 2.5 (System V). The test tubes were shaken and the compounds allowed to partition between the two phases. The results were visualized by TLC (Fig. 2.7). For each solvent system, two spots were applied, one for the upper organic phase and one for the lower aqueous phase. With solvent system I (X = 0.1 or 3.2% v/v MeOH) most of the compounds are located in the upper organic phase. With solvent

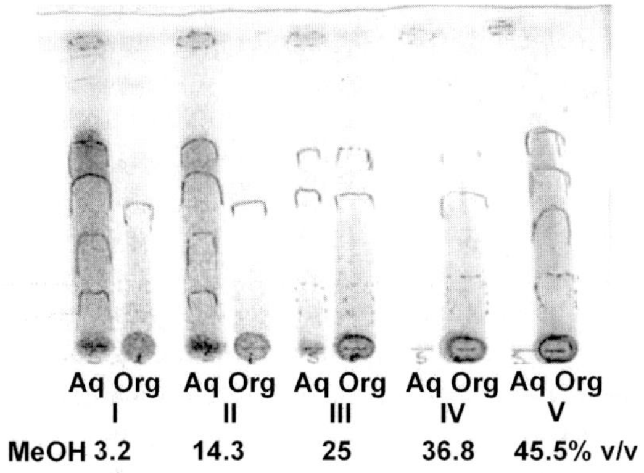

Fig. 2.7. Thin layer chromatography plate showing the partitioning of the components of a dichloromethane extract of *Lantana trifolia* between the lower aqueous phase and the upper organic phase of the hexane–ethyl acetate–methanol–water systems, 1.0:1.0:X:1.0, v/v, with X = 0.1 (System I), X = 0.5 (System II), X = 1.0 (System III), X = 1.75 (System IV) and X = 2.5 (System V). The methanol content in the system is indicated in % v/v. Silica plates, pure ethyl acetate as eluting solvent.

system V (X = 2.5 or 45.5% v/v MeOH), most of the compounds are located in the lower aqueous phase. An intermediate situation is found for solvent system III (X = 1.0 or 25% v/v MeOH), where the compounds seem to be equally partitioned between the two phases (the $K_D = 1$ situation). This study shows that, as the amount of MeOH increases in the solvent system, more compounds are extracted from the upper organic phase and driven into the lower aqueous phase. The sequence went from System I to V, and a step gradient was used. The mobile phase was the aqueous lower phase. The stationary phase was the organic upper phase of System I.

The separation: A PC Inc. hydrodynamic machine (V_T = 80 ml, 833 rpm) was used. The machine was equilibrated using System I: 65 ml of the upper organic phase was retained by the machine with a 833 rpm speed, and 2 ml/min aqueous mobile-phase flow rate in the head-to-tail direction (Sf = 81%, V_M = 15 ml). Once the machine was equilibrated, 100 mg of the dry CH_2Cl_2 extract was injected, dissolved in 5 ml of an equal amount of the two phases of System I. Figure 2.8 (top) shows the step gradient used along with the experimental time at which the mobile phases were changed. Fractions of 4 ml (2 min) were collected; the fraction numbers are indicated in Fig. 2.8. A larger number of samples was collected when the System III aqueous phase was used, since the compounds partition best in this system (Fig. 2.7). After 60 fractions were collected (2 h), the rotor rotation was stopped and all retained compounds were collected in 4-ml fractions by washing off the machine content.

After the experiment was completed, the fractions were analyzed by TLC. Figure 2.8 (bottom) summarizes the results. The first spot refers to the total extract, just for comparison. It can be seen that the step-gradient system used allowed the separation of two alkaloids (fractions 20–21, and 29–31). The latter alkaloid was in a pure form. These two compounds could not be separated by HPLC. Also, in fractions 22–23, another pure non-alkaloid compound was isolated. As expected, the more polar compounds eluted first, leaving the less polar ones to elute later.

2.6.2 From a crude plant-extract to pure flavonoids in two CCC steps

Flavonoids are phenolic compounds from plants that can be found in free and glycosylated forms. Plant extracts containing this kind of compound

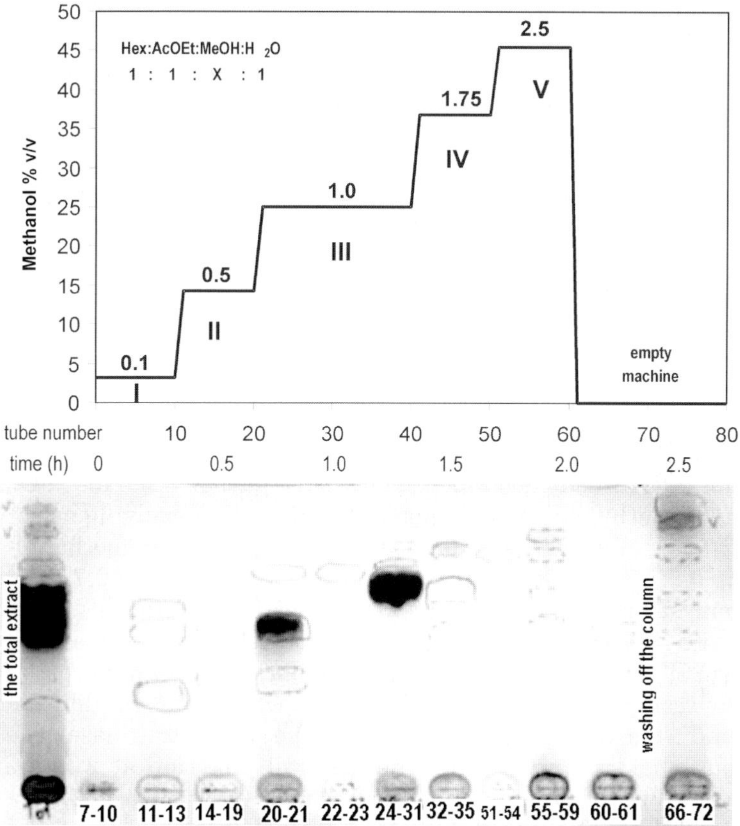

Fig. 2.8. CCC Separation of a dichloromethane extract of *Lantana trifolia*. Silica plate, pure ethyl acetate eluting solvent. Top: the step gradient used. Bottom: TLC analyses of the aqueous mobile phase fractions. PC Inc machine, 80 ml, 833 rpm. The spots were revealed by spraying with Dragendorff's reagent.

have great importance since this class of secondary metabolites displays many biological and pharmacological properties. Extracts that contain free and glycosylated flavonoids have a wide range of polarity, thus perfectly suiting them to the gradient elution. Previous investigation on the chemistry of the genus *Siparuna* (Siparunaceae) showed that these plants are rich in free and glycosylated flavonoids [16]. A crude ethyl acetate extract from the leaves of *Siparuna guianensis* was prepared.

First step: the CCC gradient separation

Search for the liquid system: The extract was submitted to the acetonitrile HPLC screening test. The result indicated that the compounds in this extract were eluted by mobile phases containing between 25 and 75% acetonitrile. This is the region of medium to high polarity range, suggesting to us the use of a System B solvent mixture such as the quaternary solvent system hexane–ethyl acetate–methanol–water. The HPLC chromatogram showed that the extract is composed mainly of three flavonoids that were eventually identified as quercetin, a free flavonol (3,5,7,3′,4′-penta-hydroxyflavone), and two quercetin derivatives: 3-*O*-β-D-glucopyranosyl (6″→1‴)-rhamnosyl-, and 7-*O*-β-D-glucopyranosyl (6″→1‴)-rhamnosyl-. Traces of kaempferol monoglucosides were also identified.

The amounts of the three solvents, hexane, ethyl acetate and methanol in the X:Y:Z:1 systems had to be optimized using the test tube + TLC method already described. The results gave $X = 0.6$ for hexane and $Y = 4.0$ for ethyl acetate. Compound partitioning evolving as shown in Fig. 2.7 was obtained with the hexane–ethyl acetate–methanol–water system 0.6:4:Z:1 v/v/v/v, with Z varying between 0.05 and 0.7. A two-step gradient starting with the 0.6:4:0.05:1 system (System I′) and ending with the 0.6:4:0.7:1 hexane–ethyl acetate–methanol–water system (System II′) should separate the components of the sample.

The separation: The 80 ml PC Inc machine was equilibrated with System I′. The upper organic phase was the stationary phase. The lower aqueous phase pumped in the head-to-tail direction at 2 ml/min displaced 13.5 ml of organic phase at 850 rpm (V_S = 66.5 ml, Sf = 83%). About 200 mg of the ethyl acetate extract from leaves of *Siparuna guianensis* were injected, dissolved in 2.5 ml of each of the two phases of System I′ (5 ml injected). Fractions of 4 ml were collected. The lower phase of System I′ was pumped for 20 min (tubes 1 to 10), next that of System II′ was pumped for 30 min (tubes 11 to 26). Then the rotor rotation was stopped and the retained compounds were also collected in 4 ml fractions ("wash-off").

Figure 2.9 (top) shows the content of the fractions after TLC analysis. It can be seen that the flavonoid-rich extract was separated into two main different-polarity fractions, one composed of the mixture of more polar derivatives (glycosylated flavonoids, tubes 10–12) and the other composed mainly of pure quercetin (tubes 34–44). As a reversed-phase two-step gradient was run, the more polar compounds elute first, whereas the least polar compounds remained on the organic stationary

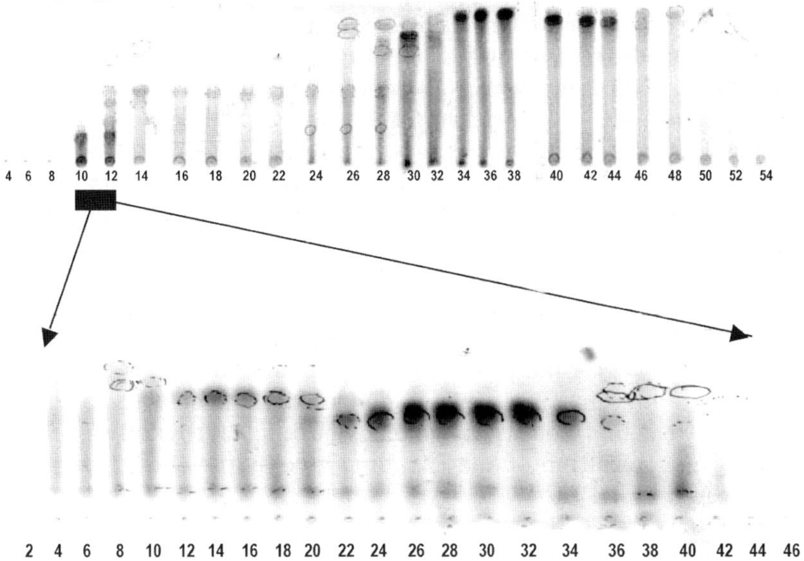

Fig. 2.9. TLC plates of the CCC fractionation of an ethyl acetate extract of *Siparuna guianensis* leaves. Top: TLC analyses of the fractions obtained after the two-step gradient CCC separation of the extract. Bottom: TLC analyses of the fractions obtained by isocratic CCC separation of the content of Tubes 10–12 of the first separation. Silica plates: eluent; top plates, ethyl acetate–acetone–water, 25:10:5, v/v; bottom plates, butanol–acetonitrile–water, 4:1:5 v/v. All plates were revealed by spraying with the Folin–Ciocalteu reagent for phenolic compounds.

phase. The compounds were identified by 1-H NMR. Small amounts of kaempferol were found in tubes 46–50.

Second step: isocratic CCC separation

Search for the liquid system: The gradient separation eliminated the apolar components of the extract. However, it did not separate the polar glycosylated flavonoids. Butanol is a polar solvent partly soluble with water. It improves the selectivity when polar compounds have to be separated by CCC [6,12]. Butanol was added to System I′. Test tubes containing the liquid systems hexane–ethyl acetate–butanol–methanol–water 0.6:4:X:0.05:1, v/v/v/v/v, were used to optimize the amount of butanol. With X = 1, the components of tubes 10–12 partitioned evenly between the two liquid phases. The normal-phase mode (mobile phase = upper organic phase) was selected to increase the resolution between the polar compounds.

The separation: The 80 ml PC Inc. machine was filled with the lower aqueous phase of the system hexane–ethyl acetate–butanol–methanol–water 0.6:4:1:0.05:1, v/v/v/v/v. The rotor was set at 850 rpm and the upper aqueous phase was pumped in the tail-to-head direction at 2 ml/min: 11 ml of aqueous phase was displaced (V_S = 69 ml, *Sf* = 86%). The separation was done for 52 min, collecting 26 fractions (4 ml each). Next, the rotor was stopped and the machine content was washed off in tubes 27–50 (4 ml each).

Figure 2.9 (bottom) shows the TLC plates obtained with the collected tube contents. Pure rhamnosyl-kaempferol was found in tubes 12 to 20, pure 7-*O*-β-D-glucopyranosyl (6″→1‴)-rhamnosyl-quercetin was found in tubes 24 to 31 and pure 3-*O*-β-D-glucopyranosyl (6″→1‴)-α-D-rhamnosyl–quercetin (rutin) was identified in tubes 32 to 35. The results show the high selectivity given by butanol on this solvent system, allowing the separation of the two di-*O*-glycosylflavonols, which were not separated by HPLC.

2.7 CONCLUSION

The most important thing to keep in mind when operating a CCC machine is that the maximum volume of stationary phase should be held by the "column" in order to obtain a good resolution power. The good results obtained in the two practical examples detailed were due to the wise choice of the solvent system and polarity adjustment, but also to the high *Sf* factors, higher than 80%, that were obtained with the liquid systems selected.

If the rotor rotation speed is increased, the stationary phase is held more tightly and *Sf* increases. This solution is only mentioned at the conclusion of this chapter because it is our experience that the occurrence of technical failures increases exponentially with the rotor rotation speed. Also many hydrodynamic machines generate high noise levels when the rpm value is increased. The technical problems are of greater and greater concern to the CCC industrial makers. Modern and recent machines are much more reliable than earlier ones.

REFERENCES

1 A. Berthod, *J. Chromatogr.*, 550 (1991) 677.
2 F. Abe and T. Yamauchi, *Chem. Pharm. Bull.*, 27 (1979) 1604.

3 K. Hostettmann, M. Hostettmann-Kaldas and O. Sticher, *Helv. Chim. Acta,* 62 (1979) 2079.
4 A. Berthod and M. Bully, *Anal. Chem.*, 63 (1991) 2508.
5 A. Berthod and B. Billardello, *J. Chromatogr. A*, 902 (2000) 323.
6 W.D. Conway, *Countercurrent Chromatography, Apparatus, Theory & Applications*. VCH Publishers, Weinheim, 1990.
7 A. Berthod, B. Billardello and S. Geoffroy, *Analusis*, 27 (1999) 750.
8 A. Berthod, C.D. Chang and D.W. Armstrong, In: A.P. Foucault (Ed.), *Centrifugal Partition Chromatography*, Chromatogr. Science Series, Vol. 68. Marcel Dekker, New York, 1995, Ch. 1, pp. 1–23.
9 R.A. Menges, G.L. Bertrand and D.W. Armstrong, *J. Liq. Chromatogr.*, 13 (1990) 3061.
10 S.J. Gluck and E.J. Martin, *J. Liq. Chromatogr.*, 13 (1990) 3559.
11 A. Berthod and B. Billardello, In: P. Brown and E. Grushka (Eds.), *Advances in Chromatography*, Vol. 40. Marcel Dekker, New York, 2000, Ch. 8, pp. 503–538,
12 L. Brown, D.E. Games, D. Wheatley and R.A. Whiteside, Lecture 776, presented at the Pittsburgh Conference in Orlando, FL, March 1999.
13 A. Berthod, S. Carda-Broch and C. Garcia-Alvarez-Coque, *Anal. Chem.*, 71 (1999) 879.
14 R.G. Whiteside, D.E. Games, C. Strawson, A.S. Graham and L. Brown, *J. Liq. Chromatogr. Rel. Technol.*, 24 (2001) 1801.
15 L.L. Mensor, F.S. Menezes, G.G. Leitão, A.S. Reis, C.S. Coube, T.C. Santos and S.G. Leitão, *Phytotherap. Res.*, 15 (2001) 127.
16 G.G. Leitão, S.S.V. Soares, T.B. Machado-Brito and F. Delle Monache, *Phytochemistry*, 55 (2000) 679.

Chapter 3

Solvent systems

Jean-Hugues Renault, Jean-Marc Nuzillard, Olivier Intes and Alexandre Maciuk

3.1 INTRODUCTION

In countercurrent chromatography (CCC), both stationary and mobile phases are liquids. The stationary phase is maintained by centrifugal fields, arranged in coiled tubes (Ito machines), or channels and ducts (centrifugal partition chromatography, CPC). CCC columns are parts of classical preparative chromatographic systems that include a preparative pump, an injection system (injection valve with preparative loop, or syringe pump for very large injection volumes), one or more detectors, and a fraction collector.

Initially, a CCC column must be considered as an "empty box" containing a long open tube (hydrodynamic apparatus) or a pack of engraved disks that form a succession of partition cells and interconnecting ducts (hydrostatic columns, CPC type). In preparative HPLC, a suitable chromatographic system must be conceived and adapted to the separation problem by selecting a stationary phase and the appropriate mobile phase. In CCC, the stationary phase and the mobile phase should be considered together. The column is filled with two equilibrated phases made of at least two solvents forming a biphasic system. Because both the stationary and mobile phases are liquids, fractionation of the sample is based on partition. The absence of a solid support, and the high diversity of available biphasic systems, make CCC a powerful and versatile tool for the separation of a wide variety of compounds such as secondary metabolites [1], peptides [2,3], lipids

Comprehensive Analytical Chemistry XXXVIII
A. Berthod (Ed.)

[4–6], cell organelles and blood cells [7], and metallic ions [8,9]. Moreover, the introduction of the displacement mode in CCC has opened new application fields for this technique, because it is very efficient for the purification of ionizable [10] or ionic molecules [11,12].

The selection of the appropriate biphasic liquid system requires consideration of various significant parameters:

- the development mode (elution or displacement) which conditions the values of suitable partition coefficients or distribution ratios (K_D) (*vide infra*);
- the solubility, the presence of ionizable functionalities (pH control) and the K_D values of the analytes;
- the rotation rate, flow rate, and pumping mode—either ascending (tail-to-head) or descending (head-to-tail). These instrumental parameters should be considered because a liquid system may be appropriate for one CCC machine and not for another.

In the preliminary optimisation step, a rational selection of suitable stationary and mobile phases is essential.

3.2 GENERALITIES ON BIPHASIC SOLVENT SYSTEMS

The problem of describing the chemical composition of the two phases of a biphasic liquid system is addressed first, in the absence of any analyte. The simplest biphasic system is made by mixing two immiscible solvents. When the equilibrium is reached, each phase is saturated with the other one and its composition fully determined by thermodynamics. When a wider choice of phase properties is needed, the presence of at least a third solvent is required.

Phase composition of ternary mixtures is described by a ternary phase diagram drawn as an equilateral triangle (Fig. 3.1). The three apexes correspond to the three pure solvents. The three sides of the triangle represent the three possible binary mixtures. A point in the inner region characterizes the global composition of the mixture. Most often, volume percentages are used, but mass percentages or molar fractions can be used, giving different diagrams for the same ternary system. Volume variations associated with mixing are ignored because they do not have a significant impact when these diagrams are exploited in the context of CCC.

Solvents are most often chosen so that two are immiscible (Solvents 1 and 3), and a third (Solvent 2) is miscible with only one (Type II

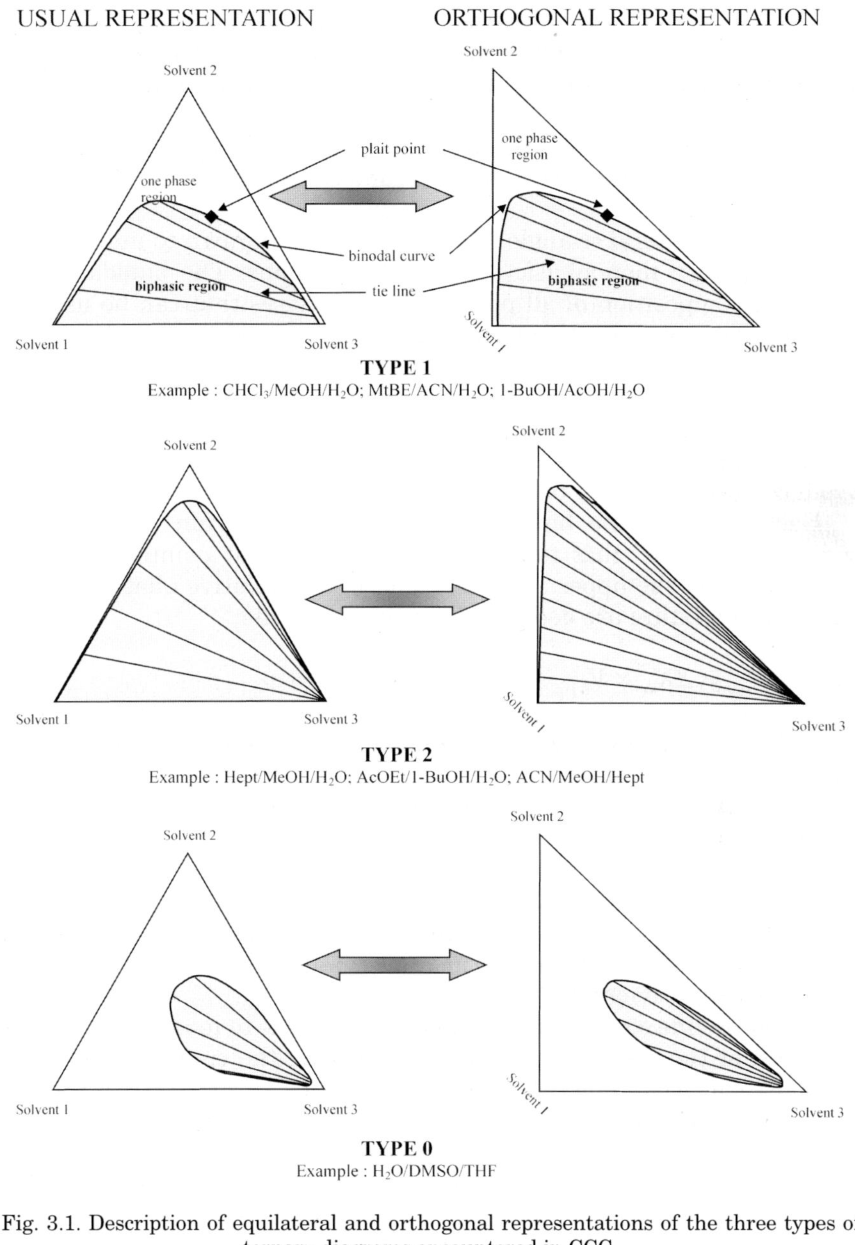

Fig. 3.1. Description of equilateral and orthogonal representations of the three types of ternary diagrams encountered in CCC.

systems) or both (Type I systems) of the other solvents (Fig. 3.1). The DMSO–THF–water system is, at the moment, the only Type 0 ternary biphasic system (Fig. 3.1) to be used in CCC [1]. In this unique system, all binary subsystems, DMSO–THF, DMSO–water, and THF–water, are monophasic. However, for certain compositions, two phases are formed, of which the lower one is rich in water, and the upper one is rich in THF and DMSO.

In the equilateral triangle, binodal curves are drawn to indicate the border between monophasic and biphasic regions. The binodal curve gives the composition of all possible liquid phases that can be used in CCC experiments with this particular ternary liquid system. A specific point of this curve, when it exists, is called the plait point. Close to this point the compositions of the two phases are nearly identical: the partition coefficient of all solutes tends toward unity. In CCC, the liquid system composition must be sufficiently far away from the plait point to avoid coalescence of the phases or mingling in the apparatus.

Each point in the biphasic region of the diagram belongs to a so-called "tie-line" that touches the binodal line at points giving the composition of the upper and lower phase. The relative phase volumes for a given mixture are deduced from the lever rule:

$$XX_{org} \times V_{org} = XX_{aq} \times V_{aq} \tag{3.1}$$

with XX_{org} being the tie-line length measured from the solvent composition to the binodal on the organic solvent side (Fig. 3.2B), V_{org}, the volume of organic phase, XX_{aq}, the length to the aqueous side (Fig. 3.2B), and V_{aq}, the volume of the aqueous phase.

The tie-lines give the following information:

- The composition of the conjugate phases, X_{aq} and X_{org} (Fig. 3.2B). Therefore, two conjugate phases can be independently prepared, provided that they belong to the same tie-line. This property is useful for solvent economy.
- When tie-line convergence occurs, it is possible to modify the composition of the other phase without changing the composition of the phase corresponding to the converging point. This property is used in the gradient elution mode (see Chapter 2 and thereafter).
- The slope of the tie-lines indicates the phase into which the blending solvent prefers to go: the tie-lines converge towards the opposite phase. This information is useful in the solvent-system-optimization step of the "best solvent strategy" (see Section 3.3.2.).

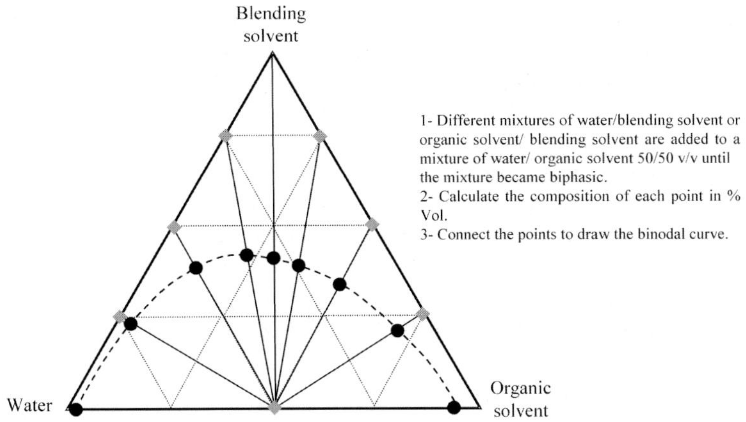

A- Binodal curve construction

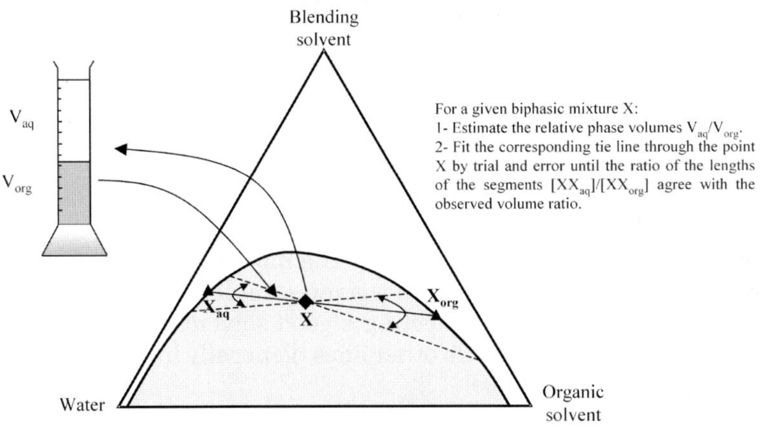

B- Tie-line construction

Fig. 3.2. A simple methodology for designing a ternary diagram.

Figure 3.2 describes an easy method to prepare and draw a ternary phase diagram. After the binodal construction, some tie-lines are drawn experimentally using a transparent template, as indicated. The volume measurement method for estimating the tie-line position is very practical since only a graduated cylinder is required. Gas chromatography or another analytical method can be used to determine more accurately the phase compositions and the tie-lines.

The representation of a ternary system by means of an equilateral triangle is somewhat difficult to manipulate. The orthogonal representation is also widely used. If the square angle represents pure solvent 1, the two orthogonal segments represent the content in solvents 2 and 3. The content of solvent 1 is then obtained by subtraction (Fig. 3.1).

3.3 STRATEGIES FOR SOLVENT-SYSTEM SELECTION

3.3.1 General comments

Numerous solvent combinations are available. Three, four, or five solvents can be mixed to modulate the upper and lower liquid-phase properties. Selectivity modulation is limitless. This is the major advantage of CCC; it can also be a disadvantage if some rules are not followed when selecting the appropriate solvent system to perform the separation. Solvent selection is rationalized in order to produce rapid results in routine laboratory experiments and the main strategies for solvent selection are reported in this chapter.

3.3.2 The "best solvent" approach

The best solvent approach is certainly the most intuitive way to find a biphasic solvent system adapted to a given sample [13,14]. The biphasic solvent system is chosen by selecting a good solvent (the so-called best solvent) for the sample and two other ones (generally immiscible) having strongly different polarities.

Type 0 and Type 1 ternary systems

In these types of ternary biphasic systems, the "best solvent" is partitioned between the most and the least polar solvents, causing the different analytes to split between the two phases. It appears that this strategy is particularly adapted to the isocratic elution mode (where the partition coefficient of the principal component in the injected sample should be close to 1), and to the dual mode.

Steps in solvent selection:

1. First gather as much information as possible on the sample. Does it contain ionizable, unstable or non-UV adsorbing compounds? If so, the pH and ionic strength will have to be adjusted. A specific detector may be needed.

2. Next, choose a solvent in which the sample is highly soluble. Considering the fact that the CCC techniques are mainly used for preparative purposes, this implies that the injected solution may contain a significant mass proportion of sample. Due to its function in the biphasic system, the best solvent is also referred to as the *blending* solvent or the *bridge* solvent.

3. Next, find two solvents (one more polar, and one less polar than the best solvent) which will form a biphasic liquid system when various amounts of the best solvent are added. Tables 3.1 and 3.2 give selected examples of mixtures that generate biphasic systems. This choice strongly influences the selectivity of the system (which cannot be predicted *a priori*).

4. Finally, the ternary diagram gives useful information for modulating the proportion of the best solvent in the system, and thus for modifying the analyte partition. Ideally, the less polar compounds of the sample should go preferentially into the less polar phase, and the more polar ones into the more polar phase. The tie-line slopes indicate the phase into which the best solvents (generally represented on the vertical axis in the orthogonal representation) will preferentially concentrate. A modulation of the best solvent proportion in the biphasic system permits one to adjust the K_D value of a compound only if the best solvent concentrates in the phase not preferred by the compound. We shall consider a ternary biphasic system constituted of three solvents, in which Solvent 2 is the best or bridge solvent. The two liquid phases are respectively referred to as 1 and 3-rich phases. If a compound prefers the 3-rich phase, its K_D coefficient will change when the amount of Solvent 2 is increased only if Solvent 2 prefers the 1-rich phase: the sample mimics the behavior of the bridge solvent (Solvent 2) (Fig. 3.3). In order to measure the K_D value of a compound, it is necessary to dissolve it in a small quantity of solvent mixture, M, belonging to a tie-line far from the plait point: the lowest one, for example. It will partition between the 1-rich and 3-rich phases. Determination of K_D can be easily monitored by thin layer chromatography (see Chapter 2).

TABLE 3.1

Liquid systems with three solvents used in CCC

Less polar solvent	Best solvent	More polar solvent
	Aqueous systems	
n-Heptane*	MeOH	Water
Toluene*		
$C_2H_4Cl_2$, $CHCl_3$		
EtOAc		
1-BuOH		
1-BuCN		
n-Heptane*	EtOH	
$CHCl_3$		
MTBE		
EtOAc		
1-BuOH, etc.		
Toluene	1-PrOH	
1-BuOH, 2-BuOH		
n-Heptane	2-PrOH	
Toluene		
$CHCl_3$		
diisopropyl ether		
Cyclohexane, *n*-Heptane*	1-BuOH	
$CHCl_3$*	2-BuOH	
MTBE*		
EtOAc*		
Toluene		
Cyclohexane, *n*-Heptane	CH_3COCH_3	Water
MTBE		
Toluene		
$CHCl_3$		
MIBK		
EtOAc		
Toluene	MeCN	
MTBE		
MIBK		
EtOAc		
n-Heptane	THF	
$C_2H_4Cl_2$, CH_2Cl_2, $CHCl_3$		
DMSO		
$C_2H_4Cl_2$*, $CHCl_3$*	HCOOH	

Less polar solvent	Best solvent	More polar solvent
Cyclohexane, *n*-heptane* Toluene MIBK MTBE EtOAc $C_2H_4Cl_2$, $CHCl_3$ iso-AmOH, 1-BuOH	Acetic acid	Water
Cyclohexane*, *n*-heptane*	MEK	
Toluene $CHCl_3$	DMF	
THF	DMSO	
Toluene	Pyridine	
	Non-aqueous systems	
Cyclohexane, *n*-heptane	Toluene	ACN
Cyclohexane, *n*-heptane	*n*-BuOH 2-BuOH	
Cyclohexane, *n*-heptane	MeOAc	
Cyclohexane, *n*-heptane	*n*-PrOH	
Cyclohexane, *n*-heptane*	EtOH	
Cyclohexane*, *n*-heptane*	MeOH	
Cyclohexane, *n*-heptane	$CHCl_3$	
Cyclohexane, *n*-heptane	Toluene	DMSO
Cyclohexane, *n*-heptane	THF 2-PrOH EtOH	MeOH
Perfluorohexane	*n*-Hexane	Toluene

*Possibility of gradient elution.

5. After step 4 is optimized, if the separation still needs to be improved, then all the classical parameters of analytical chemistry (pH, ionic strength, nature of added salts, complexing agents) can be varied, as in conventional LC chromatography. The ascending and descending modes must be tested, as well.

Type II ternary systems

In Type II systems, the best or bridge solvent notion does not exist (Fig. 3.1). Consider the biphasic ternary system in which Solvent 1 and Solvent 2 are miscible together, and Solvent 3 is not miscible with the

TABLE 3.2

Biphasic liquid systems useful in CCC, sorted by increasing polarity

Polarity	No.	Solvents system	Composition
Lipophilic	1	Hexane–ACN	1:1
Lipophilic	2	Hexane–ACN–$CHCl_3$	5:5:1
Lipophilic	3	Hexane–EtOH–H_2O	6:5:1
Semi-polar	4	Hexane–EtOAc–ACN–MeOH	5:2:5:4
Semi-polar	5	Hexane– EtOAc–MeOH–H_2O	1:1:1:1
Semi-polar	6	$CHCl_3$–MeOH–H_2O	13:7:2
Semi-polar	7	$CHCl_3$–MeOH–H_2O	1:1:1
Semi-polar	8	$CHCl_3$–MeOH–H_2O	7:13:8
Polar	9	Toluene–ACN–H_2O–EtOH	3:4:3:2
Polar	10	$CHCl_3$–MeOH–H_2O (AcOH 0.2 *M*)	1:1:1
Polar	11	EtOAc –EtOH–H_2O	2:1:2
Polar	12	1-BuOH–AcOH–H_2O	4:1:5
Polar	13	1-BuOH–EtOAc–H_2O	4:1:4

ACN = Acetonitrile, AcOH = acetic acid, EtOH = ethanol, EtOAc = ethyl acetate, MeOH = methanol, 1-BuOH = 1-butanol. Heptane, as a less toxic solvent, can replace hexane with minimum change.

other two solvents. A sample can be fractionated by this system only if its solubility in Solvent 3 is higher than that in Solvent 1 but lower than that in Solvent 2 (Fig. 3.4). If the liquid system is used in the isocratic mode, it is necessary to find the right proportion of Solvents 1 and 2 that give K_D values around unity for most compounds in the sample. The compounds partition between the 3-rich phase and the (1+2)-rich phase. Type II systems permit operation in the gradient elution mode (Table 3.3). For example, ethyl acetate–1-butanol–water was used to separate a mixture of glycosides [15–17]. These molecules prefer water rather than ethyl acetate, and *n*-butanol rather than water. The aqueous phase is used as stationary phase and the ethyl acetate-rich phase is the initial mobile phase. A polarity gradient is formed by increasing the *n*-butanol content in the mobile phase. The glycosides are successively "extracted" from the stationary phase as the concentration of the best solvent, *n*-butanol, increases in the ethyl acetate mobile phase, without

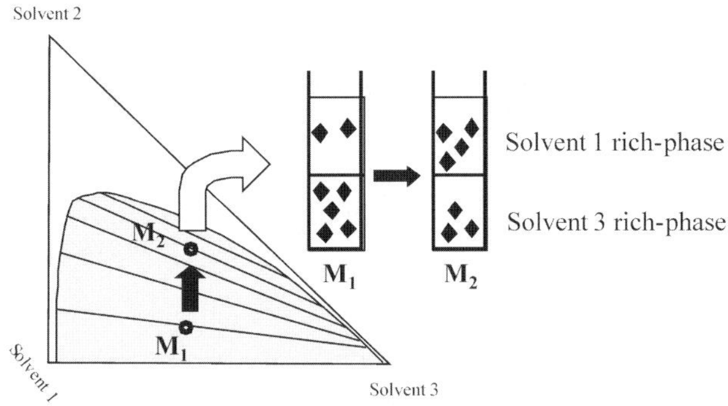

Fig. 3.3. In this scheme, the bridge solvent (Solvent 2) partitions preferentially in the Solvent 1 rich phase and the analyte prefers Solvent 3. Increasing the bridge solvent concentration (M_1 to M_2) allows one to adjust the K_D partition coefficient values.

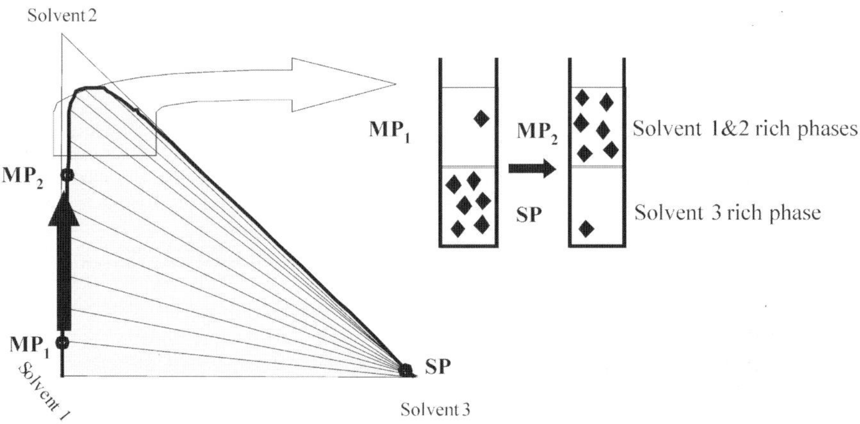

Fig. 3.4. Gradient elution run with a Type II solvent system. The analyte has more affinity for Solvent 3 than for Solvent 1, and more affinity for Solvent 2 than for Solvent 3. The Solvent 3 rich phase is used as stationary phase and the Solvent 1 rich phase as initial mobile phase (MP_1). A polarity gradient is formed by increasing the Solvent 2 proportion in the mobile phase (MP_2).

significantly changing the aqueous stationary phase. In this case, the "best" solvent is not a "bridge" solvent.

The "best solvent" approach is easy to use, but has some limitations:

- the selectivity varies only slightly for a given ternary solvent system, and thus many solvent systems must be tested;

TABLE 3.3

Examples of solvent systems used in the separation of natural molecules

Isolated compound(s)	Solvent system (v/v/v)	Development mode
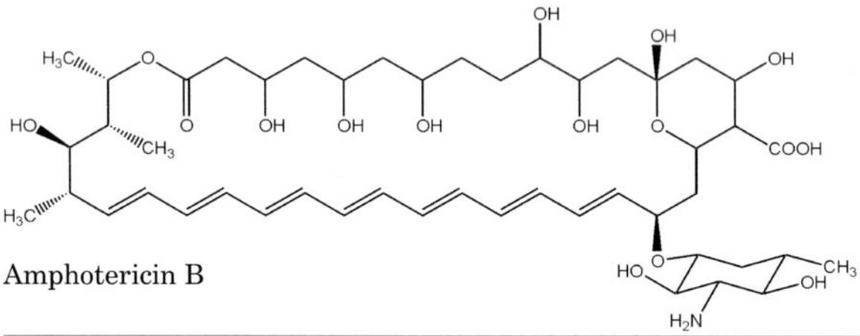 Amphotericin B	THF–DMSO–water (24.6:16.2:59.2)	Isocratic elution (ascending mode)
Fatty acids, mono-, di-, and tri-glycerides, glycosphingolipids	Hexane–ACN or hexane–MeOH–water (1:0.9:0.1)	Isocratic elution (ascending or dual mode)
Terpenoids	Hexane–ethyl acetate–nitromethane–water (9:2:2:3)	Isocratic elution (DCCC)
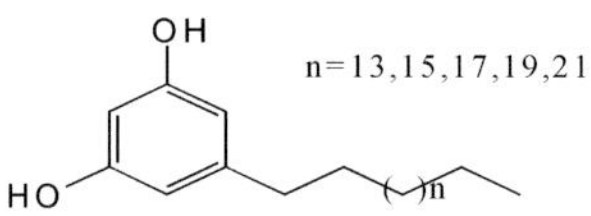	Heptane–MeOH	Isocratic elution (ascending mode)
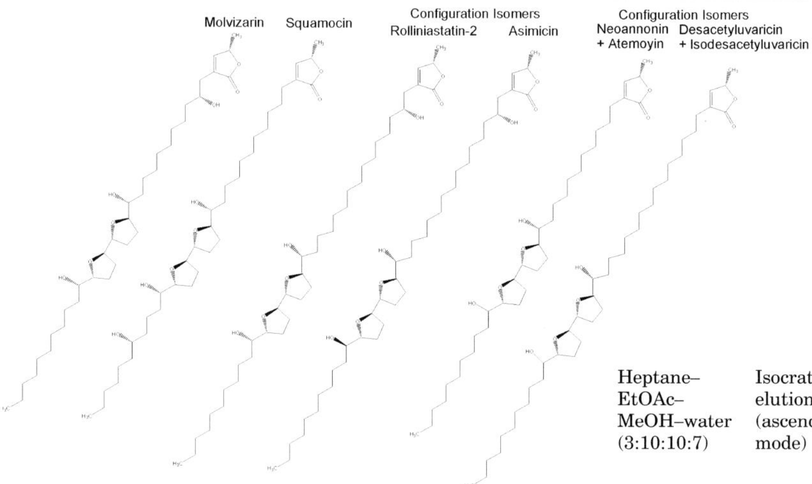	Heptane–EtOAc–MeOH–water (3:10:10:7)	Isocratic elution (ascending mode)

TABLE 3.3 (continuation)

Isolated compound(s)	Solvent system (v/v/v)	Develop-ment mode
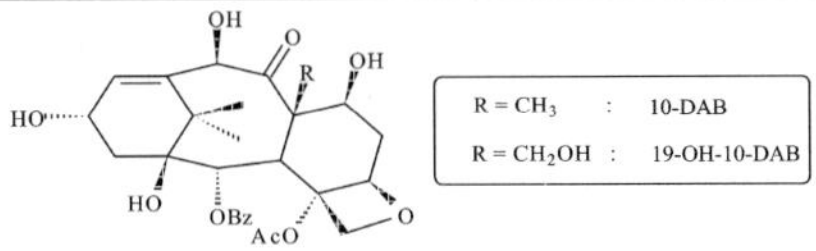	MIBK–acetone –water (2:3:2)	Dual mode elution
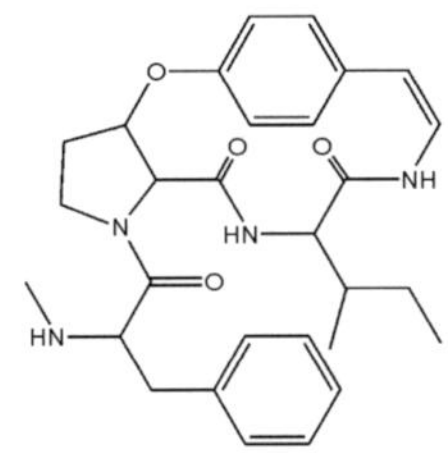 lotusine D 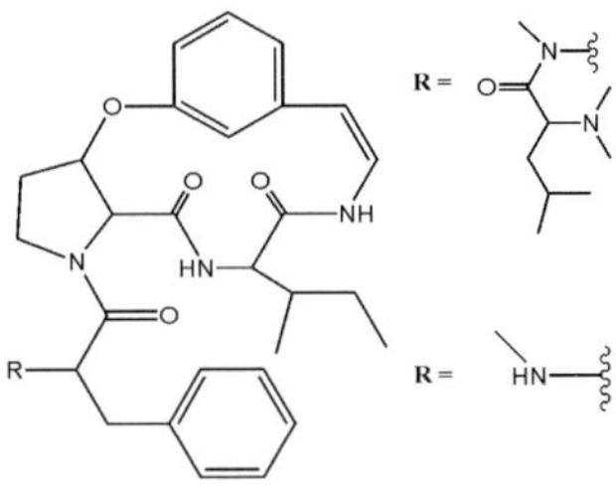 lotusine E, lotusine F	Heptane–EtOAc–MTBE –MeOH–ACN–water (1:0.5: 0.5:0.5:0.5:1)	Dual mode elution

Anthocyanins	R_1	R_2	R_3
Malvidin 3-glucoside	OCH_3	OCH_3	β-D-glucose
Peonidin 3-glucoside	OCH_3	H	β-D-glucose
Pétunidin 3-glucoside	OCH_3	OH	β-D-glucose
Cyanidin 3-glucoside	OH	H	β-D-glucose
Delphinidin 3-glucoside	OH	OH	β-D-glucose

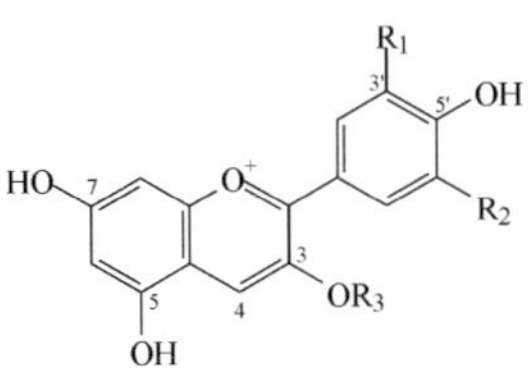	EtOAc–1-BuOH–water	Gradient elution (ascending mode)

TABLE 3.3 (continuation)

Isolated compound(s)	Solvent system (v/v/v)	Development mode
Dammarane saponins	EtOAc–1-BuOH–water	Gradient elution (ascending mode)
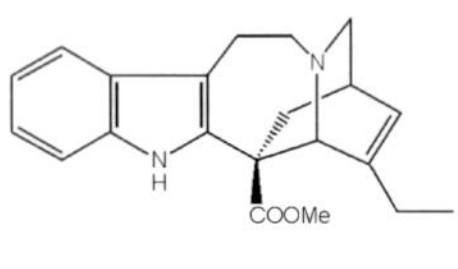 catharanthine 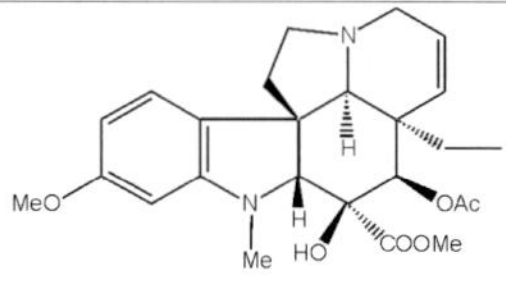 vindoline 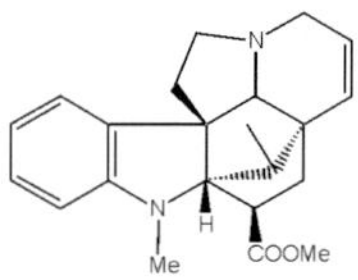 vindolinine 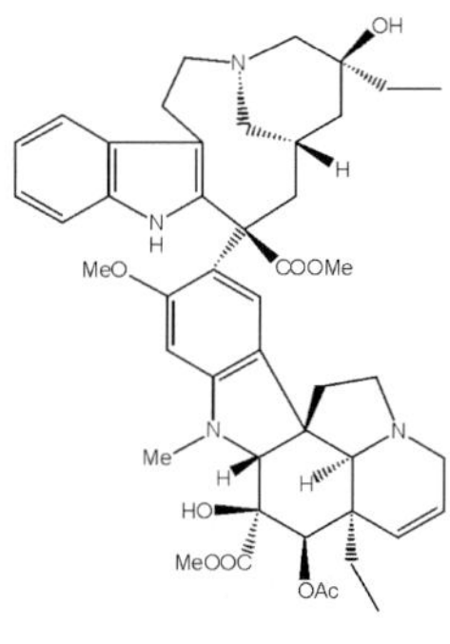 vincaleukoblastine	MtBE–ACN–water (4:1:5)	pH-zone refining
Fucans and heparins	MIBK–LA2–water (4.5:0.5:5)	Ion-exhange

– working with a high content of bridge solvent allows one to push the K_D coefficients around 1 without precipitating compounds. However, an excessive concentration of the bridge solvent means that the liquid system becomes close to the plait point, and coalescence could take place in the column. With low concentrations of bridge solvent, it is difficult to obtain practical K_D values in the elution mode. However, such low concentrations are well adapted to the displacement mode in which the solutes should be located preferentially in one phase.

3.3.3 Multisolvent system approach

Some solvents can be ingeniously combined to form a nearly continuous series which is made by gradually varying the solvent proportions. The resulting mixtures have a stepwise increasing average polarity resulting in homogenous scales. Another approach is the design of different biphasic liquid systems in a logical manner constituting heterogeneous scales. The two strategies will now be discussed.

Homogeneous scales
The Oka approach: Oka et al. designed sixteen mixtures of *n*-hexane, ethyl acetate, *n*-butanol, methanol, and water [18]. These systems are either binary or ternary, but always create nearly equal phase volumes. This solvent series covers nearly continuously a wide range of polarity, from the non-polar hexane–methanol–water, 2:1:1 system, to the polar *n*-butanol–water, 1:1 system (Fig. 3.5). This set of solvent systems is best used with rather polar samples because all the polar phases contain water. Moreover, many systems contain *n*-butanol, a solvent with a high boiling point which makes it difficult to remove by evaporation. The partition coefficients of many analytes (herbicides, indoles) were determined in these solvent systems, thus helping system selection.

The HBAW approach: In this solvent range, the binary systems *n*-heptane–acetonitrile 1:1 (non-polar), and *n*-butanol–water, 1:1 (polar), are combined, resulting in seven binary, ternary, or quaternary systems, numbered from –3 to +3 (Fig. 3.6) [13]. System 0 is tested first, indicating which systems (either positive or negative) must be further assayed. For example, if the sample prefers the polar aqueous phase, positive systems must be tested next. A useful characteristic of these solvent mixtures is UV-transparency. The HBAW system is especially useful with samples presenting a good solubility in acetonitrile or butanol. Other solvent compositions could be tested, but it should be pointed out that these solvents may produce triphasic systems.

The "ARIZONA" approach: Margraff also combined two binary systems: the ethyl acetate–water system (1:1, v/v, is System A, rather polar) and the *n*-heptane–methanol system (1:1, v/v is System Z, less polar) [19]. Then, they defined 23 liquid systems, naming them by letters, spanning the whole alphabet (except for E, I, and O) and covering the whole polarity range between System A and System Z (Fig 3.7). Margraff called his system "ARIZONA" from the abbreviation of the US State (AZ). The method is simple: two assays are done by dissolving the

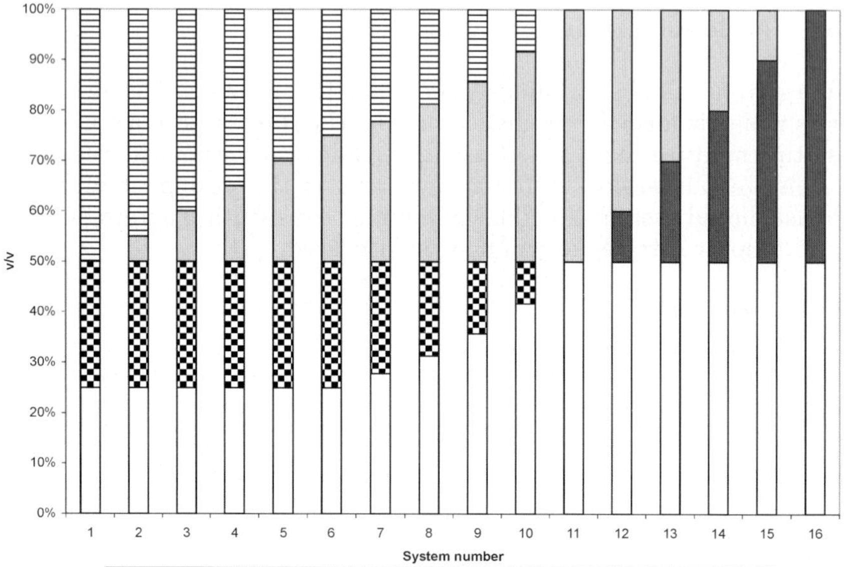

N°	n-hexane	EtOAc	n-BuOH	MeOH	Water	Upper/Lower
1	10	0	0	5	5	1,05
2	9	1	0	5	5	0,96
3	8	2	0	5	5	0,88
4	7	3	0	5	5	0,82
5	6	4	0	5	5	0,77
6	5	5	0	5	5	0,74
7	4	5	0	4	5	0,80
8	3	5	0	3	5	0,86
9	2	5	0	2	5	0,93
10	1	5	0	1	5	0,92
11	0	5	0	0	5	0,88
12	0	4	1	0	5	0,91
13	0	3	2	0	5	0,99
14	0	2	3	0	5	1,09
15	0	1	4	0	5	1,16
16	0	0	5	0	5	1,22

Fig. 3.5. The OKA scale. Solvent composition and volume ratio of the phases.

sample in the A and Z systems. If most of the sample is located in the "non-polar" ethyl acetate upper phase of A and the "polar" methanol-rich lower phase of Z, then there is necessarily a system between A and Z that will produce an equilibrated partitioning of the sample between its two phases.

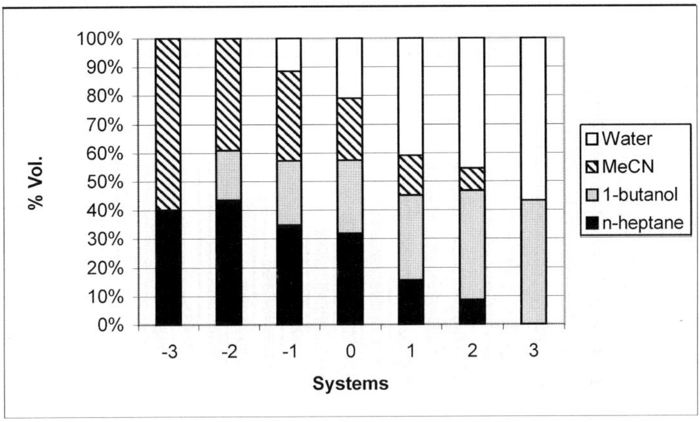

Systems	*n*-heptane	1-butanol	ACN	Water
−3	5	0	5	0
−2	5	2	3	0
−1	4	2.6	2.4	1
0	4	3.2	1.8	2
1	2	3.8	1.2	4
2	1	4.4	0.6	4
3	0	5	0	5

Fig. 3.6. The HMAW range. Solvent composition.

The sample may present a poor solubility in the mid-range systems (from J to S). Heptane and/or water represent a large part of the mixture (extreme polarity solvents), preventing the complete dissolution of the sample. The expanded Arizona approach was developed to solve this problem.

The expanded "ARIZONA" approach: Ethyl acetate can be partially or totally substituted by methyl *tert*.-butyl ether (MTBE), and methanol can be partially or totally replaced by acetonitrile [20]. Systems containing up to six solvents can also be prepared, presenting only slightly different average polarity but a greatly improved solubilization power. The intermediate systems, L, M, N, and P, may become triphasic when more acetonitrile replaces methanol.

Glyme-containing scale: In a similar approach, Foucault and Chevolot [19] combined the binary *n*-heptane–methanol, 1:1 system, with an original ternary system, MTBE–ethylene glycol dimethyl ether (glyme)–water, 1:2:1. Glyme was found to partition equally between

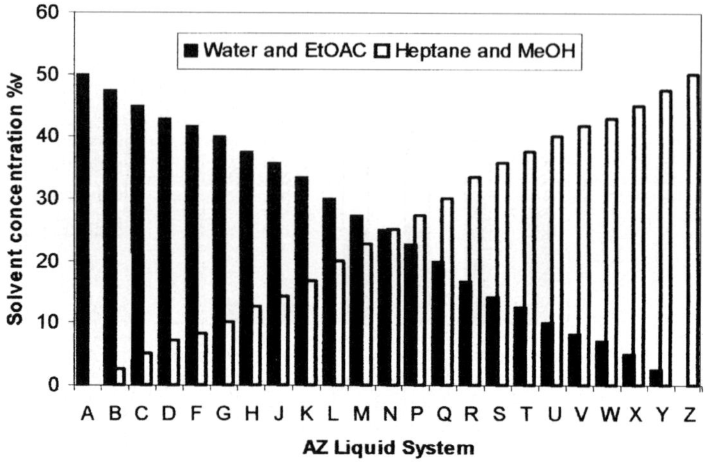

Systems	A	B	C	D	F	G	H	J	K	L	M	N	P	Q	R	S	T	U	V	W	X	Y	Z
Water	1	19	9	6	5	4	3	5	2	3	6	1	5	2	1	2	1	1	1	1	1	1	0
MeOH	0	1	1	1	1	1	1	2	1	2	5	1	6	3	2	5	3	4	5	6	9	19	1
EtOAc	1	19	9	6	5	4	3	5	2	3	6	1	5	2	1	2	1	1	1	1	1	1	0
Heptane	0	1	1	1	1	1	1	2	1	2	5	1	6	3	2	5	3	4	5	6	9	19	1

Fig. 3.7. The AZ (Arizona) system. Composition of the 23 biphasic solvent systems.

MTBE and water, where up to 60% can be added without loss of biphasic properties. Twenty-one systems were defined from the less polar heptane–methanol, 1:1, to the more polar MTBE–glyme–water, 1:2:1, with a 5% (v/v) variation between each system (Fig. 3.8). The advantage of this solvent series is the high proportion of medium-polarity strong solvents (MTBE, glyme and methanol), from 50% (System 1) to 75% (System 21) compared to the constant content of such solvent (50%) in all Arizona systems (Fig. 3.8). However, the presence of glyme imposes some experimental restrictions (high boiling point, toxicity, peroxide formation).

Acetone-based solvent scale: Another approach was developed to remedy the solubility problems encountered when trying to use the Arizona solvent scale. Acetone is a good solvent for many crude mixtures of secondary plant metabolites. It is thus highly desirable to use it as the "best" solvent in ternary systems. If A is the molecule of interest, its partitioning in a heptane–acetone–water system is in favour of the water-rich phase because acetone has more affinity for water than for heptane, as is shown by the slope of the tie lines (Fig. 3.9). Also, when

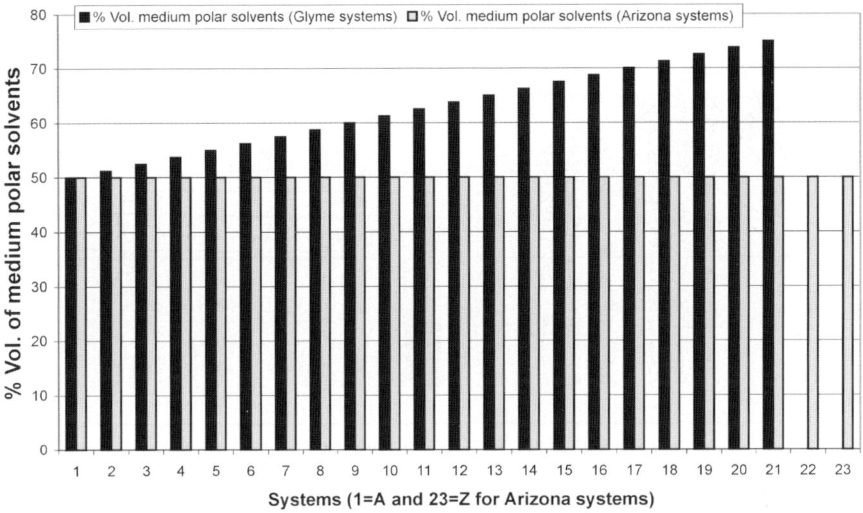

Fig. 3.8. Comparison of the ratio of medium polar solvents, the strong solvents, for Arizona and Glyme-containing systems.

dissolved in a toluene–acetone–water system, A is mainly present in the toluene-rich phase, owing to the higher affinity of acetone with toluene. A quaternary biphasic system containing toluene, heptane, acetone, and water is thus able to distribute A in nearly equivalent amounts between the two phases. The main interest in this strategy is that it permits a high acetone content in both the aqueous and the organic phases, thus allowing high concentrations of A in the system. For example, starting with two ternary systems—one based on heptane, the other based on toluene, and both containing 50% of acetone—one can build a solvent scale with the same high and constant acetone content. Following the same scheme, toluene can be replaced by methyl isobutyl ketone (MIBK) or ethyl acetate (see Ref. [14] for the corresponding ternary diagrams). All of these systems contain solvents that can be employed in industry under the usual safety conditions.

Heterogeneous scale

The Abbott approach: Abbott and Kleiman designed thirteen different biphasic systems, involving ten different solvents, which were either lipophilic (Systems 1 to 3), intermediary (Systems 4 to 8) or polar (Systems 9 to 13) (Table 3.2) [21]. This list of solvent composition scans a

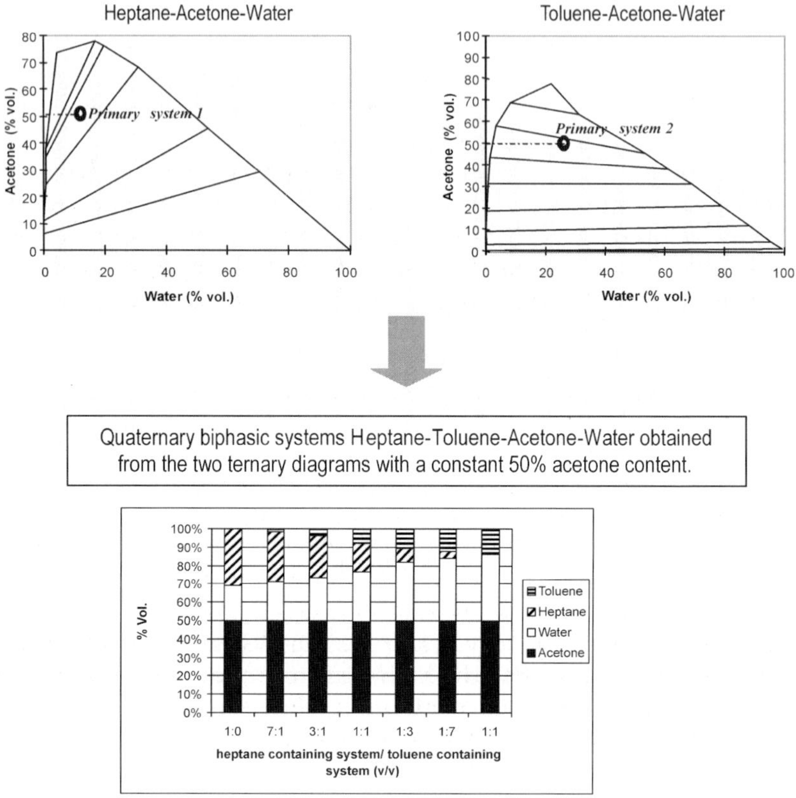

Fig. 3.9. Design of a solvent scale with a 50% acetone content using the two respective ternary phase diagrams.

large polarity range, and can deal with very diverse families of compounds.

3.3.4 Biphasic aqueous–aqueous solvent systems

Albertsson introduced aqueous–aqueous polymer-phase systems in order to fractionate or separate polar macromolecules and cell particles [23]. One phase is an aqueous solution of polyethylene glycol, while the other contains ionic (inorganic salts) or neutral species (dextrans) also in an aqueous phase. These systems have low enough viscosity that they can be used in hydrodynamic CCC as well as hydrostatic CPC machines. However, the CCC column design must be adapted somewhat, and special toroidal coils were developed for this purpose. Also, the geometry of CPC partition cells must be slightly modified to obtain a good stationary-phase retention and mass transfer.

The aqueous–aqueous polymer-phase system, polyethylene glycol 1000–potassium phosphate has been used by Shibusawa et al. [7] for the CCC separation of cytochrome-C, myoglobin, and human serum albumin. The use of these systems is delicate and somewhat specific, and will not be considered further here. Specific references should be consulted for more information [2,7,13,22,23]

3.3.5 A practical approach to the selection of solvent systems

The choice of a solvent system is mainly guided by the solubility of the analytes, which is in turn related to the non-covalent interactions developed between the solvent and solute molecules. Analyte properties may range from highly lipophilic to highly hydrophilic. For the extreme cases, we propose a single separation strategy involving only one of the solvent systems already presented. The flow-chart of Fig. 3.10 summarizes the method. For very non-polar samples, non-aqueous systems should be tried first. For intermediate samples, moderately polar aqueous systems should be tested and, if the sample is very polar, aqueous–aqueous systems can be considered.

3.4 FROM AN EMPTY BOX TO A FIT COLUMN

3.4.1 Generalities

It is quite common for CCC users to say that the selection of a biphasic solvent system corresponds to the choice of both the column and the mobile phase in HPLC. In certain respects, this remark is true. Indeed, choosing a type of solid support in HPLC (bonded silica gel or ion-exchange columns) is equivalent to defining the chromatographic development mode (elution, or displacement) in CCC. Of course, as explained below through typical examples, the choice is most often imposed by the chemical structure of the molecules to be separated.

If the molecules of interest are neutral, the elution mode is mandatory and can be applied in the isocratic or gradient mode, as in

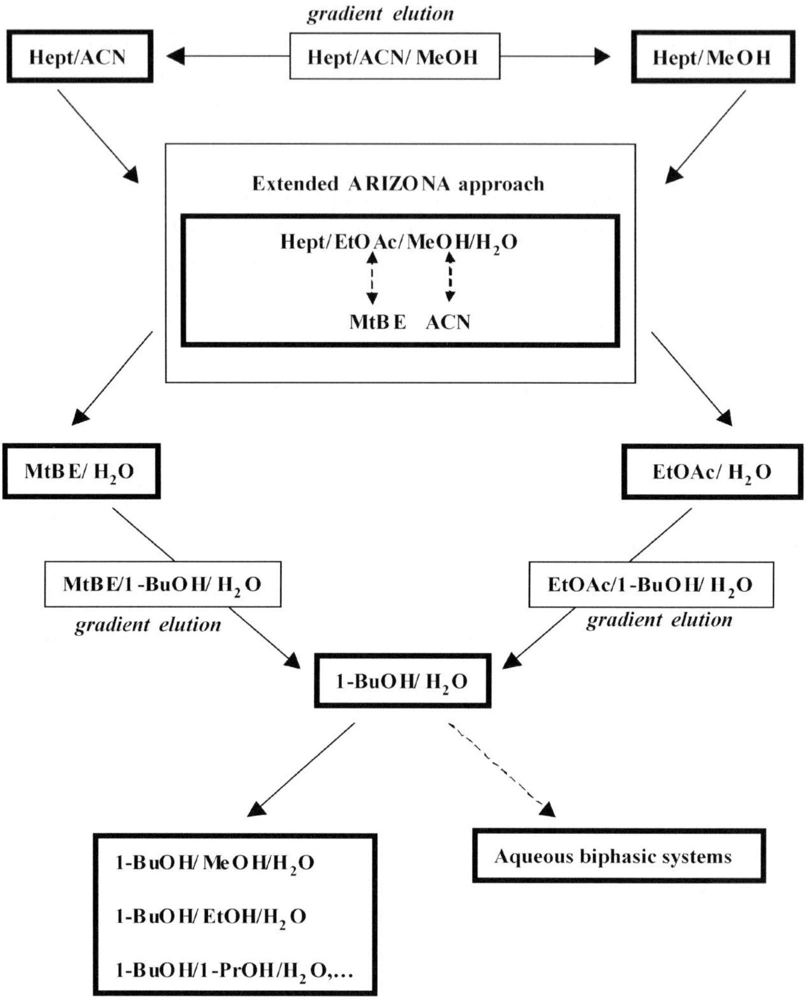

Fig. 3.10. A practical approach to the selection of solvent systems.

HPLC. Moreover, the CCC technique allows one to use an effective and original mode described below: the dual mode [24].

If the analytes are ionizable or ionic, the displacement mode would seem to be the best way to separate them. However, many charged compounds are quite well behaved in the classic or even gradient elution mode. These modes should be explored. If displacement CCC is selected, two types of mode have been developed: an original mode called pH-zone refining [10], where the ion-exchanger is in the mobile phase, and classical ion-exchange techniques [11] using a weak exchanger in the stationary phase.

After the selection of a suitable strategy (elution or displacement), the biphasic system must be chosen (*vide infra*). This is where HPLC and CCC differ. Since in CCC the mobile and the stationary phases are two liquid phases in equilibrium, they cannot be considered independently. Phase diagrams should be consulted, especially in the case of gradient elution (the tie-lines must converge). Also, the injected sample must not destroy the phase equilibrium. In HPLC, the column can be chosen from a catalog. The mobile phase will be selected using the extensive literature on HPLC separations as well as the software-assisted optimization methods. In CCC, the interdependence of the mobile and the stationary phase, the diversity of biphasic liquid systems, the different development modes, and a limited literature (e.g., for displacement mode, non-polar compounds, and peptides) may discourage the use of this technique.

Nevertheless, the CCC technique can be very efficient if a logical and rational approach is followed in the optimization of experimental conditions.

3.4.2 Chemical structures as a starting point

Introduction

Setting up a separation protocol involves selectivity optimization, and CCC is highly flexible for achieving this task [25]. The composition of both stationary and mobile phases can be finely adjusted (through the ternary diagram in the case of a ternary solvent system) to obtain the desired resolution. Figure 3.11 compares the CCC and HPLC approaches to peak resolution improvement.

As previously mentioned, in CCC the chromatographic resolution is improved by increasing selectivity, whereas in HPLC, it is improved by increasing efficiency. Different commercial chromatographic solid supports have different efficiencies. Also, the packing procedure is critical in preparative HPLC.

At this stage, it is important to point out that HPLC and CCC must not be seen as competitors, as each one presents different advantages depending on the compounds to be separated. It is almost never possible

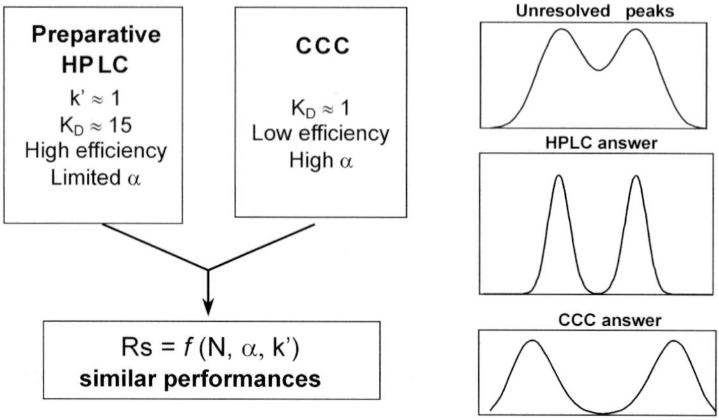

Fig. 3.11. Resolution improvements by HPLC and CCC. Left: HPLC can work with high K_D solutes because k' is low (low volume of stationary phase) and N is high. The CCC has a low number of plates but a high stationary phase volume, so it can act on the selectivity factor. Right: the different chromatograms, HPLC sharpens the peaks, CCC pulls the peaks apart.

to obtain a pure compound in a single chromatographic step. A synergistic effect may often be obtained by associating HPLC and CCC rather than carrying out repeated chromatographic steps on silica gel [26]. Combining CCC and HPLC often allows one to reduce the number of purification steps and, consequently, to increase the yields of pure isolated compounds.

The CCC techniques are most efficient in the direct fractionation of crude extracts:

– The diversity of the possible injection procedures (in the mobile phase, in the stationary phase, in a mixture of the two phases, and before- or after equilibrium of the column) allow one to dissolve a maximum amount of sample to be purified. In all cases, it is necessary to check for phase saturation. Indeed, the injection of a plug of liquid containing a high concentration of sample changes the equilibrium between the mobile and stationary liquid phases. This phenomenon can lead to the failure of the CCC run. A zone containing concentrated sample will move through the CCC column, preventing the percolation of the mobile phase and pushing the stationary phase out of the machine. To avoid this significant problem, lower sample concentrations, larger injection volumes, and additions of stationary phase in the injected mixture should be tried (see Chapter 2).

– The absence of solid support makes column clogging unlikely. However, the solvent system optimization for CCC separation requires the selection of a solvent combination in which the sample is highly soluble.

This strategy finds an interesting application in plant screening, where there is a need for the quick preparation of a small number of fractions [27]. This use of CCC, enabling the rapid screening of natural extracts, has led many to think that CCC was not suitable for the isolation of pure compounds. This wrong idea was reinforced by considering the low number of theoretical plates generated by solid-support-free chromatographic columns. The examples developed here and in other chapters of this book show that a global approach to the fractionation strategy (from the extraction broth to pure compound) allows one to take advantage of the high selectivities obtained in CCC.

Elution approach

Introduction: For historical reasons, the elution mode has been the most studied: the first separations described used droplet CCC, which was only used in the elution mode. Displacement CCC is a recent development. Most of the following examples illustrate the previously presented strategy.

The choice of a suitable biphasic system is guided by important criteria such as the sample polarity (evaluated from partition coefficient values), its solubility, ionic form, and ability to form complexes. As previously mentioned, in the elution mode it is required that the distribution constants of the sample components lie between 0.2 and 5. Different methods are available for estimating K_D values: TLC, UV–Vis spectrometry, HPLC, etc. [28].

In CCC, three types of development by elution can be applied: the classical isocratic mode, the gradient mode, and the dual mode—the latter mode being specific to countercurrent chromatographic techniques.

Isocratic mode: The isocratic mode is often used in CCC because it is easy to perform. After selecting the suitable biphasic system, the normal and the reversed-phase mode must generally be tested, i.e., ascending (tail-to-head) or descending (head-to-tail) mode. Other aspects must be considered: the stability of the biphasic system in the column, its toxicity and ease of disposal, the number of theoretical plates generated in each mode, the duration of the experiment, and the solute recovery yield (including solvent evaporation), and the final cost of the separation process.

The isocratic mode is particularly efficient and is recommended for extracts containing a mixture of homologues and/or structural analogues. It may fail when dealing with samples covering a large range of polarity.

Numerous natural extract separations using the isocratic mode are reported in the literature. Because of their versatility, the most popular strategies are the "best solvent approach" or the use of the "ARIZONA" scale.

As already mentioned, and considering the preparative character of CCC, the best-solvent strategy requires the selection of a solvent in which the sample is highly soluble. This prerequisite is well illustrated in a work by Foucault et al. [29] which reports the purification of amphotericin B, a compound that is poorly soluble in water and insoluble in heptane. It is a polyene macrolide antibiotic whose commercial form contains a mixture of homologues (tetraenes, pentaenes and heptaenes). The original biphasic system used was a Type 0 ternary liquid system. The sample was highly soluble in the DMSO–THF–water system. Thus, 500 mg of crude amphotericin B (Table 3.3), as its perchlorate salt, was purified in a single run with a 240 ml CPC column and a biphasic liquid system from the DMSO–THF–water phase diagram.

Another difficult problem is the separation of non-polar compounds such as fatty acids [30], terpenoids [31], phospholipids, or phenolic lipids [32]. Non-aqueous or biphasic systems with minimum amounts of water should be used. For this purpose, alkanes–MeOH and/or CH_3CN mixtures should be tried first. Although these systems are known to give good stationary-phase retention in blank runs, experience shows that stability problems often occur during separation of actual samples leading to large losses of stationary phase. One of the reasons for such stationary-phase instability is that these non-polar compounds often have amphiphilic properties that disturb the interaction between the solvent molecules. This problem can be partially solved by adding a small amount of water (a few percent) to the initially non-aqueous system [33].

The following examples illustrate the high selectivity that can be achieved by a suitable choice of solvent system, and how the composition of the "ARIZONA" mixtures can be modulated for a successful separation of very similar molecules. Acetogenins (characterized by a linear chain of 35–37 carbon atoms, including one or two tetrahydrofuran rings, several oxygenated functions and a γ-methyl-γ-lactone terminal group) are polyketides found only in tropical and subtropical plants of the *Annonaceae* family. Most of these compounds present interesting biological activities, but their isolation is difficult and tedious, as they require numerous chromatographic steps on solid supports (silica gel, Sephadex LH-20, or preparative TLC). Duret et al. [34] successfully used the CCC technique to shorten the isolation steps and to increase the yields. Several ARIZONA compositions (heptane–EtOAc–MeOH–water) were tested and the N-system (1:1:1:1) was retained as it provides average K_D values close to 1 for the sample. However, the solubility was very low in this solvent mixture. The sample is readily soluble in pure MeOH or EtOAc. The poor solubility in the 1:1:1:1 system is due to the high content of heptane and water (50% in volume, Fig. 3.8). The solubility of acetogenins was improved by simultaneously increasing the EtOAc and MeOH contents, starting from System N. The heptane content was reduced as well. Some fine-tuning led to the final mixture, heptane–EtOAc–MeOH–water (3:10:10:7), in which the content in EtOAc+MeOH (the best solvents) was increased to 67% (v/v). Squamocin and rolliniastatin-2 (Table 3.3), which only differ in the position of one hydroxyl group, were both obtained as pure compounds directly from a single CCC run, while six other compounds (molvizarin, asimicin, neoannonin, atemoyin, desacetyluvaricin and isodesacetyluvaricin) were purified by a subsequent HPLC run.

Dual mode: A distinctive advantage of CCC is the possibility of operating in dual mode [24]. There is no obligation to use the upper phase as the mobile phase—it can be the stationary phase as well. It is even possible to switch the role of the phases during a run. For example, the CCC separation can start in the normal-phase mode, with the apolar mobile phase and the aqueous or polar stationary phase. After a while, it can continue in the reversed-phase mode, switching the role of the liquid phases. In practice, the pumping direction of the stationary and mobile phase is reversed using a valve to switch from the ascending (tail-to-head) to the descending (head-to-tail) modes (or *vice versa*). The dual elution mode allows fractionation of molecules presenting very different polarities. Complex starting material can be completely processed, with short run-time, and without sample loss. Figure 3.12 shows the principle of dual-mode operation.

The purification of 10-deacetyl-baccatine III (10-DAB) by Margraff [35] (Table 3.3) shows the power of dual-mode operation. 10-DAB, isolated from the European yew *Taxus baccata,* is the starting material in the synthesis of docetaxel (Taxotère®), a major anti-cancer drug. The

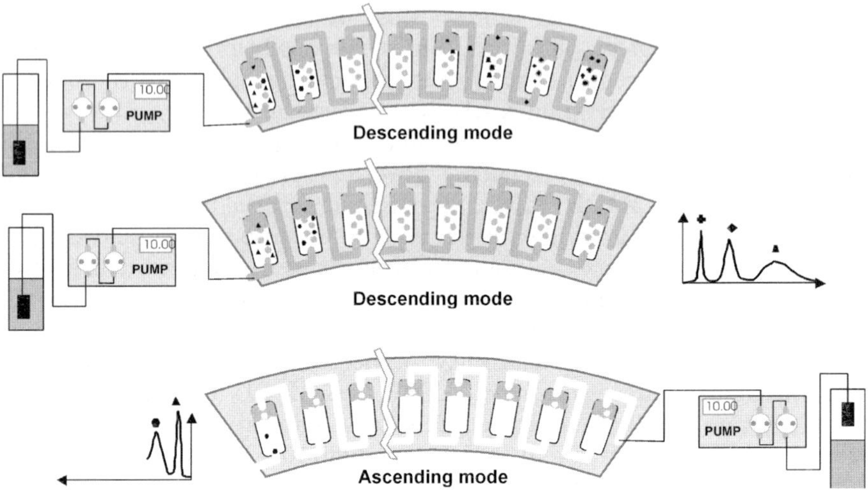

Fig. 3.12. Dual mode elution (see text for comments).

crude extract containing 10-DAB includes many undesirable products, notably the 19-hydroxy analogue (19-OH-DAB). All the ARIZONA systems were poor solvents for 19-OH-DAB and other secondary metabolites. Acetone was a much better solvent and, combined with methyl isobutyl ketone (MIBK) and water, produced biphasic liquid phases. All of these solvents are compatible with industrial safety constraints.

The MIBK–acetone–water mixture (2:3:2) led to a selectivity coefficient of 6 between 10-DAB and 19-OH-DAB. A 240-ml CPC hydrostatic machine was used. Both ascending and descending modes were checked and neither was able to isolate the molecule of interest without any contamination. In the first part of the dual-mode run (descending mode, aqueous mobile phase), one impurity and 19-OH-DAB were separated, while the second part (ascending mode, MIBK mobile phase) permitted the elimination of more impurities as well as yielding 10-DAB as a 99% pure compound. A total of 5 g of crude extract was purified using the 240 ml CPC column, in three hours, using only 0.8 l of solvent.

Similar solubility problems occurred when attempting to separate the cyclopeptide alkaloids (Table 3.3) from a crude extract of *Zizyphus lotus* (Rhamnaceae) [19]. Again, the N-system (1:1:1:1) from the ARIZONA range failed to conveniently dissolve the extract. Half of the methanol and ethyl acetate volume percentages were substituted by an equivalent amount of acetonitrile (ACN) and MTBE, respectively. The resulting complex six-solvent biphasic system heptane–EtOAc–MTBE–MeOH–ACN–Water (1:0.5:0.5:0.5:0.5:1) provided the required solubilization and selectivity properties. Three cyclopeptides (lotusines D, E and F) were isolated as pure compounds in one step using the dual elution mode (E in ascending mode, D and F in descending mode).

Gradient mode: Anthocyanins (Table 3.3) are water-soluble pigments responsible for the orange, pink, red, purple, and blue colours of higher plant parts. The purification of these compounds is made difficult by the existence of unstable forms at a pH greater than 2. Under acidic conditions anthocyanins form stable flavilium cations. These compounds have high anti-oxidant properties. A crude extract of champagne grapes by-products (*Vitis vinifera*) was prepared. It contained a rich variety of molecular species including flavonoids, sugars, and tannins. The fractionation of the extract was carried out under acidic conditions using gradient elution. The system EtOAc–BuOH–water was suitable for gradient elution since water is immiscible with both the other solvents, and the tie-lines in the ternary diagram converge in the direction of nearly pure water. Gradient elution was carried out in the normal-phase mode by varying the composition of the organic mobile phase from EtOAc–1-BuOH–water (77:15:8) (I in Fig. 3.13) to (40:46:14) (F in Fig. 3.13). The phases were acidified with 0.2% v/v trifluoroacetic acid (0.02 *M*, pH 1.7). The conditions were optimized using a 240-ml CPC column and then scaled up using a 5-l pilot-scale CPC chromatograph. Fractions of 600 ml volume were collected every ten minutes and the two major products (malvidin 3-glucoside and peonidin 3-glucoside) were isolated in one step. Additional malvidin, peonidin, cyanidin and delphinidin, as their 3-glucosides, were recovered from mixture fractions after isocratic CPC separations on the 240 ml machine using the same solvent system at intermediate compositions, depending on the polarity of the samples (A, B, and C on Fig. 3.13). Using the same strategy, four anthocyanins: cyanidin-3-glucoside, delphinidin-3-glucoside, cyanidin-3-rutinoside and delphinidin-3-rutinoside, were isolated in one step from blackcurrant using a similar gradient elution.

Another class of glycosides, namely saponins, was successfully investigated by CPC and the EtOAc–1-BuOH–water gradient system, using neutral conditions. The plants studied were *Panax quinquefolium* L. [36] and *Zizyphus lotus* (L.) Desf. [17], both containing dammarane saponins.

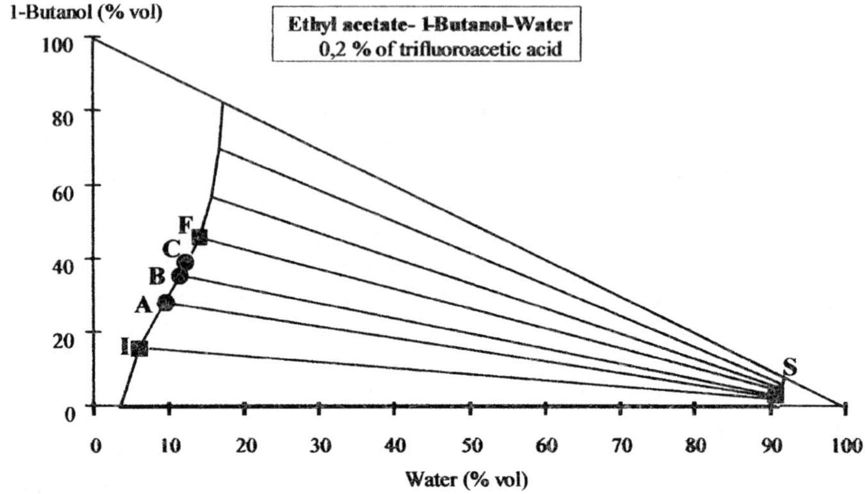

Fig. 3.13. The EtOAc–n-BuOH–water system and its use in gradient elution. A, B, C, F and I: organic upper phase with different butanol contents: S, aqueous lower phase containing more than 90% v/v water and slightly changing amounts of butanol and ethyl acetate.

Displacement approach

In displacement CCC, the chromatographic process relies on the difference in analyte solubility upon changes in solute or solvent properties. Acid–base or complexation reactions can be used. Displacement chromatography is inherently a preparative method that allows one to work with highly concentrated fractions. Displacement CCC commonly works with sample loads ten times higher than the ones used in classical CCC. The products are not eluted as separate zones each having a Gaussian-like concentration profile, but as contiguous bands of constant concentration. Unlike elution CCC, the optimization of the solvent system used in the displacement mode is not guided by the K_D partition-coefficient value of the analytes, but by the dramatic changes in solubility occurring under different pH or complexation conditions. The solutes' K_D values may be very small or very large. Two approaches have been developed. The first, called pH-zone refining, uses acid–base ionization changes and was developed by Ito [37]. It exploits the marked change in solubility observed between the neutral form of an ionizable molecule at a given pH and its charged forms at a different pH. The second approach, designed by Chevolot et al. [11], is based on ion-exchange chromatography. It has been used in the purification of water-soluble salts of strong organic acids that cannot be subjected to pH-zone refining.

pH-Zone refining (Fig. 3.14A): This technique is devoted to the purification of compounds whose electrical charge depends on the pH value. For example, a mixture of free bases is injected into the organic stationary phase that contains a base stronger than all the compounds to be separated. This stronger base is called the retainer base. The bases are moved along the column by pumping through an acidic aqueous solution called the displacer acid. Pure products are then isolated in the effluent as successive salt bands. The column is used in the descending mode, assuming that the density of the aqueous acidic mobile phase is greater than the density of the organic basic stationary phase. The ascending mode can also be used. In this case, the aqueous stationary phase will contain a strong acid called a retainer acid, and the organic mobile phase will contain a displacer base and organic acids or their salts will be separated in a similar way. The solvent system requirements are far less stringent than those for elution. The system MTBE–ACN–water (4:1:5) is adequate in most situations for analytes with moderate hydrophobicity. For the separation of hydrophobic compounds, ARIZONA systems are convenient. For the separation of polar products, the BuOH–water system may be useful. The ionic species partition preferentially in the aqueous phase, whereas neutral molecules stay in the organic phase.

pH-Zone refining is particularly well suited to the separation of alkaloids [37,38], organic acids [37] or peptides [3]. A typical example of a pH-zone refining separation is given by the isolation of the indoloterpenoid alkaloids from *Catharanthus roseus* (Apocynaceae) (Table 3.3) [39]. Up to 7 g of the tartrate salts of the crude alkaloid extract was injected and displaced by triethylamine (8 m*M*) using the above-mentioned solvent system. The aqueous stationary phase was acidified by HCl (10 m*M*) as a retainer. Pure vindoline, catharanthine, vincaleukoblastine and vindolinine were isolated in a single run.

Weak ion-exchanger (Fig. 3.14B): In 1998, Foucault and Chevolot introduced ion-exchange as a new type of displacement mode in CPC [1]. The principle of the method consists of generating lipophilic ion-pairs that can be extracted by the organic stationary phase. Amberlite-LA2, a weak anion-exchanger is used in order to significantly alter the polarity of the target ions. The LA2 is miscible with almost all common organic solvents and immiscible with water. The displacement process is carried

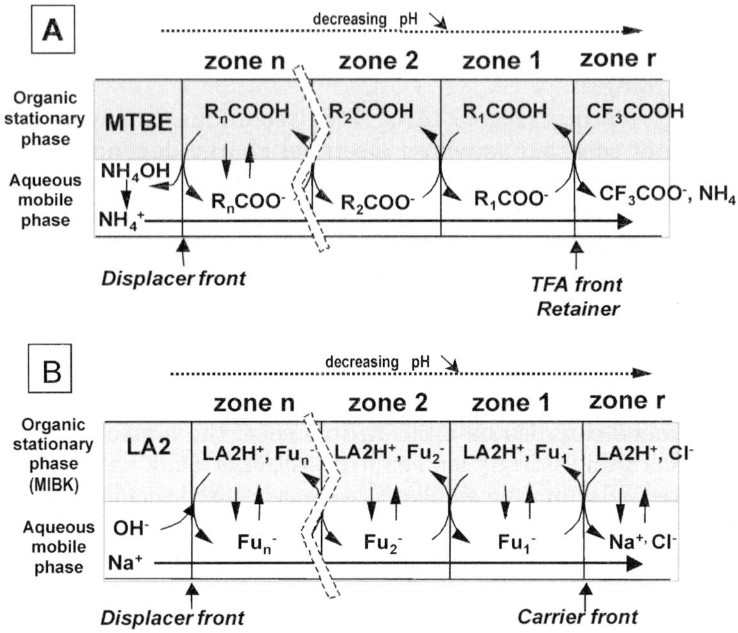

Fig. 3.14. Schematic representation of mechanism (acid–base reaction + partitioning) leading to solute band formation in pH-zone refining (A, separation of organic acids in descending mode), and in ion-exchange with LA2 (B, fractionation of fucans in descending mode).

out by using a strongly basic aqueous mobile phase. The hydroxide ions displace the anions from the organic stationary phase to the aqueous mobile phase by neutralizing the weak anion-exchanger. The differences in the association constants of LA2 and the anionic analytes should be high enough to produce separated bands. If this condition is fulfilled, the analytes emerge from the CPC column in characteristic band-like zones of homogeneous material. This development mode was successfully applied to the separation of partially depolymerized fucans (polysulfated polysaccharides) [11], and low-molecular-weight heparins [12]. The same biphasic system was used: the stationary phase comprised LA2 dispersed in MIBK and was protonated by an aqueous solution of HCl whereas the mobile phase was an aqueous solution of NaOH saturated by MIBK. Metallic ions were also separated using ion-exchange displacement CCC [8].

3.5 CONCLUSION

Solvent-system selection is an important part of the design of a CCC separation protocol. The chemical nature of the sample determines the appropriate development mode. The solvent system also determines the analyte partition-coefficient range. In the displacement mode, the main point is to obtain a strong variation of the analyte solubility upon ionization or complexation. This is relatively easy to obtain, and does not impose a strong constraint on the solvent-system composition. Conversely, the classical elution mode imposes a much finer composition adjustment. Strategies such as the "best solvent" and/or the "multi-solvent" approaches allow the rapid design of a liquid system that will ensure a high degree of selectivity and high yields minimizing the separation duration and solvent volume. The few reported examples illustrate the vast diversity of preparative separation problems that have been solved using CCC and a rational choice of solvent systems. Other developments will certainly emerge from a better understanding of column hydrodynamics and physico-chemical modelling of solvent–solvent and solvent–analyte interactions.

REFERENCES

1 A.P. Foucault and L. Chevolot, *J. Chromatogr. A,* 808 (1998) 3–22.
2 M. Knight, in N.B. Mandava and Y. Ito (Eds.), *Countercurrent Chromatography: Theory and Practice*, Chromatographic Sciences Series, Vol. 44. Marcel Dekker, New York, 1988, Chap. 10, pp. 583–616.
3 Y. Ma and Y. Ito, *Anal. Chim. Acta,* 352 (1997) 411–427.
4 J.G. Alvarez, J. Cazes, J.C. Touchstone and R.L. Grob, *J. Liq. Chromatogr.*, 13 (1990) 3603–3614.
5 H. Becker, J. Reichling and W.C. Hseih, *J. Chromatogr.,* 237 (1982) 307–310
6 G. Baudimant, M. Maurice, A. Landrein, G. Durand and P. Durand, *J. Liq. Chromatogr. Rel. Technol.*, 19 (1996) 1793–1804.
7 Y. Shibusawa, M. Yamaguchi and Y. Ito, *J. Liq. Chromatogr. Rel. Technol.*, 21 (1998) 121–133.
8 K. Talabardon, M. Gagean, J.M. Mermet and A. Berthod, *J. Liq. Chromatogr. Rel. Technol.*, 21 (1998) 231–250.
9 T.A. Maryutina, P.S. Fedotov and B.Y. Spivakov, *Zh. Anal. Khim.,* 52 (1997) 1140–1146.
10 Y. Ito and Y. Ma, *J. Chromatogr. A,* 753 (1996) 1–36.

11 L. Chevolot, S. Colliec-Jouault, A. Foucault, J. Ratskol and C. Sinquin, *J. Chromatogr. B*, 706 (1998) 43–54.
12 O. Intes, J.H. Renault, C. Sinquin, M. Zèches-Hanrot and J.M. Nuzillard, *J. Chromatogr. A*, 918 (2001) 47–57.
13 A.P. Foucault, In: A.P. Foucault (Ed.), *Centrifugal Partition Chromatography*, Chromatographic Sciences Series, Vol. 68. Marcel Dekker, New York, 1995, pp. 71–97
14 A.P. Foucault, In: A.P. Foucault (Ed.), *Centrifugal Partition Chromatography*, Chromatographic Sciences Series, Vol. 68. Marcel Dekker, New York, 1995, pp. 363–407.
15 A.P. Foucault and K. Nakanishi, *J. Liq. Chromatogr.*, 13 (1990) 3583–3602.
16 J.H. Renault, P. Thépenier, M. Zèches-Hanrot, L. Le Men-Olivier, A. Durand, A.P. Foucault and R. Margraff, *J. Chromatogr. A*, 763 (1997) 345–352.
17 J.H. Renault, K. Ghédira, P. Thépenier, C. Lavaud, M. Zèches-Hanrot and L. Le Men-Olivier, *Phytochemistry*, 44 (1997) 1321–1327.
18 F. Oka, H. Oka and Y. Ito, *J. Chromatogr.*, 538 (1991) 99–105.
19 A.P. Foucault and L. Chevolot, *J. Chromatogr. A*, 808 (1998) 3–22.
19 G. Le Crouérour, Ph.D. Thesis, Chap. 9, Reims, 2000.
21 T.P. Abbott and R. Kleiman, *J. Chromatogr.*, 538 (1991) 109–112.
22 I.A. Sutherland, In: N.B. Mandava and Y. Ito (Eds.), *Countercurrent Chromatography: Theory and Practice*, Chromatographic Sciences Series, Vol. 44. Marcel Dekker, New York, 1988, pp. 617–652.
23 P.A. Albertsson, *Partition of Cell Particles and Macromolecules*. Wiley-Interscience, New York, 1986.
24 A.P. Foucault, in A.P. Foucault (Ed.) *Centrifugal Partition Chromatography*, Chromatographic Sciences Series, Vol. 68. Marcel Dekker, New York, 1995, pp. 87–90.
25 J.-H. Renault, K. Ghédira, P. Thépenier, C. Lavaud, M. Zèches-Hanrot and L. Le Men-Olivier, *Phytochemistry*, 44 (1997) 1321–1327.
26 K. Ingkaninan, A.P. Ijzerman, T. Taesotikul, R. Verpoorte, *J. Pharm. Pharmacol.*, 51 (1999) 1441–1446.
27 K. Ingkaninan, A.C.J. Hermans-Lokkerbol and R. Verpoorte, *J. Liq. Chromatogr. Rel. Technol.*, 22 (1999) 885–896.
28 W.D. Conway (Ed.), *Countercurrent Chromatography — Apparatus, Theory and Applications*. VCH, New York, 1990, pp. 302–316.
29 A.P. Foucault, P. Durand, E. Camacho-Frias and F. Le Goffic, *Anal. Chem.*, 65 (1993) 2150.
30 J.G. Alvarez, J.C. Touchstone, R.L. Grob and J. Cazes, *J. Liq. Chromatogr.*, 13 (1990) 3603–3614.
31 R.C. Bruening, F. Derguini and K. Nakannishi, *J. Chromatogr.*, 357 (1986) 340.
32 O. Intes, J.H. Renault, P. Thépenier, M. Zèches-Hanrot, J.M. Nuzillard, B.

Laignel, J.N. Bertho and R. de Baynast, 1st Countercurrent Chromatography Symposium (CCC2000), London, 10–14 September 2000.
33 A. Berthod and M. Bully, *Anal. Chem.,* 63 (1991) 2508–2512.
34 P. Duret, A.I. Waechter, R. Margraff, A.P. Foucault, R. Hocquemiller and A. Cavé, *J. Liq. Chromatogr.* 20 (1997) 627–635.
35 R. Margraff, In: A.P. Foucault (Ed.), *Centrifugal Partition Chromatography*, Chromatographic Sciences Series, Vol. 68. Marcel Dekker, New York, 1994, Ch. 12, pp. 331–353.
36 L. Le Men-Olivier, J.H., Renault, P. Thépenier, M.J. Jacquier, M. Zèches-Hanrot and A.P. Foucault, *J. Liq. Chromatogr.*, 18 (1995) 1655–1662.
37 Y. Ito and Y. Ma, *J. Chromatogr. A*, 753 (1996) 1–36.
38 F. Yang, J. Quan, T. Zhang and Y. Ito, *J. Chromatogr. A*, 822 (1998) 316–320.
39 J.H. Renault, G. Le Crouérour, P. Thépenier, J.M. Nuzillard, M. Zèches-Hanrot and L. Le Men-Olivier, *J. Chromatogr. A*, 849 (1999) 421–431.

Chapter 4

Centrifugal partition chromatography: the story of a company

A.P. Foucault

Dedicated to the memory of Kanichi Nunogaki

4.1 INTRODUCTION

This chapter is entirely devoted to the one-axis centrifugal systems operating with two rotary seals, which have been classified by Y. Ito as hydrostatic countercurrent chromatographs. There is some controversy about the multitude of acronyms in countercurrent chromatography. The story of these one-axis instruments actually began under another name, as we will see below. In this chapter, we call this well defined category of instruments "Centrifugal Partition Chromatographs", or CPC, since this appears to be the internationally accepted term.

A CPC instrument or a CPC column is a series of channels linked in cascade by ducts, and aligned in cartridges or disks in a circle around a rotor. Setting the rotor in motion submits this assembly to a constant centrifugal field.

4.2 THE STORY OF CPC

4.2.1 The roots

Can we attribute the first countercurrent column to Cornish et al. [1]? In 1934 they constructed a countercurrent column of 210 units, each with a plate efficiency of about 50% under the operating conditions. It was used for the fractionation of fat-soluble vitamins (vitamins A, D, E, and F) between solvents such as hexane and ethanol. The authors had already clearly understood the relationship to fractional distribution

Comprehensive Analytical Chemistry XXXVIII
A. Berthod (Ed.)

(the name of counter-current distribution at that time) and chromatography since they wrote: "The column used here is analogous to Tswett's chromatoscope[1] in which the remarkable separations obtained are due to the use of the column idea, but where a solid absorbing substance replaces the polar solvent."

In 1941, A.J.P. Martin and R.L.M. Synge had two historical papers published in the *Biochemical Journal* [2,3]. In the first paper, a 40-stage countercurrent liquid–liquid extraction machine was described and its efficiency tested in separating acetamino acids. However, these scientists were not fully satisfied: "we wish to emphasize that the procedure is still in a crude state and capable of technical development." This is where CPC could have been invented since the second of these papers was entitled, "A new form of chromatogram employing two liquid phases", and since the first sentence reads, "In most forms of counter-current extraction column the very small drop required for the rapid attainment of equilibrium, and hence for high efficiency, cannot be used owing to the difficulty of ...". Martin and Synge opted for immobilizing one liquid phase onto small particles of a solid matrix. This was the invention of liquid–liquid partition chromatography.

Thirty years later (1970) the same idea led Y. Ito et al. [4] to disperse a liquid phase into another one by means of a centrifugal field. They described the helix countercurrent chromatograph, a prototype later abandoned by these scientists, who developed various types of CCC apparatus (see other chapters in this book). Eventually, after another twelve years (1982), the same idea led W. Murayama et al. [5] to design and build the first instrument at Sanki Engineering Ltd. in Kyoto (whose president was the late Kanichi Nunogaki). This instrument is currently known as the Centrifugal Partition Chromatograph, or CPC.

– *Remark*: It is important to avoid any confusion with the coil planet centrifuge apparatuses. The coil planet centrifuge, or CPC analyzer, invented by Y. Ito et al. [6,7], and first constructed by Sanki Engineering Ltd. in 1966, is not a chromatographic instrument, and does not belong to the story of liquid–liquid chromatography. But these two articles [6,7] are of the greatest interest since they prove that the starting point was the same for Y. Ito, inventor of many counter-current chromatographs (CCC), and K. Nunogaki, the president of Sanki at that time and with his family and collaborators the inventor

1 The authors referred to Palmer, "Carotenoids and related pigments, the chromolipoids", Chemical Catalog Company, 1922, while Tswett's idea dates back to 1903.

of the centrifugal partition chromatograph (CPC). They later terminated their relationship, for reasons unknown to us.

4.2.2 The first paper

The first paper describing the CPC prototype manufactured by Sanki [5] appeared in April 1982. The paper corresponded to a talk given at the 18th International Symposium on Advances in Chromatography, Tokyo, Japan, April 15–18, 1982. The paper was entitled "A new Centrifugal Counter-Current Chromatograph and its application", and in the introduction one may read: "we have developed another kind of Centrifugal Partition Chromatography which employs a rotary seal joint (CCCC-RJ)", and then: "the key point of the apparatus is the construction of the rotary seal joint...". In a similar way, the legend of Fig. 3 of this document refers to the "Centrifugal Partition Chromatograph" while that of Fig. 5 refers to the "Centrifugal Counter-Current Chromatograph system".

It appears from this first paper that the Nunogaki family had not yet decided on one name over the other. The first acronym for the instrument was CCCC-RJ, for "Centrifugal Counter-Current Chromatography with Rotary Seal Joints", probably to point to the difference from the CCC instruments already manufactured according to Y. Ito's designs. A photograph of the first Sanki prototype is shown in Fig. 4.1. The separation column consists of twelve cartridges arranged around the rotor of a centrifuge, the inner volume of each cartridge is ca. 15 ml for 50 channels. This results in a CPC column of around 190 ml for 600 channels. The rotary seal joint consists of two disks (one made from graphite and the other from ceramics, both highly polished and pressed against each other by a spring), one of which is fixed while the other rotates with the axis of the centrifuge. This first generation of rotary seal joints was rather fragile and one of the weak spots of the instrument. The other weak point was the manual speed control whereby a belt had to be moved by hand from one pulley to the other.

Despite these early drawbacks, encouraging results were obtained with this prototype, and two important laws were apparent from this very first paper:

- The resolution between peaks varies little, or not at all, with the flow-rate over a wide range.
- Higher rotational speeds give better peak resolution.

These two laws stay true for all subsequent CPC instruments.

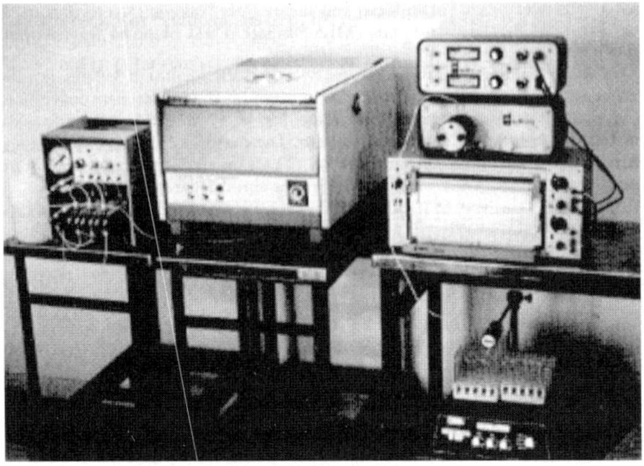

Fig. 4.1. The first Sanki prototype called the CCCC-RJ.

4.2.3 Change of name from CCCC-RJ to CPC

During the period 1983–1986, six papers appeared in the literature, three from Japan and three from the USA, and from them we can select 1986 as the date for the adoption of the CPC acronym [8–13]. In the first paper by T. Okuda et al. [8], tannins had been isolated by Centrifugal Counter-Current Chromatography, with no details given but with reference to the first Sanki paper [5]. During the summer of 1983, Kanichi Nunogaki sent a prototype instrument to his friend Koji Nakanishi, professor of chemistry at Columbia University, New York. At that time, Koji was confronted with the problem of isolating Tunichromes from Tunicate blood, and Dr. Reimar Bruening was heading the project. Tunichromes are extremely labile compounds and decompose extensively during adsorption chromatography, but they could finally be separated, thanks to CCCC [9,10]. In these two papers, Reimar Bruening called his Sanki instrument the 'CCCC' or 'C^4'.

In February 1984, Dr. Fadila Derguini, on leave of absence from the Centre National de la Recherche Scientifique (CNRS), the French state research institute, joined Nakanishi's group, and she and Reimar became very familiar with the Sanki machine. In 1985 they acquired a new instrument; the accompanying manual (Fig. 4.2) stated that the final choice between 'Centrifugal Counter-Current Chromatography' and 'Centrifugal Partition Chromatography' had not yet been made. At that time, Sanki had an informal agreement with E.J. Tolle, president of

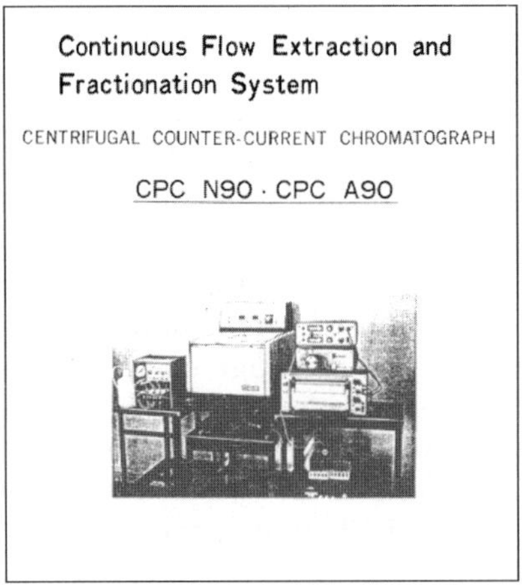
Continuous Flow Extraction and
Fractionation System

CENTRIFUGAL COUNTER-CURRENT CHROMATOGRAPH

CPC N90 · CPC A90

Fig. 4.2. Facsimile of the cover of the first manual of the CPC N90 (later named CPC-LLN), highlighting Sanki's hesitation to name the instrument CCCC or CPC.

Imass, to sell their instrument in the USA. Some visitors came to Nakanishi's laboratory to see demonstrations performed by Reimar and Fadila. However, Imass was not able to sell any instruments manufactured by Sanki [14].

The decision to keep "CPC" as a generic name for the Sanki instruments seems to have occurred during the first symposium organized by Sanki in Kyoto, in early Spring 1986. At that time Sanki was also manufacturing refractive index detectors for Showa Denki (Shodex detectors) which sold quite well. Kanichi Nunogaki, the president of Sanki, was "brilliant, a strong drinker, and probably not organized", as his close friend Koji Nakanishi liked to joke. Let us now quote Reimar Bruening: "I remember discussing the name CCCC with the Nunogakis during the first symposium in Kyoto. I felt the acronym was too long and silly sounding, but more importantly, I pointed out that it was actually not a real counter-current situation, since one phase was kept stationary by centrifugal force. Out of that discussion came the more appropriate name, CPC".

After this memorable first symposium, the acronym CPC and the name "Centrifugal Partition Chromatography" were adopted by all scientists using the Sanki instruments. A few years later it became a trademark. The term 'counter-current' was left to the numerous designs of two-axis instruments invented by Y. Ito and described in other chapters of this book.

Two anecdotes may be given before leaving the story of the name. In October 1990 we had to withdraw an article from the Counter-Current Chromatography thematic issue of J. Chromatogr., 538, No. 1 (1991), with Y. Ito, M. Knight, Y.W. Lee and W.D. Conway as guest editors, because the two referees for our paper did not accept our calling the Sanki instrument a 'CPC'. On the other hand, in the same issue was a paper by A. Weisz et al. [15] entitled, "Preparative purification ... by Centrifugal Counter-Current Chromatography", and using an Ito CCC instrument (a horizontal flow-through coil planet centrifuge) instead of the early CCCC instrument made by Sanki. Joking apart, all these early hesitations between acronyms and the excessive use of them we find in several books, did not make it any easier for those who wanted to get involved in Countercurrent Chromatography.

4.2.4 The CPC-LLN

Description of the instrument
This instrument was introduced in 1984 and sold until 1992; it has now been replaced by the HPCPC (*vide infra*). Figure 4.3 shows the instrument with its accessories, a power supply, a pump, a valve module with

Fig. 4.3. CPC-LLN with its accessories.

Fig. 4.4. Upper view of the rotor of the CPC-LLN arranged with six 250 W cartridges.

the injection valve and the elution mode valve, a UV detector, and a fraction collector. The CPC main frame is a 44×67×70 cm centrifuge operating in the 200–2000 rpm range. Its temperature can be regulated from 15 to 35°C in an ambient temperature of 25°C. Both the power supply and the centrifuge are equipped with a 24-hour timer for automatic unattended operation and shutdown. The rotor is a 30 cm (diameter) by 15 cm (height) cylinder made with compartments that receive cartridges of two kinds (12-cartridge capacity). The cartridges consist of engraved polychlorotrifluoroethylene (PCTFE) plates separated by Teflon seal sheets and maintained in position by two stainless steel plates.

Partition cartridge-type 250W: There are 400 channels per 18-ml cartridge. Each cartridge consists of PCTFE plates, with 50 channels on each side of each PCTFE plate. A channel is 12.4 mm long by 2.4 mm wide, and a duct is 12.4 mm long by 0.9 mm wide; they are both 1.1 mm deep. The total internal volume is then the sum of 13 ml for the 400 channels, and 5 ml for the ducts, that is 73% and 27% of the total cartridge volume, respectively. Up to 12 cartridges can be connected in series, but most of the applications found in the literature were performed using three or six cartridges. Figure 4.4 is an upper view of the rotor equipped with six cartridges.

Partition cartridge-type 1000E: There are 40 channels per cartridge. The cartridges are similar to the type 250W, except that each channel is larger, which results in fewer channels per cartridge. They are called preparative cartridges, with a total nominal volume of 75 ml each, or 900 ml for an entire CPC column of 12 cartridges.

The rotary seal: The rotary seal of the CPC-LLN consists of two drilled rods terminated with carefully polished disks, which are maintained in close contact by a spring. One rod is made of a hard ceramic while the other is made of soft graphite and acts as a dry lubricating face. This assembly is able to maintain a leak-free connection up to 60 bars, provided the two facing disks are extremely clean. Special attention must be paid if the liquid phases, such as buffers, contain some dissolved solids, as the evaporating solvent could leave crystals on the sealing surfaces. They are, however, easy to dismantle and wash by sonication in hot water and solvents.

The late eighties

The first appearance of the instrument can be found in Refs. [11–13], but the first full description of the CPC-LLN appeared in *American Laboratory* in February 1987 and was written by Dr. J. Cazes and K. Nunogaki, to promote the instrument. The article appeared after Dr. Cazes had joined Sanki Laboratories to market the CPC-LLN in the United States and to support Mr. Hiroshi Nunogaki, the son of Kanichi Nunogaki, who was now in charge of the USA distribution. The headquarters of Sanki Lab. were at Sharon Hill, Pennsylvania, for a few years, then at Mount Laurel, New Jersey, until January 1994. Jack Cazes was a specialist in HPLC, having worked for several industrial companies (e.g., Waters, and Varex), and had a thorough knowledge of the industrial branch of chromatography. He was a brilliant businessman and the story of the CPC-LLN outside Japan is intimately linked to him. He is now retired but is still the editor of *Journal of Liquid Chromatography & Related Technology* (Marcel Dekker), and we would guess that 99% of all chromatographers probably know of him.

The beginning of 1988 was marked by the publication of the first book on Countercurrent Chromatography edited by N.B. Mandava and Y. Ito [16], in a series supervised by Jack Cazes and published by Marcel Dekker (*Chromatographic Science Series*). Chapter 7 written by R.C. Bruening et al. was devoted to CPC, with a few examples from Nakanishi's laboratory in New York and from the Sanki laboratory in Kyoto. During the same period, two papers appeared by H. Terada et al., dealing with the determination of partition coefficients by CPC [17,18], and prefiguring several important contributions from around the world.

At that time Dr. A. Berthod joined Professor D.W. Armstrong at the University of Missouri-Rolla just before I joined Koji Nakanishi at Columbia University in New York. Alain Berthod and I did not know each other and it was probably a coincidence that the French CNRS, our common institution, allowed us to go to the USA to do research. Moreover, while Berthod and I were sure to do research in Countercurrent Chromatography, we did not know the difference between CCC and CPC, our knowledge of CCC being based on reading Ito's articles. That is why I think both Berthod and I had, by pure coincidence, the extraordinary fortune of joining two laboratories where CPC was a hot topic, and were thus able to bring back our knowledge to France. This period produced a wealth of articles by Berthod [19–25] and by myself [26–30].

The same period was also the starting point of the extensive use of the CPC-LLN by Professor K. Hostettmann and Dr. A. Marston who acquired an instrument in August 1986 in Lausanne, and for Dr. R. Verpoorte who acquired one in 1988 in Leiden. These scientists were aware of CPC because in 1985 Dr. Verpoorte had met K. Nunogaki in Japan, while K. Hostettmann had been a post-doc in Nakanishi's laboratory in New York. Those were the first of many publications by these authors with this instrument [31,32].

The 2nd CPC colloquium, Kyoto, 4 June 1988

I was fortunate to be invited by Sanki Engineering for a week in Kyoto to give a 50-minute talk on my work at Nakanishi's laboratory. Let me take this opportunity to express my gratitude to Sanki for their generosity towards me: they attended to all my needs during my stay at Nakanishi's laboratory, with no obligation but to understand better how to get the best out of a CPC instrument.

This second symposium was the first where non-Japanese scientists could talk on their use of the CPC[2]. D.W. Armstrong reported his work with A. Berthod [33]. R.G. Salomon talked about the purification of a water-sensitive natural product with an aprotic solvent system [34], and I talked about the separation of two large and highly hydrophobic peptides [26]. The presence of three (out of nine) foreign (i.e., non-Japanese) speakers, along with the arrival in Europe of R. Verpoorte's and K. Hostettmann's CPC instruments, and the marketing activities of Jack Cazes in the USA, definitely marked the beginning of the widespread use of CPC outside Japan in 1987–1988.

Figure 4.5 shows the Sanki Staff and speakers of the second colloquium. Everybody is there except K. Nakanishi, who gave the general

2 Except for Reimar Bruening, who gave a talk at the first meeting in Kyoto.

Fig. 4.5. The second CPC colloquium, Kyoto, 4 June 1988. Kanichi Nunogaki (president of Sanki Engineering Ltd.) is in the centre, between Fadila Derguini and Yoshiaki Nunogaki. Behind him is Prof. D.W. Armstrong.

introductory talk ("The Significance of Isolation"). He was always rushing between the hotel, congress hall and another hall as, in addition to the Sanki Conference, he was giving talks at the 16th IUPAC Natural Products Symposium and the 3rd International Conference on Retinal Proteins, all at overlapping times in Kyoto. Fortunately, I was able to get a picture of him while he was giving a talk in a restaurant, highlighting the merits of natural products such as wines and sushi (Fig. 4.6). The articles corresponding to the talks given at the 2nd CPC colloquium are grouped in *J. Liq. Chromatogr.*, Vol. 11, No. 12, 1988.

Fig. 4.6. Prof. Koji Nankanishi talking about natural products. He was attending the 2nd CPC Colloquium, the 16th IUPAC Natural Products Symposium and the 3rd International Conference on Retinal Proteins, all at the same time.

The 3rd CPC colloquium, San Mateo, California, 21–22 June 1990
While the 2nd CPC colloquium was characterized by having three out of nine non-Japanese speakers, the 3rd CPC colloquium had only two Japanese out of eleven (Professor Araki from Kyoto and Professor Okuda from Okatama). Seven speakers were from the USA, highlighting the work accomplished by Jack Cazes in the USA from 1987 to 1990. The CPC-LLN was by now very well known outside Japan, mainly in the USA.

Unfortunately, the CPC-LLN suffered from very fragile rotary seals that quickly started to leak if not kept extremely clean. It was not possible to repair a leaking rotary seal on site, which resulted in time lost while the rotary seal had to go to Japan and back to the USA. Sanki was aware of this problem since 1985. In 1986, Reimar Bruening, while travailing with Hiroshi Nunogaki on the Bullet train (Sanki had asked him to give seven seminars to various corporations in Japan), had the idea for a new rotary seal design, based on piston seals like the ones on HPLC pumps, and a rotating hollow sapphire rod. He drew the design on a napkin for Hiroshi to show to his uncle (Yoshiaki Nunogaki) who was the chief engineer at Sanki. In 1991 these new rotary seals appeared with the HPCPC series 1000 (HP for high performance, *vide infra*). They were clearly a lot more reliable than those mounted on the CPC-LLN.

4.2.5 The HPCPC series 1000

The body of this instrument was shown at the 2nd CPC Colloquium in Kyoto, June 1988, but it was still an empty box, the redesign of the rotor being in progress. It was introduced at the 3rd CPC Colloquium in San Mateo, June 1990, and presented for the first time at the 1991 PittCon in Chicago. In October/November of 1991 I and one of my students had the privilege of being the first scientists to test an HPCPC. This was by courtesy of Jack Cazes who graciously loaned us the first unit received in the USA. Needless to say, we were enthusiastic about our results [35].

Description of the instrument
The HPCPC is a bench-top CPC (31×47×50 cm) weighing around 43 kg, and operating in the range 200–2000 rpm. Its temperature cannot be regulated, and it has no automatic shutdown facilities. Figure 4.7 shows the instrument. The four-way valve, which controls the elution mode, is mounted on the HPCPC so that the instrument can be connected

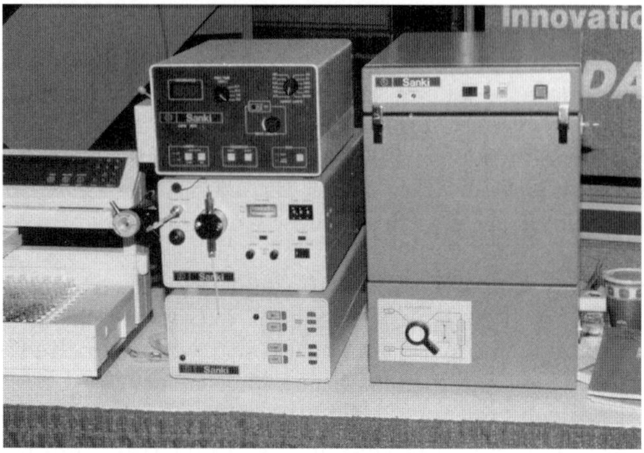

Fig. 4.7. The HPCPC.

Fig. 4.8. Upper view of the HPCPC, with the pack of disks and the upper rotary seal.

between the outlet of a sample injector and the inlet of a detector, in a manner similar to a standard HPLC column. The rotor consists of one pack of twelve disks (two packs of six disks each, for the early instruments), connected through 1/16″ tubing, and is easily removable. The upper side of the pack is shown in Figure 4.8. A disk is made of polyphenylene sulfide (PPS) and consists of 89 channels, and ducts on each side which have been obtained by injection moulding. A detail of a disk is shown in Fig. 4.9.

A channel measures 15×2.8×2.1 mm and a duct is 15×1×1 mm, which results in an internal volume of ca. 88 μl per channel; the total

Fig. 4.9. Detail of a disk, with channels and ducts.

volume for the pack is around 220 ml — 85% of that volume being for the channels, and 15% for the ducts. The rotary seals (the upper one can be seen in Fig. 4.8) are made of a drilled sapphire rod going through two toroidal seals similar to those used around HPLC pump pistons. They need much less maintenance than the previous seals used with the CPC-LLN. Only one toroidal seal moves relative to the sapphire rod, and it is very easy to replace with a seal insertion tool comparable to that used for HPLC pumps.

Figure 4.10 shows both the CPC-LLN (right) and the HPCPC (left). The smaller size of the HPCPC is mainly due to the absence of

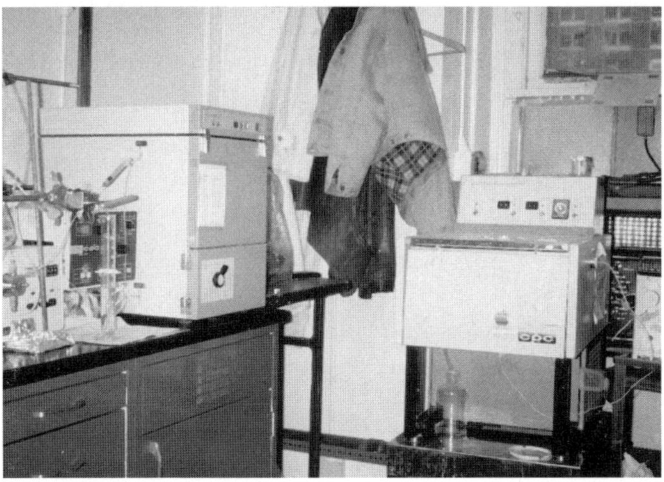

Fig. 4.10. An HPCPC unit (left) and a CPC-LLN in the Nakanishi laboratory (1993).

temperature control on this instrument. Intensive use of a HPCPC has proved that there is only a small increase in the temperature of the rotor during its rotation. In most cases this has no effect on the liquid–liquid equilibrium between the two phases, unless the chosen biphasic system is too close to the binodal (see Chapter 3 by J.H. Renault).

Sanki at the Pittsburgh Conference
Sanki US exhibited at six PittCon meetings, from 1988 (New Orleans) to 1993 (Atlanta). I had the good fortune to attend all these meetings. For my first PittCon, I was invited by the Chromatography Forum of the Delaware Valley for the Del Nogare Award of Professor Harold Walton, a wonderful scientist who had invented Ligand Exchange Chromatography in 1954 [36]. Sanki supported me partly or fully for the other PittCon meetings, where I gave talks on CPC at the CCC sessions, always chaired by Dr. N.B. Mandava. There I met Dr. Ito, Professor Conway, Dr. Chou (Pharmatec), Mrs. Carmeci (P.C. Inc.), and many other scientists, most of them working with CCC instruments, while only a few of them had a CPC: Dr. S. Gluck (Dow Chemical), Dr. S. Muralidharan (University of Arizona) and D.W. Armstrong (University of Missouri-Rolla).

The Sanki booth was small but attractive. I remember that Jack Cazes anticipated the coming of the HPCPC, as shown in Fig. 4.11 where we see the 1990 PittCon in New York. The HPCPC acronym is here but a CPC-LLN is on the table. Two years later, at the 1992 PittCon in New

Fig. 4.11. Sanki booth at the 1990 PittCon, New York. Y. Kosuge (left), H. Nunogaki (centre) and J. Cazes (right). The HPCPC acronym is there, but the instrument has not yet reached the USA, and the CPC-LLN is shown.

Fig. 4.12. Sanki booth at the 1992 PittCon, New Orleans. From left to right: Dr. J. Cazes, W. Murayama and H. Nunogaki. The HPCPC is available.

Orleans (Fig. 4.12), the HPCPC had arrived, here with H. Nunogaki, W. Murayama and J. Cazes (right to left). The expression of Jack Cazes was "Tomorrow's innovations ... today!": it was a great idea, and he was always pushing Sanki Japan to go faster, but complained that the Nunogaki family was too slow. Jack Cazes always had good stories to make you happy, and all these PittCons with him constitute a wonderful part of my life.

From Sanki US to EverSeiko Corp.

In 1992, the need for better distribution of the series 1000 HPCPC in Europe being obvious, Sanki US started collaboration with Varex in order to sell the HPCPC through Varex representatives. To my knowledge, the first laboratories to acquire an HPCPC in Europe were French. They were the Laboratory of Pharmacognosy of Professor Le Men-Olivier and Dr. G. Massiot at the University of Reims and the Laboratory of Analytical Science (A. Berthod) at the University of Lyon 1. Collaboration between Varex and Sanki US became very efficient, and the Varex representatives were trained to understand fully what they were selling. The HPCPC was widely accepted in universities and industries, and this golden age lasted for about a year, when quite suddenly J. Cazes was forced to stop doing business with Sanki in January 1994. Figure 4.13 is the facsimile of the letter received by users of Sanki CPC equipment. This letter was abrupt, with no explanation at all for this amazing and damaging decision. We tried to obtain more

SANKI LABORATORIES, INC.
FELLOWSHIP BUSINESS CENTER
520 FELLOWSHIP ROAD
SUITE D-406
MOUNT LAUREL N J 08054
(609) 231-4496 (609) 231-4599 (FAX)

January 10, 1994

TO ALL USERS OF SANKI CPC EQUIPMENT

I must regrettably inform you I am leaving Sanki Laboratories, Inc. at the end of this week. I know you have relied upon me and Sanki to provide technical support and replacement parts for your CPC instruments, and you have a right to expect this support. Unfortunately, I do not know what will happen to the Sanki office in the U.S., but I will be willing to answer your technical questions at my office at home. I will keep a limited stock of replacement parts and can supply them to you directly until my stock is depleted.

If you need replacement parts, order them directly from The Cazes Group, P. O. Box 2180, Cherry Hill, NJ 08034-0162. I can be reached by phone at (609) 424-3505 or by FAX at (609) 751-8724. I will also be available, on a consulting basis, to provide more extensive technical support to you. Call me to discuss consulting service.

I apologize for any inconvenience this situation will cause you. Thank you for your support and for your understanding. I look forward to being helpful to you in the future.

Sincerely,

Dr. Jack Cazes

P.S.: If you would want to contact Sanki in Japan, you may write to Mr. Kanichi Nunogaki, President, Sanki Laboratories, Inc., 2-16-10 Imazato, Nagaokakyo, Kyoto 617, Japan (FAX: 81-75-954-3540).

TOMORROW'S INNOVATIONS...TODAY!

Fig. 4.13. Facsimile of J. Cazes' letter informing people of the end of his collaboration with Sanki.

information on the reasons leading J. Cazes to end this collaboration, but received no comments.[3]

---

3 Editor's comment: As mentioned in the text, Jack Cazes had an extensive knowledge of the industry's needs. He always thought that CPC would be an excellent tool in industry. In 1993, he was able to involve a major pharmaceutical company in CPC after proving the capability of the HPCPC machine in their particular purification requirement. This company was willing to order very large-scale CPC units. The contract was ready and in the 7-figure range (in US $!). Due to an economic crisis in Japan and probably Kanichi Nunogaki's health problems, Sanki rejected the order. This veto ruined all Cazes' efforts to promote the CPC technique in the US industry. It was considered by him to be a disapproval of his hard work implying his resignation.

After J. Cazes' departure the situation remained very confused for several months while we were finishing a book devoted to Centrifugal Partition Chromatography [37], published by Marcel Dekker in the Chromatographic Science Series and edited by Jack Cazes (so the story goes). The book was scheduled for late 1994, and we were all close to thinking that it would be useless, because of the disappearance of the technique described in it!

Despite some major problems in Japan, Sanki wanted to pursue the manufacture of the HPCPC (and of larger units). They were lucky to find in Mr. Hideaki Matsuoka, CEO of EverSeiko Corp. in Tokyo, the right businessman to become the worldwide sole distributor of the CPC instrument. Mr. Matsuoka was a young, brilliant and very nice businessman, knowledgeable about chromatographic equipment. He was already doing business mainly in Asian Countries. Starting in June 1994, it took him only a few months to prepare a new brochure for the various CPC models, and to start some new collaborations, such as that with JM Sciences in the USA.

After the death of Dr. Kanichi Nunogaki, the manufacture of the CPC was entrusted to EverSeiko and System Instrument Corp. (SIC) in late 1999 [38].

4.2.6 Large-scale instruments

Very little detailed information is available concerning larger instruments. This is because they are custom-designed for specific separation processes. Their basic design and operation are similar to the smaller laboratory preparative units. The lack of detailed information concerning these instruments is probably due to the proprietary separations they perform. Table 4.1 shows the main characteristics of large-scale CPC. The interested reader will have to contact EverSeiko for additional detailed information.

A CPC LLI-7 (formerly LLI-005) is shown in Fig. 4.14. For comparison there is an HPCPC on the left; the control box of the large-scale unit is on the upper side of the picture. This 5-L CPC is supplied with two preparative pumps (Shimadzu), an additional syringe pump (Isco) for injection of large volumes, and a fraction collector. The upper side of the 5-L rotor is shown in Fig. 4.15; it looks similar to the bench-top HPCPC, except for the connections which are made with 1/8″ tubing. The rotary seals are in proportion, too. Figure 4.16 shows industrial-size equipment in Japan; we have no precise information

TABLE 4.1

Specifications of large-scale centrifugal partition chromatographs

Feature	LLL-1.5	LLI-1.7	LLI-30
Internal volume (litres)	1.36	5.4	27–30
Capacity (max) (g)	50	200	1000
Flow rate (approx.) (ml/min)	20–40	50–140	400–700
Max rpm	1500	1500	1000
Rotor radius (cm)	15	15	25
Number of channels	1296	1040	792
Channel length (cm)	2.9	4.6	8
Number of rotor packs	1	2	2
Number of disks per rotor pack	12	12	11

Fig. 4.14. General view of the CPC model LLI-7 (5 L) and its accessories. An HPCPC unit can be seen on the left.

Fig. 4.15. Upper view of the LLI-7, where we can admire the quality of the finish of the instrument. Compare with Fig. 4.8.

Fig. 4.16. Industrial equipment in Japan. This could be a 30 L CPC supplied with a large three-head pump.

concerning this installation, but this could probably be a 30-L CPC supplied with very large three-head pumps.

There are not many examples of separations with large-scale CPC in the literature. Figure 4.17 shows the preparative separation of unsaturated fatty acid ethyl esters using a CPC LLI-7 [39]: 300 g of a crude mixture was injected, and the dual mode elution used to perform the final purification of C18:1, C18:2, C18:3 ethyl esters. Considering the

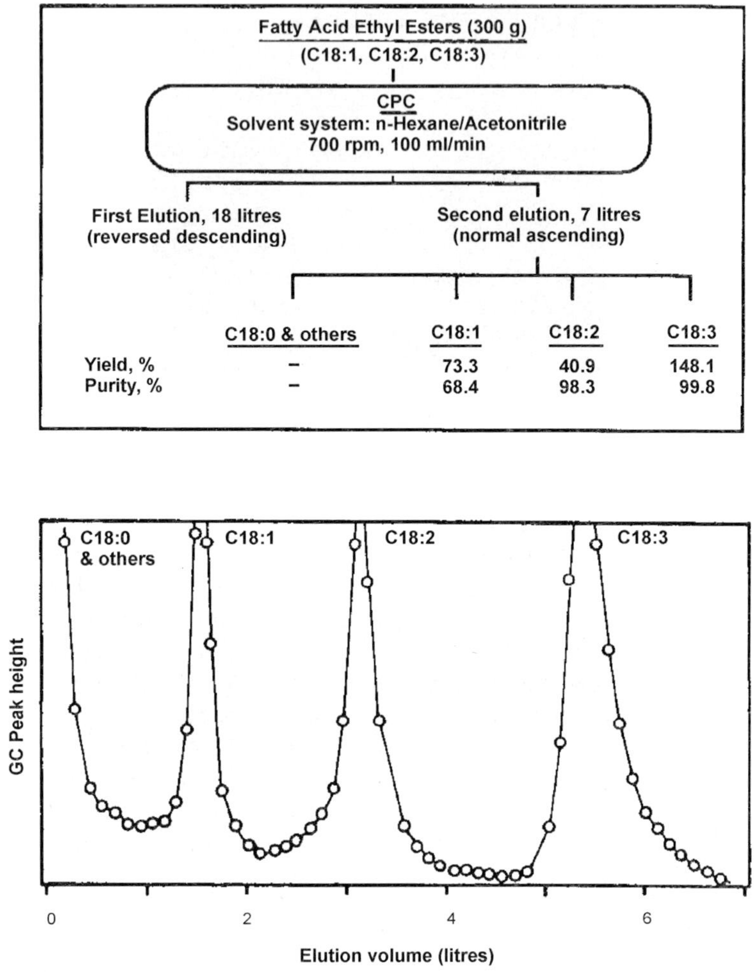

Fig. 4.17. Large-scale separation of linolenic, linoleic and oleic acid ethyl esters with a CPC LLI-7 (5 L). Operating parameters are shown on the flow sheet.

flow-sheet, it needs 25 l of solvent to purify 262 g, i.e. 95 ml/g of purified product, *which is around ten times less solvent than needed for the same purification by preparative HPLC* [40]! Another example can be found in [30], with the purification of di-homo-γ-linolenic acid ethyl ester (C20:3) after injection of 130 g of a mixture in a CPC LLI-7.

4.3 NEWCOMERS AND NEW APPROACHES

4.3.1 The Dutch School

In July 1991 a collaboration was started between Rob Verpoorte, who was already aware of CPC, at the Leiden University, Gorlaeus Laboratories (Drug Research), and Luuk van der Wielen and Karel Luyben, working at the Kluyver Institute for Biotechnology, Delft University of Technology, who had a thorough background in chemical engineering. They had a need to scale up the CPC instrumentation and wanted to understand better all the phenomena involved in CPC, but were hampered by a lack of knowledge concerning the flow behaviour of the liquid–liquid two-phase system.

At first, CPC was used essentially by natural-product chemists who where interested in their molecules and not in the technique itself. To a lesser extent, CPC was also used by chromatographers who had an analytical approach with the use of 'plate height' or 'resolution', but who were generally unaware of hydrodynamics and chemical engineering. Armstrong et al. [25] developed a mass-transfer model to describe the efficiency of a CPC column as a function of the flow-rate of the mobile phase. They assumed that the bulk of the mass transfer occurs while the mobile phase is present as an emulsified band between the two phases. However, no visual evidence of the presence of such an emulsified band was presented. Foucault et al. [41] introduced the 'Stokes' model which describes the flow of the mobile phase in a channel as droplets with an average radius corresponding to Stokes' Law. The average radius is calculated from the average linear velocity of the mobile phase in a channel, which in turn is calculated from stationary-phase hold-up experiments [42]. The 'Stokes' model is used to explain the stationary phase hold-up of various two-phase systems with different properties, and the influence of rotational speed on efficiency. It was then suggested that a direct visualization of the mobile-phase flow through the stationary phase would be very useful.

Luuk van der Wielen and Karel Luyben therefore undertook the realization of a transparent CPC cartridge and a rotor, with the channel and duct dimensions and geometries equal to those of the commercial CPC-LLN. They interested Michel van Buel, who was beginning a Ph.D. on the project, and this group of scientists then achieved a milestone in the story of CPC, as evidenced by their publications [43–48]. They presented photographs of the flow-states inside a CPC column, and

Fig. 4.18. Michel van Buel's defense, 13 October 1997, at Delft University of Technology (The Netherlands). Michel is in the centre, on his right is Prof. L.A.M. van der Wielen; on his left is Prof. K.Ch.A.M. Luyben and Dr. A. Foucault. Left of the picture is Prof. R. Verpoorte.

through a chemical engineering approach they were able to model the effluent concentration profiles [46], and gradient-elution profiles [45], and to adapt the model to the case of large injections of weak acids (hop acids) [49,50]. More details of their investigations are found in Chapter 5 of this book, by L. Marchal et al. I had the great pleasure of being a member of the committee examining Michel van Buel's thesis, on October 13, 1997. Figure 4.18 shows all the committee members surrounding Michel. Behind the apparent formal solemnity of this ceremony was an extremely warm relationships between professors and students. The closing speech by Prof. K.Ch.A.M. Luyben seems to have been especially cordial, according to Michel's and the audience's faces, while I was a little frustrated since he was speaking Dutch (half of the ceremony was in English, half in Dutch).

Unfortunately, it does not seem that all these fundamental studies induced the realization of new commercially available CPC instruments. I was told that a prototype of a 5-L CPC has been built at Delft University, but four years later I still have no details.

4.3.2 The French School

By courtesy of CNRS, I was allowed to move to Nantes in June 1996, and started collaborating with several scientists in Nantes and Saint Nazaire (western France) with the aim of developing a new CPC apparatus. Largely helped by the previous works of the Dutch school, we set up a visualization system to film flow patterns in a single channel of a transparent CPC prototype at any rotational speed. Our observations not only confirmed those of Michel van Buel and his colleagues [48], but also supplemented them. We were then lucky to welcome Luc Marchal, a young graduate student starting a Ph.D.; Luc knew how to give to our project a larger impact than was initially foreseen. Details of our observations are given by Luc himself in Chapter 5 of this book.

It must be pointed out that in channels in rotation, the trajectories of jets and droplets of mobile phase into the stationary phase are not radial at all, despite the strong centrifugal field generated by the rotation [51,52]. This is owing to the Coriolis acceleration, $\vec{\Gamma}_{Cor} = -2\vec{\Omega}\Lambda\vec{V}_R$, where $\vec{\Omega}$ is the rotation vector and $\vec{V}_R$ the relative velocity of the mobile phase in a channel. $\vec{\Gamma}_{Cor}$ deviates any moving object towards the right (relative to $\vec{V}_R$), when subjected to a trigonometric rotation. Our fruitful observations proved that the action of the Coriolis acceleration was more powerful than previously though, and that it was very important to take it into account for future design of CPC channels. A complete hydrodynamic study will be found in Chapter 5.

Our collaboration with ANVAR (French agency for the development of industrial innovation), IFREMER (French national research institute of exploitation of the sea), and the French company Kromaton (formerly SEAB) [53], resulted in a first series of CPC instruments called FCPC, for Fast Centrifugal Partition Chromatographs.

The FCPC

This first French instrument is a bench-top model (58×36×61 cm, 80 kg) equipped with a pack of 20 disks for 1300 channels and a column volume of around 200 ml. Its originality lies in the geometry of the channels, which are still confidential, with patent pending [The patent was granted on Feb. 9, 2000, ref. FR 2791 578]. The FCPC is characterized by a very good retention of the stationary phase, which allows high flow-rates and short analysis times, and a good efficiency, when used at adequate rotational speed (ca. 1200–2000 rpm) and flow-rate (5–20

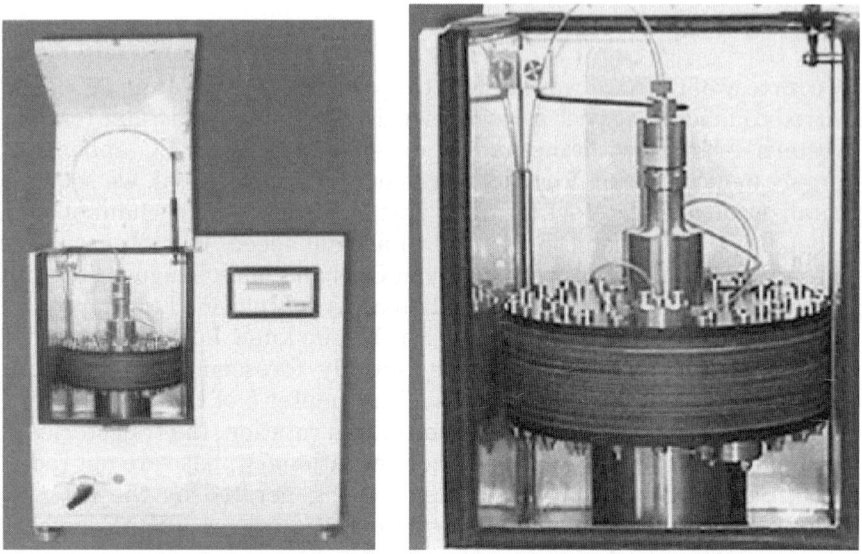

Fig. 4.19. General view of a FCPC (left); close view of the rotor (right).

ml/min). Figure 4.19 shows a general view of a FCPC. The fast performance of the FCPC will be highlighted by two examples.

Figure 4.20 compares the performance of the instrument in the elution mode when used in the wrong way, i.e., with slow rotational speed and low flow-rate, and in the right way, i.e., high rotational speed and high flow-rate. Increase of the efficiency, owing to a better mixing of the mobile and the stationary phases in each channel (see Chapter 5) greatly compensates for the light loss of stationary phase owing to a higher flow-rate. For this example, the increase in efficiency is by a factor of 8 (or $\sqrt{N}$ varies by a factor 2.8), while the stationary-phase ratio varies by a factor only 1.3. Consequently, with proper operating conditions, the resolution is greatly enhanced and the duration of analysis is drastically shortened.

Figure 4.21 is an example of ion-exchange displacement chromatography of Z-dipeptides (Z = benzyloxycarbonyl), and already obtained by Ma and Ito [54] using a commercial model of high-speed counter-current chromatograph (HSCCC, P.C. Inc., Potomac, Maryland, USA). A complete description of ion-exchange displacement chromatography (also called pH-zone refining), will be found in Chapter 7 by A. Berthod. One characteristic of ion-exchange displacement chromatography is the

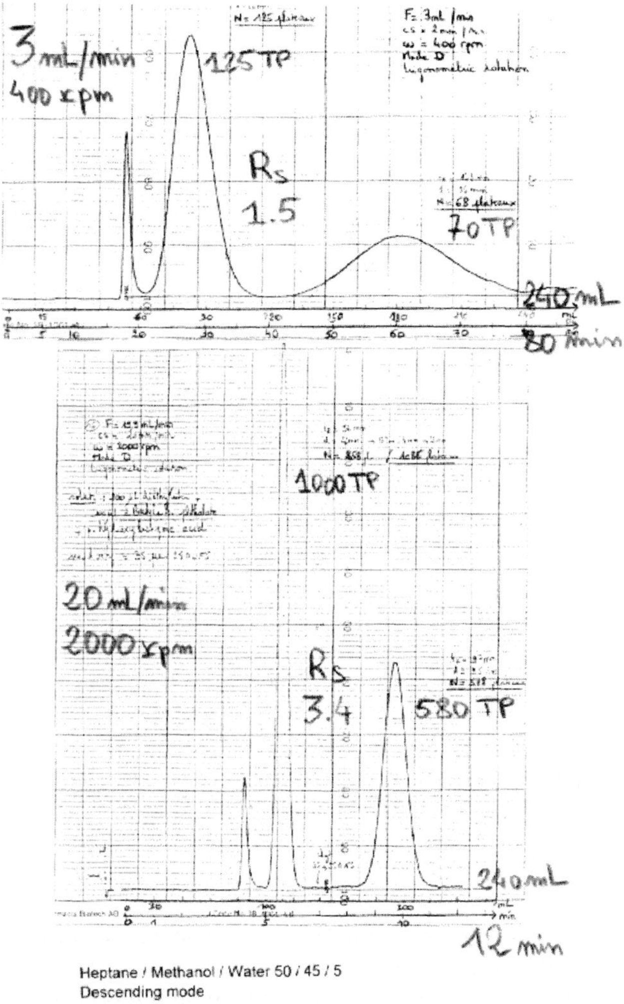

Fig. 4.20. Influence of both flow-rate and rotational speed upon efficiency of the FCPC. Heptane/methanol/water (50/45/5) biphasic system, descending mode; F and ω as indicated; sample: unweighted but identical mixture of *p*-hydroxybenzoic acid (a marker for t_0) diethylesterphthalate and dibutylesterphthalate.

release of products from the column in continuous bands arranged according to their pK_a values and partition coefficients. For this chromatogram, the dark line indicates the pH of the effluent, which is controlled

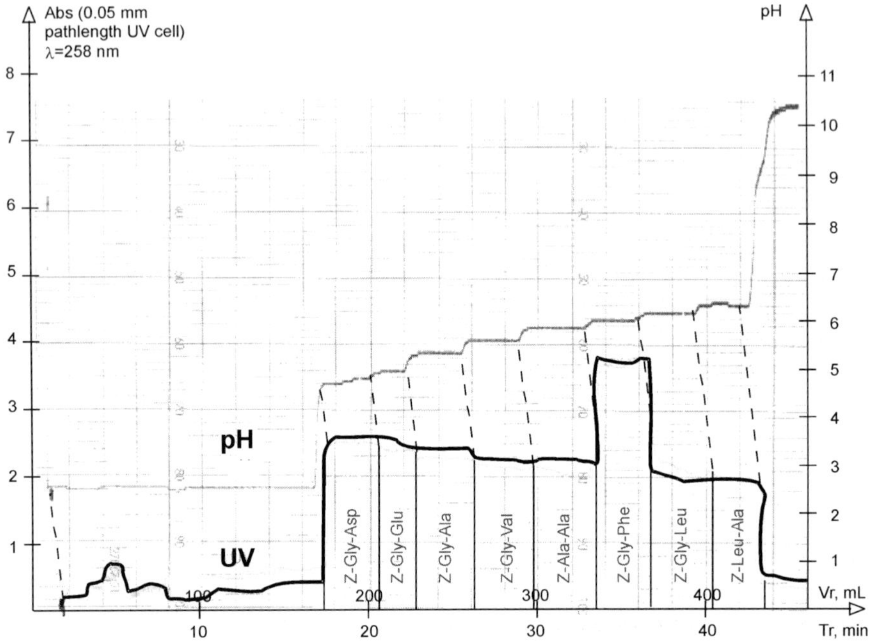

Fig. 4.21. Ion-exchange displacement chromatography (also called pH-zone refining) of Z-dipeptides using the FCPC. Methyl *tert*-butyl ether/acetonitrile/water 2/2/3 biphasic system. Stationary phase: upper phase with trifluoroacetic acid 16 mM. Mobile phase: lower phase with ammonia 11 mM. Descending mode, F = 10 ml/min, stationary phase ratio from 67% to 53%, rotation: 1200 rpm. Injection: 100 mg of each Z-dipeptide in 5.5 ml of stationary phase.

by the pK_a of the analyte, and the grey line corresponds to the UV absorbance at 258 nm, recorded with a prep-scale UV monitor (0.05 mm pathlength). It can be seen that Z-Gly-Phe, which has two aromatic rings instead of one for the other Z-dipeptides, is clearly defined here. This evidences the quality of the borderlines between zones generated by the displacement mode. For a similar retention of the stationary phase (≈ 65% for the HSCCC, ca. 60% for the FCPC), but different flow-rates (3.3 ml/min for the HSCCC, 10 ml/min for the FCPC), the time for the separation is only ca. 50 min with the FCPC compared to 230 min when using the HSCCC. This time difference is mainly due to the very good retention of the stationary phase in the FCPC at higher flow-rates.

4.4 FUTURE TRENDS

The story of CPC instruments has for almost twenty years been the story of a family, the Nunogakis of Kyoto, and of a company, Sanki Engineering Ltd. Through success and ordeals, they gave the CPC technique (i.e., the one-axis centrifugal chromatographic system with two rotary-seal joints) a place among the large panel of CCC instruments (i.e., the two-axis centrifugal chromatographic systems with no rotary-seal joint). CPC apparatus is easier to scale up than the corresponding CCC instruments, and they allow almost any imaginable liquid–liquid biphasic system to be operated with no problem of stationary phase loss: these could be the two most important differences between CPC and CCC. However, a CPC apparatus is still much more expensive than a CCC machine. It is desirable for the scientific community that other companies should develop new and more affordable instruments. Imagine what HPLC might be today if only one company were to make HPLC instruments! It would also be desirable for new research groups to be involved in centrifugal partition chromatography. It is now time for CPC to grow and to prove its usefulness. At the time of writing this chapter, a small French company has started manufacturing FCPC instruments and we expect a great future for it. May this example prompt other manufacturers in the world to follow suit.

REFERENCES

1 R.E. Cornish, R.C. Archibald, E.A. Murphy and H.M. Evans, *J. Ind. Eng. Chem.*, 26 (1934) 397.
2 A.J.P. Martin and R.L.M. Synge, *Biochem. J.*, 35 (1941) 91.
3 A.J.P. Martin and R.L.M. Synge, *Biochem. J.*, 35 (1941) 1358.
4 Y. Ito and R.L. Bowman, *Science*, 167 (1970) 281.
5 W. Murayama, T. Kobayashi, Y. Kosuge, H. Yano, Y. Nunogaki and K. Nunogaki, *J. Chromatogr.*, 239 (1982) 643.
6 Y. Ito, M. Weinstein, I. Aoki, R. Harada, E. Kimura and K. Nunogaki, *Nature* 212 (1966) 985.
7 Y. Ito, I. Aoki, E. Kimura, K. Nunogaki and Y. Nunogaki, *Anal. Chem.*, 41 (1969) 1579.
8 T. Okuda, T. Hatano and K. Yazaki, *Chem. Pharm. Bull.*, 31 (1983) 333.
9 R.C. Bruening, E.M. Oltz, J. Furukawa, K. Nakanishi and K. Kustin, *J. Am. Chem. Soc.,* 107 (1985) 5298.
10 R.C. Bruening, E.M. Oltz, J. Furukawa, K. Nakanishi and K. Kustin, *J. Nat. Prod.,* 49 (1986) 193.

11 R.C. Bruening, F. Derguini and K. Nakanishi, *J. Chromatogr.,* 357 (1986) 193.
12 T. Okuda, T. Yoshida, T. Hatano, K. Yazaki, R. Kira and Y. Ikeda, *J. Chromatogr.*, 362 (1986) 375.
13 T. Okuda, T. Yoshida, T. Hatano, Y. Ikeda, T. Shingu and T. Inoue, *Chem. Pharm. Bull.*, 34 (1986) 4075.
14 E. Fusoni, Imass, Accord, MA, USA, personal communication.
15 A. Weisz, A.J. Langovski, M.B. Meyers, M.A. Thicken and Y. Ito, *J. Chromatogr.*, 538 (1991) 157.
16 N.B. Mandava and Y. Ito, *Countercurrent Chromatography, Theory and Practice*, Chromatographic Sciences Series, Vol. 44. Marcel Dekker, Inc., New York, May 1988.
17 H. Terada, Y. Kosuge, W. Murayama, N. Nakaya, Y. Nunogaki and K. Nunogaki, *J. Chromatogr.*, 400 (1987) 343.
18 H. Terada, Y. Kosuge, W. Murayama, N. Nakaya, Y. Nunogaki and K. Nunogaki, *Chem. Pharm. Bull.*, 35 (1987) 5010.
19 A. Berthod and D.W. Armstrong, *J. Liq. Chromatogr.*, 11 (1988) 547.
20 A. Berthod and D.W. Armstrong, *J. Liq. Chromatogr.*, 11 (1988) 567.
21 A. Berthod, J.D. Duncan and D.W. Armstrong, *J. Liq. Chromatogr.*, 11 (1988) 1171.
22 A. Berthod and D.W. Armstrong, *J. Liq. Chromatogr.*, 11 (1988) 1187.
23 A. Berthod, Y.I. Han and D.W. Armstrong, *J. Liq. Chromatogr.*, 11 (1988) 1441.
24 A. Berthod and D.W. Armstrong, *J. Liq. Chromatogr.*, 11 (1988) 1457.
25 D.W. Armstrong, G.L. Bertrand and A. Berthod, *Anal. Chem.*, 60 (1988) 2513.
26 A. Foucault and K. Nakanishi, *J. Liq. Chromatogr.*, 11 (1988) 2455.
27 A. Foucault and K. Nakanishi, *J. Liq. Chromatogr.*, 12 (1989) 2587.
28 A. Foucault and K. Nakanishi, *J. Liq. Chromatogr.*, 13 (1990) 2421.
29 A. Foucault and K. Nakanishi, *J. Liq. Chromatogr.*, 13 (1990) 2583.
30 A. Foucault, *Anal. Chem.*, 62 (1991) 569A.
31 A. Marston, C. Borel and K. Hostettmann, *J. Chromatogr.*, 450 (1988) 91.
32 A. Hermans-Lokkerbol, R. van der Heijden and R. Verpoorte, *J. Liq. Chromatogr.*, 16 (1993) 1433.
33 D.W. Armstrong, *J. Liq. Chromatogr.*, 11 (1988) 2433.
34 M.R. Jiroussek and R.G. Salomon, *J. Liq. Chromatogr.*, 11 (1988) 2507.
35 A.P. Foucault, O. Bousquet, F. Le Goffic and J. Cazes, *J. Liq. Chromatogr.*, 15 (1992) 2721.
36 H.F. Walton and R.H. Stokes, *J. Am. Chem. Soc.*, 76 (1954) 3327.
37 A.P Foucault (Ed.), *Centrifugal Partition Chromatography*, Chromatographic Science Series, Vol. 68. Marcel Dekker, New York, 1994.
38 EverSeiko Corp., 4F Takagi Bldg, 4-36-17, Ikebukuro, Toshima-ku, Tokyo 171-0014, Japan, matsuoka@everseiko.co.jp

39 W. Murayama, Y. Kosuge, N. Nakaya, Y. Nunogaki and K. Nunogaki, *J. Liq. Chromatogr.*, 11 (1988) 283.

40 C. Viron, R. Pennanec, P. André and M. Lafosse, *J. Liq. Chromatogr. Rel. Technol.*, 23 (2000) 1681.

41 A.P. Foucault, E. Camacho-Frias, C.G. Bordier and F. Le Goffic, *J. Liq. Chromatogr.* 17 (1994) 1.

42 A.P. Foucault, O. Bousquet and F. Le Goffic, *J. Liq. Chromatogr.*, 15 (1992) 2691.

43 M.J. van Buel, L.A.M. van der Wielen and K.Ch.A.M. Luyben, In: A.P. Foucault (Ed.), *Centrifugal Partition Chromatography*, Chromatographic Science Series, Vol. 68. Marcel Dekker, New York, 1994, Ch. 3, p. 51.

44 M.J. van Buel, L.A.M. van der Wielen and K.Ch.A.M. Luyben, *J. Chromatogr. A*, 773 (1997) 1.

45 M.J. van Buel, L.A.M. van der Wielen and K.Ch.A.M. Luyben, *J. Chromatogr. A*, 773 (1997) 13.

46 M.J. van Buel, L.A.M. van der Wielen and K.Ch.A.M. Luyben, *AIChE J.*, 43 (1997) 693.

47 M.J. van Buel, Thesis, Delft University of Technology, The Netherlands, 1997.

48 M.J. van Buel, F.E.D. van Halsema, L.A.M. van der Wielen and K.Ch.A.M. Luyben, *AIChE J.*, 44 (1998) 1356.

49 A.C.J. Hermans-Lokkerbol and R. Verpoorte, *J. Chromatogr. A*, 664 (1994) 45.

50 A.C.J. Hermans-Lokkerbol, A.C. Hoek and R. Verpoorte, *J. Chromatogr. A*, 771 (1997) 71.

51 A.P. Foucault, L. Marchal, J.M. Rosant, S. Farias Neto, J. Legrand and P. Durand, Presented at the Pittsburgh Conference, New Orleans, 1998, Abstract 878.

52 L. Marchal, A. Foucault, G. Patissier, J.M. Rosant and J. Legrand, *J. Chromatogr. A*, 869 (2000) 339.

53 Kromaton Technologies, 64 rue Pasteur, 94807 Villejuif, France, http://www.kromaton.com/

54 Y. Ma and Y. Ito, *J. Chromatogr. A*, 702 (1995) 197.

Chapter 5

Centrifugal partition chromatography: an engineering approach

Luc Marchal, Alain P. Foucault, Gerard Patissier, Jean-Michel Rosant and Jack Legrand

5.1 INTRODUCTION

The first CPC apparatus was introduced in 1982 [1] by scientists working for Sanki Engineering Ltd., and for almost 20 years this company remained the only one manufacturing such apparatus. By definition we call a Centrifugal Partition Chromatograph (CPC) a one-axis centrifugal system operating with two rotary seals. A CPC column is a series of cells linked in cascade by ducts and regularly disposed around a rotor as cartridges or disks, and submitted to a constant centrifugal acceleration generated by a rotation. Figure 5.1 gives a typical view of the upper side of a commercial instrument with its characteristic pack of disks and the upper rotary seal joint.

The first photographs of flow patterns in a CPC column were published in 1994 [2], as part of the main contribution of the Dutch School who adopted an innovative approach to studying criteria of performance of a CPC apparatus. Between these 1982 and 1994, scientists were working "blind" with these instruments, making assumptions about the behaviour of two non-miscible liquid phases subjected to a strong centrifugal acceleration, from studies of chromatographic profiles coming from them.

– Remark: in a sense CPC has followed a story opposite to that of column chromatography on solid support, which started with a glass column and coloured mixtures (Michael Tswett [3]) before disappearing in UV and in steel columns (HPLC).

Comprehensive Analytical Chemistry XXXVIII
A. Berthod (Ed.)

Fig. 5.1. Upper side of an HPCPC (Sanki Engineering Ltd.), with the pack of disks and the upper rotary seal.

We report here the first assumptions on flow patterns found in the literature, then we will highlight the contribution of the Dutch School before studying thoroughly some characteristic examples using our own results.

5.2 ASSUMPTIONS AND FIRST VISUALIZATIONS

5.2.1 The first image of CPC

The first image of CPC was given in 1982 by W. Murayama et al. [1] as shown in Fig. 5.2, a drawing seen in numerous publications. The idea is that while the rotor of the centrifuge is in motion, the mobile liquid phase is pumped into the cells and "bubbles down" if heavier, or "bubbles up" if lighter, through the stationary phase by the action of centrifugal force. The first experiments [1] showed that the resolution, R_s, between peaks was not significantly changed when the flow rate was increased. The resolution did improve with higher rotational speeds. If no explanation was given for the action of rotational speed, the apparent independence of R_s with the flow rate was explained as follows: "These results indicate that the 'true' flow rate of the bubbles of the mobile phase through the stationary phase is determined by the strength of the applied centrifugal force. As long as the pumping rate (the apparent flow rate) does not exceed the 'true' flow rate, it does not affect the resolution of the peaks. This means that the separation can be completed in a short time provided a sufficient centrifugal force is applied" [1].

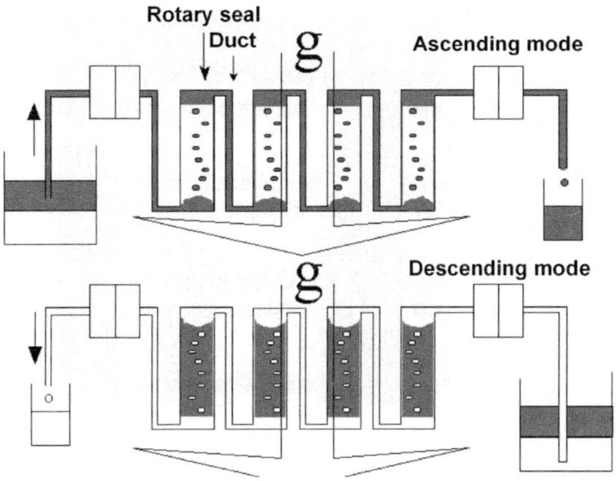

Fig. 5.2. Simple scheme used by the pioneers of CPC to represent the arrangement between cells and ducts, with the droplets of mobile phase flowing in the stationary phase.

It is remarkable that, by the study of only one chromatographic parameter, i.e. R_s, the authors developed the idea of apparent flow rate and 'true' flow rate, which is not really true but not far from the truth. They should better have discoursed on the linear velocity of the mobile phase. In fact, when steady-state equilibrium is reached in the CPC column, the volumetric flow rate is constant whatever the pattern of the mobile phase, but the linear velocity is dependent on the flow structure. However, the fundamental difference for the mobility of this mobile phase between ducts (pump driven) and cells (not pump driven) makes CPC very different from its parent CCC.

5.2.2 Van Deemter-type plots and the three-stage model

In the course of their studies of the effects of various parameters upon various chromatographic performances of a CPC-LLN (see the complete description of this apparatus in Chapter 4), A. Berthod and D.W. Armstrong discovered that the plots of efficiency versus flow rate showed a behaviour roughly opposite to that of modern liquid chromatography (LC) or gas chromatography (GC) [4,5]. Indeed, van Deemter-type plots have maxima rather than minima. In view of these results, the question arose as to whether or not traditional non-equilibrium treatment for chromatography is applicable to CPC techniques. The evidence of its

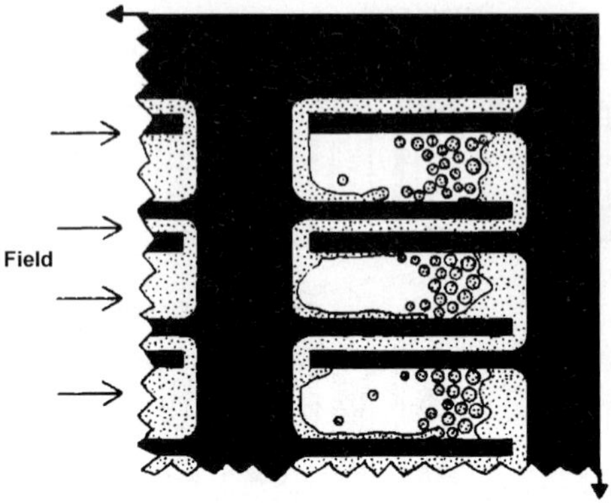

Fig. 5.3. Berthod's 1987 scheme of the three stages in a CPC channel, here in the descending mode: first a fast fall of droplets or jet, then an emulsified layer that forms at the phase boundary, and finally a pool stirred but non-turbulent. Adapted from Refs. [4,5].

failure led them to formulate a model to account for the observed flow behaviour toward efficiency.

The cell is divided conceptually into three stages (Fig. 5.3):

- Stage 1: the mobile phase is introduced at the top of the stationary phase as droplets. These droplets fall through the stationary phase for a period of time which is normally very short, particularly at higher flow rates where streaming occurs.
- Stage 2: for a second period of time the droplets are in an emulsified band between the two bulk phases.
- Stage 3: the reminder of the residence time is spent in a fairly quiescent state with stirred but non-turbulent exchange across the cross-section area.

A mathematical treatment with few approximations leads to the equation:

$$-\ln(1-E) = A / Q + BQ^{b} \tag{5.1}$$

where E corresponds to the fraction of theoretical plates per cell ($E = 1$ for optimum efficiency, $E < 1$ in all other case); A and B are constants depending on the solute as well as the solvent system and the condition of the experiment (rotational speed, cell shape), but independent of Q, the flow rate; b and A/B should be independent of the solute, provided it has no surfactant effect on the system.

From Eq. (5.1) it arises that the efficiency is maximal at very low flow rates (A/Q). This is not really useful for practical applications because of the extremely long time required for a separation. At higher flow rates the BQ^b term makes CPC well suited for both preparative and high-velocity separations [5].

As already stated in Section 5.2.1, it is remarkable that, by studying only one chromatographic parameter (the efficiency, N), the authors developed the idea of a three-stage cell. Even if it does not correspond exactly to the three stages seen by further visual observations (injection zone, curvilinear moving zone, coalescence zone, but no emulsified band), it led the authors to right conclusions, i.e., the flow rate partly promotes the mixing of the mobile phase with the stationary phase.

5.2.3 The Stokes model

Foucault and colleagues [6] introduced the Stokes model to explain their experimental results concerning the influence of the rotational speed, the flow rate, and the physical properties of biphasic systems on the stationary phase holdup and the chromatographic efficiency. In this model, they figured the mobile phase in a cell as droplets with an average radius corresponding to Stokes law. At that time it was uncertain how the mobile phase really flows through the stationary phase in CPC. The mobile phase travels from cell to cell through the ducts, where it is a continuous phase. While in the cell, it may be continuous, like a film against one of the walls, flowing very quickly then accumulating as in a pool and waiting to be transferred in the next cell through the duct. The mobile phase may be also discontinuous, like clouds or droplets going their own way under the influence of the gravitational field and being transferred when they reach the outlet of the cell. Whatever the pattern, when the steady state is reached, with the cells having their volume shared by the stationary and the mobile phases, the governing law is always the same: what comes in, goes out, directed by the flow rate.

Foucault et al. [7] introduced first a very simple equation based on the steady-state equilibrium actually encountered in a CPC cell submitted to a constant flow rate, F, and rotational speed, ω, and corresponding to the schematic shown in Fig. 5.4:

$$\frac{V_{\mathrm{m}}^{\min}}{V_{\mathrm{c}}} = \frac{(1-d)}{\bar{u}}\frac{F}{S} + d \tag{5.2}$$

where $V_{\mathrm{m}}^{\min}$ is the volume of mobile phase corresponding to the steady-state equilibrium between all the parameters acting upon the hydrodynamics of the two phases in the column, V_{c} is the volume of the CPC

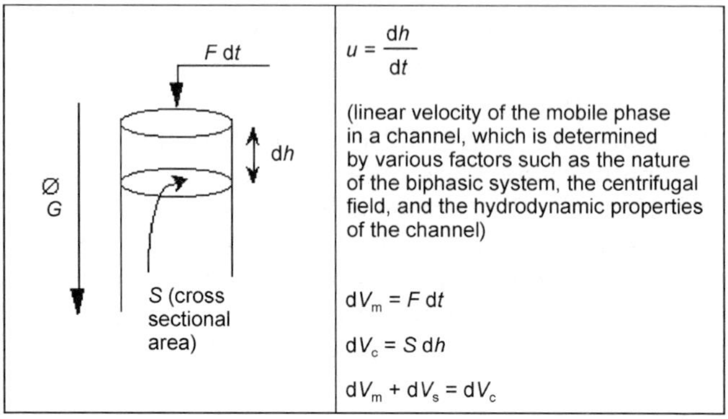

Fig. 5.4. Steady-state equilibrium in a CPC channel, leading to Eq. (5.2) (see Ref. [8]).

column, $\bar{u}$ is the average velocity of the mobile phase in a cell, which may be calculated through its residence time in a cell (not in a duct), d is the ratio of the column volume corresponding to the ducts ($V_{\text{ducts}} = dV_c$), and S is the cross section of a cell.

Equation (5.2) is valid whatever the experimental conditions are and does not need any assumption. The first important experimental result (based on data for 12 solvent systems in both descending and ascending mode) was that there is a linear relationship between V_m^{min} / V_c and F/S over a broad range of values of F. This means that $\bar{u}$ is independent of F in CPC, a conclusion close to that made by W. Murayama et al. [1].

As is customary in other fields (e.g., sedimentation of proteins in a centrifuge), the authors adopted a very simple model to describe the hydrodynamics of the flow in a cell. They imagined the mobile phase in a cell as droplets with an average radius a flowing in the stationary phase at constant linear velocity $\bar{u}$, previously defined, and corresponding to the radius of rigid spheres, which will follow the Stokes law:

$$\bar{u} = \frac{2a^2 \Delta\rho g}{9\eta_{\text{SP}}} \tag{5.3}$$

where $g = \omega^2 R$, R being the average radius of the centrifuge and ω the rotational speed, $\Delta\rho$ is the density difference between the two phases, and η_{SP} is the viscosity of the stationary phase, which can be the upper or the lower one, depending on the chosen mode of operation.

The model transforms something discontinuous flowing with an average linear velocity $\bar{u}$ into regular droplets with average radius a, flowing with a constant linear velocity $\overline{u}$. As a model, it is not necessary to know if the mobile phase is actually flowing as droplets, but the model must be compatible with experimental results.

The second important experimental result was that for higher rotational speed the retention of the stationary phase was independent of it. From Eqs. (5.2) and (5.3) this meant that the higher the acceleration field, the smaller the radius of the "droplets" of mobile phase in a cell.

As a consequence of the Stokes model, and if it is assumed that the chromatographic efficiency, N, is linearly related to the interfacial area between the two phases, the authors concluded that a linear relationship between N and ω should be observed, as well as a linear relationship between N and V_m^{min}. The authors verified these relationships using the hexane/methanol–4% water biphasic system in the descending mode [6,8]. They concluded that, whatever the system, the best results will be obtained for higher rotational speed not because the retention of the stationary phase is higher, but because the dispersion of the mobile phase in the stationary phase becomes better. The authors suggested that direct visualization of the mobile phase through the stationary phase would be very useful.

5.2.4 The first visual and chemical engineering approach: the Dutch school

Van Buel et al. [2] were the first to present photographs of the typical flow states of a CPC column for the hexane/water two-phase system at various flow rates and rotational speeds. This work opened the way to a major contribution of the Delft University of Technology for a better understanding of CPC [9–13]. Four main types of flow states were observed as a function of rotational speed and flow rate: large droplets, a jet stuck along the cell walls, a broken jet, and atomization. The authors applied to CPC the 'plug flow with axial dispersion' model, a chromatographic modelling introduced by Villermaux et al. [14]. This model allows one to describe the effluent concentration profile as a function of indebted physical and chemical parameters. It was shown theoretically and experimentally that the inverse overall mass transfer coefficient, k_0, is a linear function of the partition ratio, K_D, according to:

$$\frac{1}{k_0} = \frac{K_D}{k_m} + \frac{1}{k_s} \qquad (5.4)$$

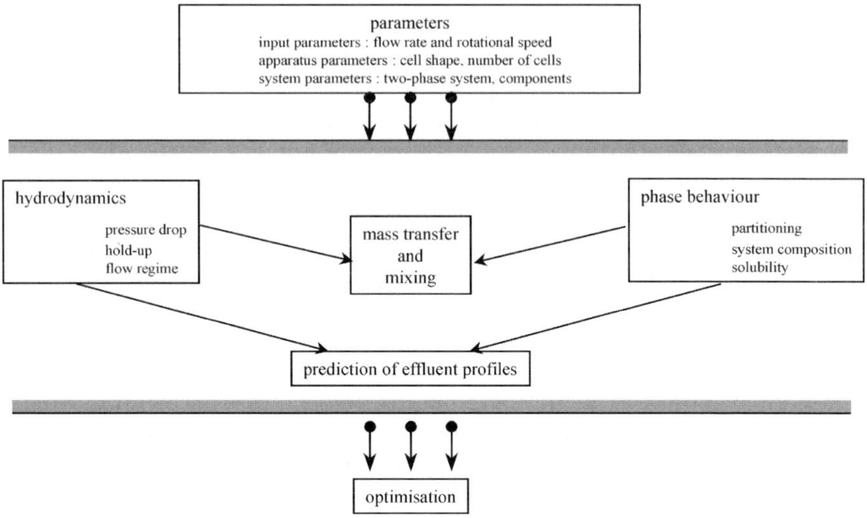

Fig. 5.5. M. van Buel's approach for optimizing CPC separations given the various parameters available.

where k_m and k_s are the local mass transfer coefficients in the mobile and stationary phases. Furthermore, it was shown that the overall volumetric mass transfer coefficient, k_0a (a being the specific interfacial area between the stationary and mobile phases) depends on cell geometry.

According to van Buel [12], for the optimization of CPC separations (with respect to productivity and component purity) in a short time and with minimal experimental work needed, it is necessary to understand how various phenomena govern a CPC separation. Figure 5.5 gives an overview of the approach the author has developed for optimizing CPC separations given the various parameters available.

5.3 CHROMATOGRAPHY CURVES MODEL

5.3.1 Chromatography theory

A transient and a reversible exchange between a mobile phase and a stationary phase characterizes chromatographic processes. Due to the differences in affinity of the solutes to the stationary phase, a certain degree of separation is achieved. For example, if a mixture of two

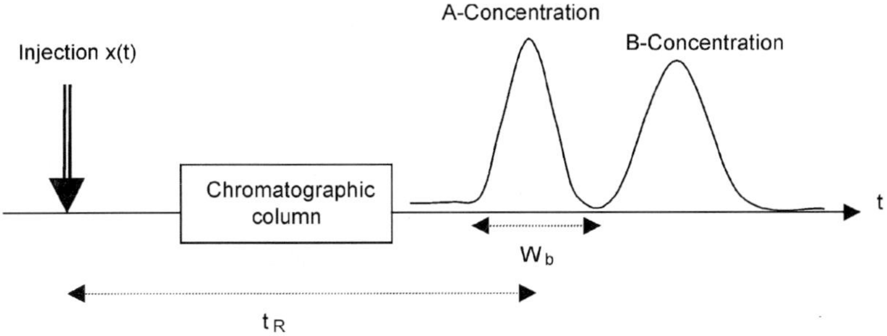

Fig. 5.6. Chromatographic separation of two components A and B.

components—one of which stays preferably in the stationary phase while the other prefers the mobile phase—is injected into the chromatographic column, the two components do not have the same residence time in the column and consequently leave the column separately (Fig. 5.6).

Each peak at the column exit can be characterized by the retention time, t_R, which corresponds to the residence time of a given component in the column, and by the peak width, W_b, at the baseline. Retention time is a function of the partition ratio of the component between the phases, $K_D = C_s/C_m$, and of the residence time distribution (RTD) of the mobile phase in the column. Bandwidth, W_b, is also dependent on the RTD and on the mass transfer in the two phases. The chromatographic efficiency can be evaluated from the experimental values of t_R and W_b. The resolution of a separation can be expressed by the factor, R_S:

$$R_S = \frac{t_{R_B} - t_{R_A}}{\frac{1}{2}(W_{b_A} + W_{b_B})} \tag{5.5}$$

The subscripts A and B refer to components A and B.

The chromatographic system can be considered as a linear dynamic system. The problem is to determine the relationship existing between the output signal $y(t)$ and the input signal $x(t)$. The transfer function, $G(s)$, of the chromatographic system is defined by:

$$G(s) = \frac{Y(s)}{X(s)} \tag{5.6}$$

where $X(s)$ and $Y(s)$ are the Laplace transforms of the signals. $G(s)$ is independent of the shape of the input signal and characterizes the

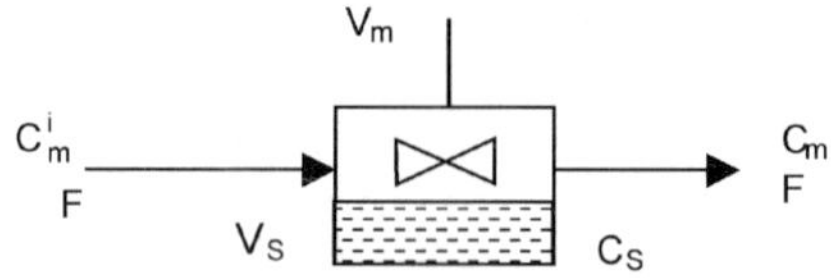

Fig. 5.7. Mixing-cell model for chromatographic separation.

transient behaviour of the chromatographic system. In the real domain, the inverse Laplace transform corresponds to the impulse response $E(t)$ of the system to a Dirac delta function input, $x(t) = \delta(t)$. The impulse response $E(t)$ can be characterized by its statistical moments:

$$\mu_k = \int_0^\infty t^k E(t)\mathrm{d}t \tag{5.7}$$

or by the central moments:

$$\mu'_k = \int_0^\infty (t - \mu_i)^k E(t)\mathrm{d}t \tag{5.8}$$

The first moment, μ_1, corresponds to the mean residence time, $\bar{t}$, while the second central moment, μ'_2, is the variance, σ^2, and measures the amount of dispersion in the column.The chromatographic column can be regarded [15] as an association of N_c mixing-cells (or plates) in series containing mobile and stationary phases (Fig. 5.7).

A reversible mass exchange exists between the mobile phase (volume, V_M, solute concentration, C_m) and the stationary phase (volume, V_S, solute concentration, C_s). The mass balance in a mixing-cell is written as:

$$F\,C^i_m = F\,C_m + V_m \frac{\mathrm{d}C_m}{\mathrm{d}t} + V_s \frac{\mathrm{d}C_s}{\mathrm{d}t} \tag{5.9}$$

In linear chromatography, the equilibrium concentrations in the two phases are proportional. By defining the partition ratio of the solute between the phases, $K_D = C_s / C_m$, and the mean residence time of the mobile phase in the column, $\bar{t} = N_c V_m / Q$, we obtain:

$$C_{\mathrm{m}}^{\mathrm{i}} = C_{\mathrm{m}} + \frac{\bar{t}}{N_{\mathrm{c}}}\left[1 + \frac{(1-\varepsilon_{\mathrm{m}})K_{\mathrm{D}}}{\varepsilon_{\mathrm{m}}}\right]\frac{\mathrm{d}C_{\mathrm{m}}}{\mathrm{d}t} \tag{5.10}$$

where ε_{m} is the volume fraction occupied by the mobile phase. Then, the impulse concentration response is given by:

$$E(t) = \frac{t^{N_{\mathrm{c}}-1}}{t_{\mathrm{R}}^{N_{\mathrm{c}}}}\frac{N_{\mathrm{c}}^{N_{\mathrm{c}}}}{(N_{\mathrm{c}}-1)!}\exp\left(-\frac{N_{\mathrm{c}}t}{t_{\mathrm{R}}}\right) \tag{5.11}$$

where t_{R} is the retention time:

$$t_{\mathrm{R}} = \bar{t}\left(1 + \frac{(1-\varepsilon_{\mathrm{m}})K_{\mathrm{D}}}{\varepsilon_{\mathrm{m}}}\right) \tag{5.12}$$

$E(t)$ is an ideal chromatographic peak, i.e. without axial dispersion and no mass transfer resistance. The variance is given by:

$$\sigma^2 = \frac{t_{\mathrm{R}}^2}{N_{\mathrm{c}}} \tag{5.13}$$

It is interesting to note that the residence time, t_{R}, is proportional to the number of mixing-cells, N_{c} and that the resolution, i.e. the separation time between two peaks which is quantified by $1/\sigma$, is proportional to $\sqrt{N_{\mathrm{c}}}$.

In case of finite mass transfer rate between mobile and stationary phases, the mass balance in the mixing cell is written [14] as:

in the mobile phase:

$$FC_{\mathrm{m}}^{\mathrm{i}} = FC_{\mathrm{m}} + V_{\mathrm{m}}\frac{\mathrm{d}C_{\mathrm{m}}}{\mathrm{d}t} + k_0 a\left(C_{\mathrm{m}} - \frac{C_{\mathrm{s}}}{K_{\mathrm{D}}}\right) \tag{5.14}$$

in the stationary phase:

$$k_0 a\left(C_{\mathrm{m}} - \frac{C_{\mathrm{s}}}{K_{\mathrm{D}}}\right) = V_{\mathrm{s}}\frac{\mathrm{d}C_{\mathrm{s}}}{\mathrm{d}t} \tag{5.15}$$

where k_0 is the overall mass transfer and a the interfacial specific area. The mass transfer time constant, T_{m}, is defined as:

$$T_{\mathrm{m}} = \frac{K_{\mathrm{D}}(1-\varepsilon_{\mathrm{m}})}{k_0 a} \tag{5.16}$$

In the Laplace domain, the transfer function,

$$G(s) = \frac{L(C_{\mathrm{m}})}{L(C_{\mathrm{m}}^{\mathrm{i}})},$$

corresponding to the resolution of the differential equations (5.14) and (5.15), is the following:

$$G_i(s) = \left[\frac{1+sV_{\mathrm{m}}}{F}\left(1+\frac{K_{\mathrm{D}}(t-\varepsilon_{\mathrm{m}})}{\varepsilon_{\mathrm{m}}(1-T_{\mathrm{m}}s)}\right)\right]^{-1} \tag{5.17}$$

where $\varepsilon_{\mathrm{m}} = V_{\mathrm{m}}/(V_{\mathrm{m}} + V_{\mathrm{s}}) = 1 - Sf$, and for a N_{c}-plate chromatographic column:

$$G(s) = \left[\frac{1+sV_{\mathrm{m}}}{F}\left(1+\frac{K_{\mathrm{D}}(t-\varepsilon_{\mathrm{m}})}{\varepsilon_{\mathrm{m}}(1-T_{\mathrm{m}}s)}\right)\right]^{N_{\mathrm{c}}} \tag{5.18}$$

The impulse concentration response is obtained by numerical inversion of Eq. (5.18). The moments of $E(t)$ can be determined using Van der Laan's theorem:

$$\mu_{\mathrm{k}} = (-1)^{\mathrm{k}} \frac{\partial^{\mathrm{k}} G(s)}{\partial s^{\mathrm{k}}} \tag{5.19}$$

The first moment gives:

$$t_{\mathrm{R}} = \bar{t}\left(1+\frac{(1-\varepsilon_{\mathrm{m}})K_{\mathrm{D}}}{\varepsilon_{\mathrm{m}}}\right) \tag{5.20}$$

This expression is similar to Eq. (5.12), meaning that the position of the peak depends only on the partition coefficient. The broadening of the peak is conversely due to both mass transfer rate and axial dispersion, as can be seen on the second-order moment:

$$\frac{\sigma^2}{t_R^2} = \frac{1}{N_c} + \frac{T_m}{t_R} \frac{2}{\dfrac{\varepsilon_m}{(1-\varepsilon_m)K_D} + 1} \tag{5.21}$$

The term $1/N_c$ is related to axial dispersion, the peak becomes sharper with an increase of mixing cells in series. The second term of Eq. (5.21) characterizes the effect of mass transfer resistance on peak broadening, which is sensitive to the parameter T_m/t_R.

An alternative theory was proposed by Van Deemter et al. [16], who considered a continuous chromatographic column. The mobile phase occupies the fraction ε_m of the column volume and the stationary phase the fraction $1 - \varepsilon_m$. Lapidus and Amundson first derived the mathematical model, so-called plug flow with axial dispersion and mass exchange [17]. The assumptions of the model are:

- the retention of a component is only due to its distribution between the two phases;
- the composition of the injected sample does not influence the partition coefficient of the individual components;
- the mobile phase flow is modeled by an axial dispersion plug flow (axial dispersion coefficient, D_{ax}, axial position in the column, z);
- the stationary phase is considered as a dead zone.

In the column, the mass transfer flux is superimposed to the purely convective flow (linear velocity, U) and the axial dispersion flux model. It gives for the mobile phase:

$$\varepsilon_m u \frac{\partial C_m}{\partial z} + \varepsilon_m \frac{\partial C_m}{\partial t} + k_0 a \left(C_m - \frac{C_s}{K_D} \right) = \varepsilon_m D_{ax} \frac{\partial^2 C_m}{\partial z^2} \tag{5.22}$$

and for the stationary phase:

$$(1-\varepsilon_m) \frac{\partial C_s}{\partial t} - k_0 a \left(C_m - \frac{C_s}{K_D} \right) = 0 \tag{5.23}$$

Equation (5.23) is equivalent to:

$$K_D C_m = T_m \frac{\partial C_s}{\partial t} + C_s \tag{5.24}$$

where T_m is a mass transfer time constant. In the Laplace domain, where $L(C)$ is the Laplace transform of C, Eqs. (5.22) and (5.24) become:

$$-D_{\mathrm{ax}}\frac{\partial^2 L(C_{\mathrm{m}})}{\partial z^2}+U\frac{\partial L(C_{\mathrm{m}})}{\partial z}+s\left(1+\frac{k'}{T_{\mathrm{m}}s+1}\right)L(C_{\mathrm{m}})=0 \tag{5.25}$$

$$K_{\mathrm{D}}L(C_{\mathrm{m}})=(T_{\mathrm{m}}s+1)L(C_{\mathrm{s}}) \tag{5.26}$$

with $k'=K_{\mathrm{d}}\dfrac{1-\varepsilon_{\mathrm{m}}}{\varepsilon_{\mathrm{m}}}$ the retention factor.

The resolution of these equations leads to the transfer function $G(s)$ such as:

$$L(C_{\mathrm{m}})(z=l)=G(s)L(C_{\mathrm{m}})(z=0) \tag{5.27}$$

where

$$G(s)=\mathrm{e}^{\frac{Pe}{2}\left(1-\sqrt{1+\frac{4\bar{t}s\left(1+\frac{k'}{T_{\mathrm{m}}s+1}\right)}{Pe}}\right)} \tag{5.28}$$

with the Péclet number, $Pe=\dfrac{Ul}{D_{\mathrm{ax}}}$ and l the column length, $l=\bar{t}U$. There is no analytical solution in the time domain of Eq. (5.28). The two first moments of the impulse response can be calculated from the Van der Laan's theorem:

$$t_{\mathrm{R}}=\bar{t}(1+k') \tag{5.29}$$

which is similar to Eq. (5.12) and:

$$\frac{\sigma^2}{t_{\mathrm{R}}^2}=\frac{2}{Pe}+\frac{2k'}{1+k'}\frac{T_{\mathrm{m}}}{t_{\mathrm{R}}} \tag{5.30}$$

Equation (5.30) is identical to Eq. (5.21) if:

$$Pe=2N_{\mathrm{c}} \tag{5.31}$$

This condition is realized [18] for flow with low axial dispersion, i.e. with large values of N_{c} or Pe.

5.3.2 The CPC model

Van Buel et al. [9–13] have developed a model for describing the effluent concentration profiles in CPC. The peak retention is caused by partitioning over the two phases, while peak broadening is caused by mass transfer limitation, axial dispersion in column volume and off-column mixing. The mobile phase in the off-column is modeled by the axial dispersed plug flow model, which is described by the following convection–diffusion equation:

$$U\frac{\partial C}{\partial z}+\frac{\partial C}{\partial t}=D_{\mathrm{ax}}\frac{\partial^2 C}{\partial z^2} \tag{5.32}$$

The transfer function, G_{off}, corresponding to the resolution of Eq. (5.32) with the boundary conditions given by Eq. (5.27) is written as:

$$G_{\mathrm{off}}=\frac{e^{Pe^*/2}}{\cosh\left[\frac{Pe^*}{2}\beta\right]+0.5\left(\beta+\frac{1}{\beta}\right)\sinh\frac{Pe^*\beta}{2}} \tag{5.33}$$

with

$\beta=\left(\frac{1+4\bar{t}^*s}{Pe^*}\right)^{1/2}$, $\bar{t}^*$ and $Pe^*=\frac{Ul}{D_{\mathrm{off}}}$ the mean residence time and the Péclet number related the off-column mixing, respectively.

The effluent concentration profiles in the chromatographic column are obtained from the set of differential equations (5.22) and (5.23). Van Buel et al. [13] solved this problem in the Fourier domain assuming that the sample injection could be modeled by a Dirac pulse. We propose here a different approach based on a two-point measurement associated with curve fitting in time domain [19,20]. The CPC apparatus is modeled in Fig. 5.8. C_{i} and C_{o} are the effluent concentration at the inlet and outlet of the CPC apparatus, respectively. The curve fitting in time domain method is based on the following steps:

(1) transformation of the experimental concentrations, C_{i} and C_{o}, in Fourier series, $\widetilde{C}_{\mathrm{i}}$ and $\widetilde{C}_{\mathrm{o}}$.

(2) determination of the calculated outlet concentration in the Fourier domain:

$$L(\widetilde{C}\widetilde{C}_{\mathrm{o}})=G(s)\cdot L(\widetilde{C}_{\mathrm{i}}\widetilde{C}) \tag{5.34}$$

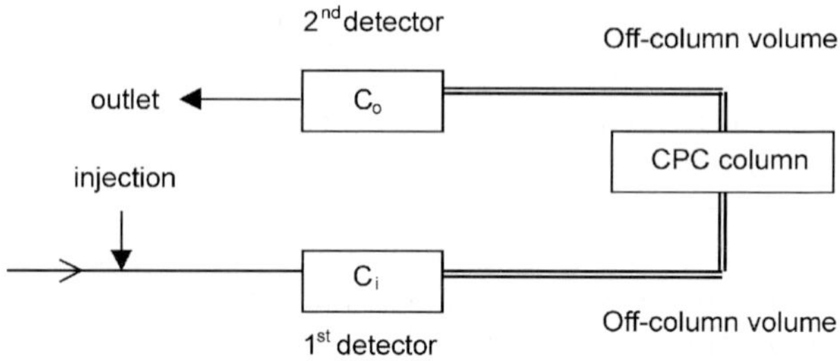

Fig. 5.8. Sketch of CPC apparatus.

(3) determination of the parameters of the transfer function by minimization of the error between $\widetilde{C}_{\mathrm{i}}$ and $\widetilde{C}\widetilde{C}_{\mathrm{o}}$:

$$\mathrm{Err} = \frac{\int (\widetilde{C}_{\mathrm{o}} - \widetilde{C}\widetilde{C}_{\mathrm{o}})^2 \, \mathrm{d}t}{\int \widetilde{C}_{\mathrm{o}}^2 \, \mathrm{d}t} \tag{5.35}$$

The application of this method to CPC apparatus is made by modelling the experimental set-up sketched in Fig. 5.8 by the process scheme given in Fig. 5.9. The CPC model is characterized by six parameters: the mean residence time in the off-column, $\bar{t}\,^*$, and in the CPC column, $\bar{t}$, the axial dispersion coefficient, or the Péclet number, in the off-column, Pe^*, and in the CPC column, Pe, the partition ratio, K_{D}, and the mass transfer time constant, T_{m}.

The determination of these six parameters requires three independent residence time distribution RTD experiments.

Experiment 1: The CPC column is short-circuited in order to determine $\bar{t}\,^*$ and Pe^* from:

$$G_{\mathrm{off}}(t^*, Pe^*) = \frac{L(C_{\mathrm{o}})}{L(C_{\mathrm{i}})} \tag{5.36}$$

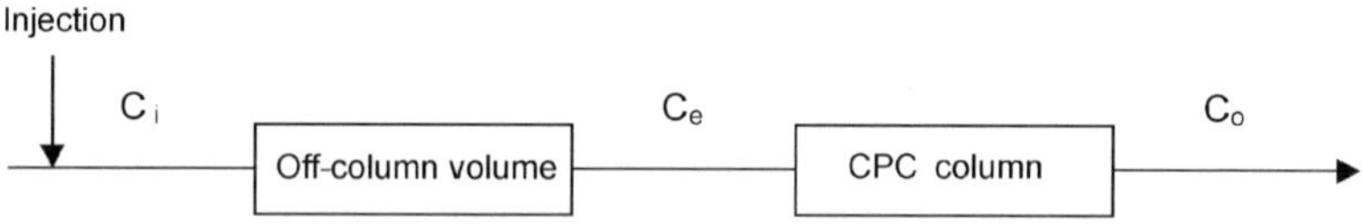

Fig. 5.9. Process scheme corresponding to the experimental set-up sketch in Fig. 5.8 and used for the CPC model.

Experiment 2: The CPC column is connected to the circuit and a tracer with no affinity for the stationary phase ($K_D = 0$) is used for the RTD experiment. The transfer function, G_T, is:

$$G_T = G_{off} \cdot G = \frac{L(C_o)}{L(C_i)} \quad (5.37)$$

From the knowledge of G_{off}, G (Eq. (5.28)) can be calculated and then the parameters $\bar{t}$ and Pe are easily determined.

Experiment 3: Determination of RTD by using a tracer with a given partition ratio between mobile and stationary phases. Equation (5.37) is also used to determine K_D and T_m from the $\bar{t}$ *, $\bar{t}$, Pe* and Pe values.

5.4 FLOW PATTERNS, RESIDENCE TIME DISTRIBUTION AND MASS TRANSFER

5.4.1 Experimental

Visualization techniques

The experimental set-up of van Buel et al. [13]: For their first visualizations, M.J. van Buel and colleagues used a transparent cartridge with cells and ducts equal to those of the commercial CPC-LLN 250W cartridges [2]. The cartridge was positioned vertically, and by using an optical system of two mirrors and a stroboscope, the images were frozen and photographs were taken using a Pentax Z-20 camera equipped with a 70-210 mm zoom lens and a 50 mm extension tube (Fig. 5.10). All experiments were performed with an aqueous mobile phase in the descending mode, except for the system chloroform/water where it was in the ascending mode.

Experimental set-up in our laboratory [21]: An overview of our experimental set-up is shown in Fig. 5.11. The visual CPC consists of a body (steel) supporting an electrically driven asynchronous motor (Leroy Somer, Angoulême, France) and a vertical shaft bearing the CPC rotor with the upper side transparent. CPC cells were engraved in a 5-mm thick steel ring clamped between a steel plate and a glass ring of optical quality (Ediver, Rebais, France). A 0.05-mm-thick Teflon sheet (Angst+Pfister, Roissy-Charles de Gaulle, France) was tightened on each side of the engraved ring to ensure a leak-free rotor. The engraved ring ($\varnothing_{int}$ = 175 mm, $\varnothing_{ext}$ = 240 mm), had 66 parallelepiped CPC cells of 10×4×5 mm connected by ducts 0.8 mm in width. The rotor was

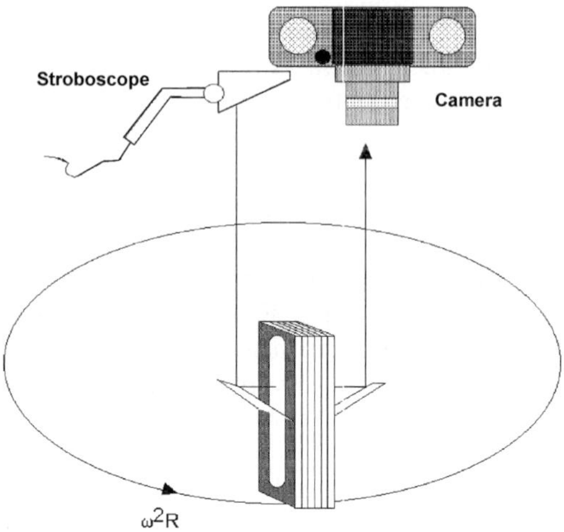

Fig. 5.10. van Buel's experimental set-up for the visual experiments. The various chromatographic apparatuses are not shown [9–13].

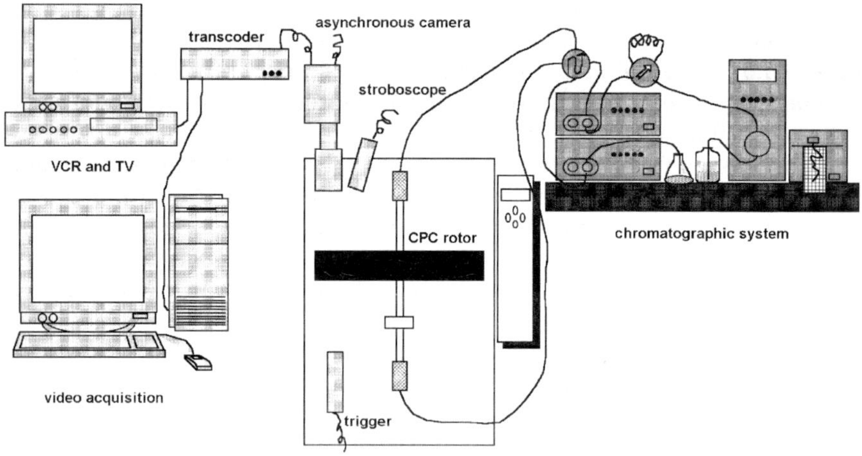

Fig. 5.11. Overview of our experimental set-up with the visualization system on the left, the visual CPC in the centre and the chromatographic system on the right.

connected to the chromatographic system through 1/16″ PEEK tubing (Upchurch Scientific, Oak Harbor, WA, USA.) and two rotary seals (Tecmeca, Epernay, France). The rotational speed could be varied from 0 to 3000 rpm.

A slightly modified Videostrobe system (Sysmat Industrie, St Thibault des Vignes, France) was used. It consisted of a TMC-9700 progressive scan CCD digital colour camera (Pulnix, Sunnyvale, CA, USA) with an asynchronous shutter, two Phylec stroboscopic units (Sysmat) and a VLS7T optical speed sensor (Compact, Bolton, UK) which triggered both the stroboscopes and the camera. The camera was equipped with an 18-108 mm F 2.5 TV Zoom lens and connected to a CDM-810 multi-system converter (World Advanced Technologies, Toulouse, France), allowing NTSC/PAL/SECAM transcoding and 60/50 Hz field conversion. Images were recorded on an HR-DD946MS videorecorder (JVC, Japan). They were displayed with a CPM1404 monitor (Hitachi, Japan), and/or on a computer equipped with a Pentium 233 MHz processor and a Matrox Millennium II AGP power desk card (Matrox, Montreal, Canada) driven by MiroVideo DC30 Plus software (Pinnacle Systems, Mountain View, CA, USA). Films were stored using Adobe Premiere v. 4 (Adobe Systems Inc., San Jose, CA, USA).

Chromatographic equipment
A P4000 quaternary gradient pump equipped with inert assemblies was used (ThermoQuest, San Jose, CA, USA), which delivered constant flow rates from 0.1 to 30 ml/min. A 6-port injection valve and a 4-port switching valve, both in PEEK material (Upchurch), were used for injection and selection of either ascending (upper phase mobile) or descending (lower phase mobile) mode. Two Spectra 100 UV detectors equipped with 0–3 mm, 0–4.6 μl, variable path-length preparative scale LC flow cells were used for chromatographic data acquisition (ThermoQuest). The first UV detector was positioned just after the injection valve and used to record the input profile of the injected sample. The second detector was conventionally positioned after the CPC column and used to record the output profile of the injected sample. The two profiles were acquired on a SA32 32-channel data acquisition system (AOIP, Evry, France), then exported onto a computer for further calculations.

Chemicals
Heptane, methanol, *n*-butanol and chloroform (puro) were from Carlo Erba (Rodano, Italy); and 2-propanol and dipotassium phosphate from Prolabo (Paris, France). Polyethylene glycol 3350 (PEG) and the various solutes used to generate chromatographic peaks were from Sigma (St. Quentin Fallavier, France). Water was deionized using a Milli-Ro®4 water purification system (Millipore, Bedford, MA, USA).

Two-phase systems and tracers

The two-phase system constituted of two solvents, *n*-heptane and methanol, denoted HM (a well-known system for the separation of fatty acids and lipids) was mainly used. The physical properties of each phase are summarized in Table 5.1. As heptane is partially miscible with methanol (miscibility of 30% w/w), the two mutually saturated phases have similar properties and low interfacial tension (measured at 23°C by Foucault et al. [6] by the spinning drop method). The upper phase (UP) is the heptane-rich phase and lower phase (LP) the methanol-rich one. Viscosity was calculated by pressure drop measurements in a 4-m capillary tube (Upchurch, internal diameter of 0.254 mm), preliminary calibrated with pure solvents. The method gives accurate results (± 5%) according to literature values (Handbook of Chemistry and Physics, 67th ed.). Densities of saturated phases were measured by weighting a flask of well-known volume, calibrated with pure water at 22°C.

The study was done in the descending mode (LP is the mobile phase) for rotational speed values from 400 up to 2400 rpm (± 1 rpm) and flow rates from 3 up to 24 ml/min (± 0.1 ml/min). Potassium nitrate KNO_3 (R.P. Normapur® from Prolabo, Paris, France) was used as a tracer for RTD determination in the mobile phase, because it has no affinity with heptane ($K_D = 0$). KNO_3 has a maximum UV absorption at 220 nm (the spectrum was achieved with a UV/Vis spectrometer (ATI UNICAM UV2 series spectrometer, Cambridge, UK)). Phenyl salicylate used for the determination of mass transfer coefficients has a partition ratio around 0.7 in HM and is detected at 280 nm.

The four other two-phase systems studied in our laboratory have the properties given in Table 5.1.

Operating the visual CPC

The temperature of the laboratory was fixed at 22°C (± 0.5°C) during all experiments. The solvents, stocked at 22°C too, were mechanically stirred together during few minutes which, after few minutes of settling time, resulted in two perfectly mutually saturated phases. Phases were let together in the flask, to be sure they are continuously equilibrated.

The following *modus operandi* is typical and often described [5–7, 13]. The four-port valve is set on the chosen mode position, ascending or descending. The injection loop is filled with the sample in mobile phase. The column is filled with the stationary phase. The rotor is driven at the given rotational speed value and then the mobile phase elution is started at the selected flow rate. The stationary phase volume displaced during

TABLE 5.1

Physical properties of the two-phase systems at 22°C (except γ, at 23°C)

System	ρ UP kg/m^3	ρ LP kg/m^3	$\Delta\rho$ kg/m^3	γ mN/m	η UP cP	η LP cP
HM	684	756	72	1.16	0.47	0.65
HMW	684	924	240	13.90	0.42	1.68
HCPMW	939	1037	98	1.47	1.46	0.60
BAW	891	992	101	1.21	2.48	1.51
ATPS	1076	1163	87	0.93	19.96	1.23

Abbreviations: HM: hexane-methanol; HMW: *n*-heptane–methanol–water (50–25–25); HCPMW: *n*-heptane–chloroform–*n*-propanol–methanol–water (20–23.2–5.2–31–20.6); BAW: *n*-butanol–acetic acid–water (40–10–50); ATPS: aqueous two-phase system made of polyethylene glycol 3550–dipotassium phosphate–water (17.4–10.6–72, w/w/w).

the equilibrium phenomena is measured with a 25 ml graduated cylinder. If no more stationary phase is driven and if the two UV detector responses are stabilized (constant and noiseless baseline), the equilibrium is reached. The six-port injection valve is switched to the "inject" position and simultaneously data acquisition is started. When the solute elution is finished, the data acquisition is stopped, the injection valve is returned to the "load" position and the column is refilled with stationary phase at a rotation speed of about 300 rpm. The two-phase system is stirred again, and so on for the next experiment.

Video acquisition (15 s) is done during the solute elution. The observed flow regimes for heptane–methanol two-phase system, according to rotational speed and flow rate, will be first presented. Then the RTD results and the fits obtained with our model will be given and analyzed with respect to the flow patterns. Lastly, the effects of the physical properties of various two-phase systems will be presented.

5.4.2 Review of the different types of flow-regimes

The flow in a CPC cell has to be considered as a vertical flow under g-gravity due to centrifugal acceleration. When a two-phase mixture is flowing in a conduit, several spatial distributions of the phases can exist, depending on the flow rates, the fluid properties, the conduit geometry and the external forces. Each configuration called flow pattern characterizes similar interfacial geometries. Visual observations or photographic and video techniques are the usual methods for recognizing flow patterns through transparent test-section walls. Knowledge of the flow pattern is important in modelling the flow (holdup, pressure drop) and the transfer processes. For almost all applications, normal-gravity is the only acting force, so that horizontal and vertical flows are commonly distinguished. In vertical pipes different flow situations are possible: upward or downward flows, co-current or counter-current flows.

Although there is a considerable number of published studies on liquid/gas two-phase flows, there are relatively few articles on liquid/liquid two-phase flows. Numerous flow pattern maps propose transition boundaries for liquid/gas mixtures in dimensional or dimensionless coordinates [22]. The extrapolation of these predictive models to 0-gravity conditions for space application indicates that the transitions are quite different [23]. This is moreover confirmed by experiments performed at reduced-gravity [24]. No study or data on two-phase flow patterns in pipes at high-gravity is available in the literature.

In liquid/liquid systems the interfacial tension is generally lower than in liquid/gas systems, the interface is more distortable and the dispersion is favoured. In many chemical applications such as extraction equipment the physical parameters of the operating liquids—density difference, viscosity ratio, and interfacial tension—can vary in a wide range. Because the objective of the process is generally to produce a dispersion of drops or droplets, the flow conditions not of interest are not studied, so no further attention is focused on the other flow patterns [25]. Because of their major interest for petroleum engineering, many works have been concerned with water/oil mixtures in pipes. With reference to liquid/gas flows, water/oil pattern maps have been proposed. In vertical tubes four basic patterns are observed [26]: bubble (oil dispersed), slug (oil dispersed with large drops), froth (both phases dispersed) and mist (water dispersed). Despite some obvious similarities, water/oil flows show important differences from liquid/gas flows. In addition, Angeli and Hewitt pointed out that the nature of the wall material and especially its wetting properties can strongly modify the flow patterns in liquid/liquid systems [27].

A CPC cell is a relatively short conduit and one can assume that the flow inside is greatly controlled by the injection conditions of the mobile phase. The hydrodynamics of liquid break-up in an immiscible liquid is of great interest in many industrial applications. The conditions of vertical injection in a large volume from a horizontal orifice have been fully described by Clift et al. [28]. Three flow regimes can be distinguished according to the orifice velocity: drops, jet, atomization (cloud of droplets). At low velocity individual drops are formed close to the orifice. By increasing the velocity, at a critical value, a jet occurs and drops form by breakup through axi-symmetric instabilities (varicose breakup). Then the jet length increases to a maximum, before it decreases again due to asymmetric instabilities (sinuous breakup) and when the jet length finally becomes zero, atomization occurs at the orifice. Due to the size of the CPC cell, the wall and its wetting properties can alter the hydrodynamics of the breakup [13].

The above-mentioned approach of the Dutch school is really a pioneer work for the understanding of the liquid/liquid flow patterns in a centrifugal system.

5.4.3 Flow patterns in CPC

Description of flow patterns

The two-phase flow behaviour in CPC depends on the liquid's physical properties (viscosity and interfacial tension) and the flow characteristics in the injecting orifice of each cell. The velocity mean value and the velocity profile (turbulent flat one or fully developed parabolic one) in the orifice are deciding factors [29,30], as well as the wettability of the orifice material for expansion-contraction of the flow.

As the cross-sectional dimensions of the duct are 0.8×5 mm, it can be considered as a "flat nozzle". For a flat nozzle, regimes are quite different from those obtained with circular nozzles, and the mobile phase does not form a jet, but rather a flat sheet. Fraser [31] defined three modes of liquid sheet disintegration in gas: rim, wave and perforated-sheet. Rim mode occurs when both viscosity and interfacial tension are high. Surface forces then cause the free edge of the liquid sheet to contract into a thick rim, which then breaks up like a jet. In perforated-sheet mode, holes appear in the sheet, delimited by rims formed by the liquid present inside. The holes grow rapidly until the rims of adjacent holes coalesce to produce ligaments that finally break up into non-uniform size droplets. Wave motion, amplified in the flow direction, can also appear leading to torn away waves. The resulting areas rapidly contract under the action of surface tension to form ligaments and then droplets.

In CPC, mobile phase is pumped in the ducts whereas it falls or settles in the cells under centrifugal acceleration effect. In the cell part, mobile phase flow can be conceptually divided into three sequences: injection, free fall and coalescence.

Evidence of the Coriolis acceleration in CPC cells. The stroboscopic visualization allows us to see a cell as if we were taken on rotor board. The observed flow paths are then considered in a rotating non-Galilean system. The resulting flow paths are submitted to two complementary accelerations resulting from the composition law of acceleration [32]: centrifugal acceleration Γ_c and Coriolis acceleration Γ_{cor}

$$\Gamma_c = R_d \omega^2 \tag{5.38}$$

$$\vec{\Gamma}_{Cor} = -2\vec{\Omega} \wedge \vec{V}_R \tag{5.39}$$

where R_d is the disk radius at the cell's centre, ω the angular velocity, $\vec{V}_R$ the relative velocity vector, and $\vec{\Omega}$ the rotation vector.

Our observations of flow-patterns confirm those of van Buel et al. [13] and also supplement them. Van Buel's observations were made in CPC cells parallel to the rotation axis $\vec{\Omega}$, so that flow patterns are described as seen in a plane parallel to $\vec{\Omega}$ (Fig. 5.12). Our observations were made in a plane perpendicular to $\vec{\Omega}$. Then, in addition to the flow patterns described by van Buel, we observed deviations of jets or droplets from the radial direction, caused by an acceleration perpendicular to $\vec{\Omega}$ and $\vec{V}_R$, the relative velocity of the mobile phase in the cell, i.e., the Coriolis acceleration $\vec{\Gamma}_{Cor}$ [32,33]. Further evidence of complementarity of our observations is shown on Fig. 5.13. Let us draw similar cells for van Buel's observations (Fig. 5.13a and b) and ours (Fig. 5.13c). The rotation axis $\vec{\Omega}$ in Fig. 5.13a and c are perpendicular to each other, so

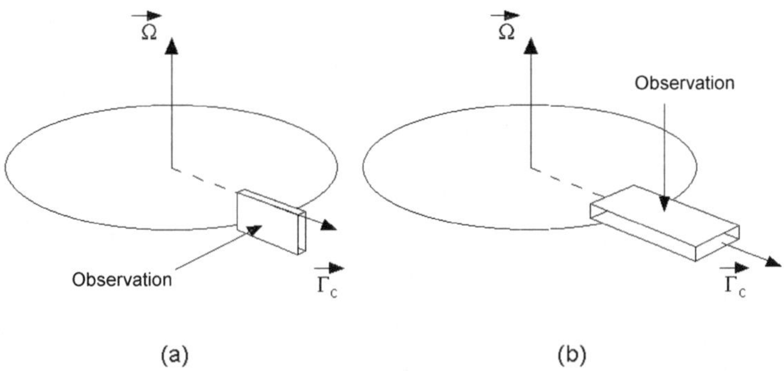

Fig. 5.12. Observation of flow patterns in CPC. (a) Observation in a plane parallel to the rotation axis; (b) observation in a plane perpendicular to the rotation axis.

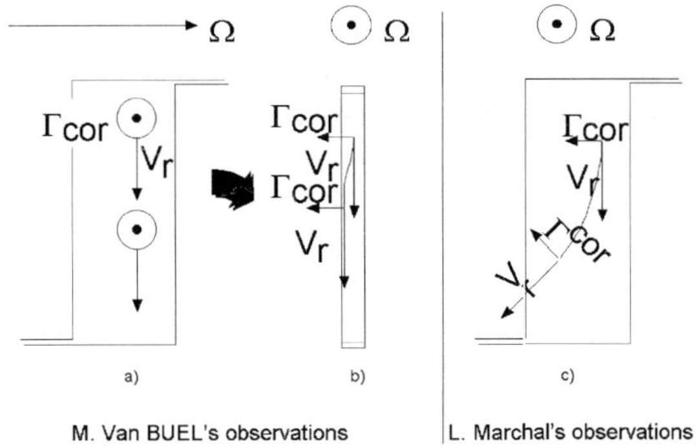

Fig. 5.13. Scheme explaining how the action of Coriolis acceleration can deviate the flow from the radial direction when there is no wall counteracting its effect.

that in Fig. 5.13a the Coriolis acceleration, $\vec{\Gamma}_{Cor}$, is directed toward the observer, who cannot really see its effect on 'droplet' trajectory. Let us now rotate this rotation axis $\vec{\Omega}$ so that it becomes parallel to that of Fig. 5.13c, thus giving van Buel's cell the look of Fig. 5.13b. For Figs. 5.13b and c, the Coriolis accelerations are then clearly directed in the same way, and corresponding to the vector $\vec{\Gamma}_{Cor}$ (Eq. (5.39)).

However, in the case of Fig. 5.13b, the action of $\vec{\Gamma}_{Cor}$ is counterbalanced by the close proximity of the left wall of the cell, which forces the 'droplet' to fall vertically. In the case of Fig. 5.13c there is room, and the deviation from the vertical fall of the 'droplet' can be observed. An example of such deviation is shown in Fig. 5.14, in the descending mode, i.e., with the flow going from the top to the bottom of the photographs.

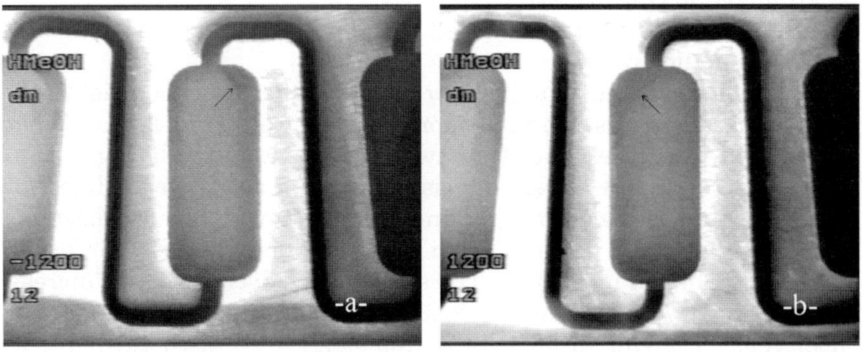

Fig. 5.14. Evidence of the Coriolis acceleration effect: descending mode, 1200 rpm, 12 ml/min, (a) trigonometric rotation; b) clockwise rotation.

For Fig. 5.14a the rotation is trigonometric and the jet deviates as soon as it leaves the top duct, while for Fig. 5.14b the rotation is clockwise, and the jet deviates in the opposite direction, according to Eq. (5.39).

Injection: Determination of the cell injecting conditions is not obvious in CPC. As a result of the first visualizations, it appears that the duct is not perfectly wetted by the mobile phase [2]. Part of the stationary phase is forced into the duct. The same phenomenon is also observed with our apparatus and the mobile phase inlet velocity may depend on either the flow rate and/or the centrifugal acceleration. For high rotational speeds, the mobile phase spreads and forms a film in the non-radial part of the duct. The film thickness depends on both rotational speed and flow rate. For lower rotational speed values, gravity appears to have a significant effect on the two-phase flow in the duct end. A kind of meniscus is observed; mobile phase is then supposed to flow along the bottom wall of the cell in descending mode. Wettability of the disk material and interfacial tension are expected to have a particular influence. In particular in descending mode, for the heptane–methanol two-phase system, the mobile phase (methanol-rich phase) is a polar one, and so it leads to wet the cell steel side surfaces.

Free fall: This corresponds to the duration of the interfacial area life. Created during the injection step, the interface path is then correlated to the initial conditions and the hydrodynamics of both phases in the cells under the influence of the external forces. Owing to the stroboscopic visualizations, three main flow regimes are observed:

- film stuck along the cell side walls (Fig. 5.15a);
- curvilinear sheet or film with oscillating instabilities (Fig. 5.15c);
- atomization (Fig. 5.15d).

No single droplets or emulsified bands were observed with this particular liquid system.

For each flow rate value, at low centrifugal acceleration fields, the mobile phase flows as two films along the cell walls (Fig. 5.15a). When the flow rate increases, one of the two films leaves the wall (Fig. 5.15b). When the rotational speed increases, there is a change in injection conditions and the two films become unstuck, giving one oscillating or wavy sheet, strongly deviated from the radial direction by the Coriolis force (Fig. 5.14 and Fig. 5.15c). The sheet length decreases with increasing centrifugal acceleration, due to interfacial instabilities dependent on sheet velocity, giving droplets often too small to be distinguishable on the video screen. For lower flow rates, the sheet length decreases to 0, and a cloud of droplets is formed near the duct outlet. For higher flow

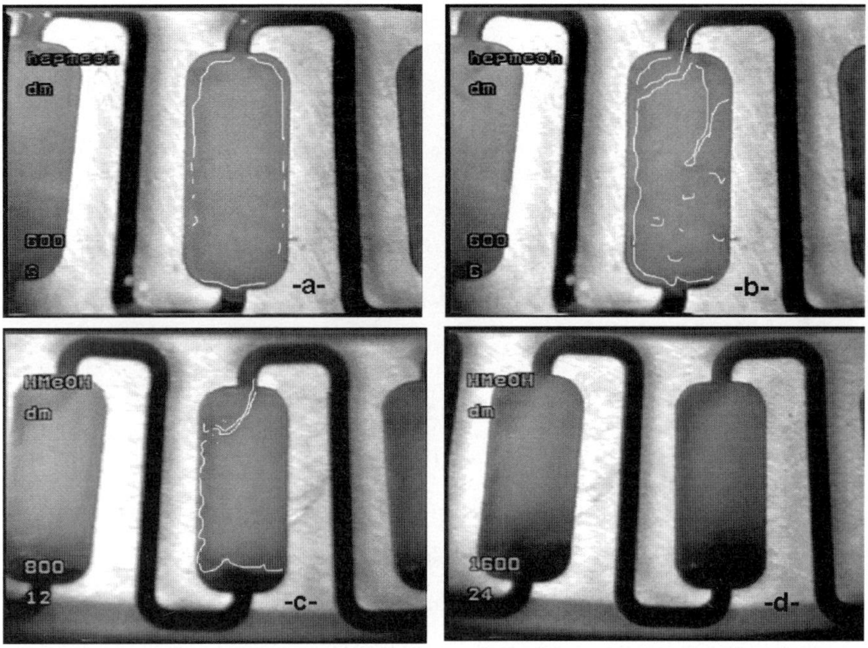

Fig. 5.15. Flow regimes (methyl purple dye in methanol-rich mobile phase). (a) Stuck film; (b) stuck film and oscillating sheet; (c) oscillating sheet; (d) atomization.

rates ($F \geq 15$ ml/min), instabilities appear on mobile phase film in the duct, leading suddenly to an atomization shape (Fig. 5.15d). Those favourable instabilities bound to flow rate value result in the detachment of mobile phase film in the duct, maybe under the influence of Coriolis acceleration and stationary phase flow structure. In spite of the Coriolis force effect, in atomization flow regime the mobile phase is well distributed all over the stationary phase volume. It can be assumed that a strong re-circulation of the flow occurs in the stationary phase volume of the cell. The flow chart of the flow regimes observed at each couple of operating conditions is given in Fig. 5.16 and shows the beneficial effects of both centrifugal acceleration and flow rate on the mobile phase dispersion in the CPC cells.

The two-phase behaviour in CPC cells is in agreement with previous studies, even if it is difficult to analyze the evolution of the flow regime of mobile phase flow within the cell thickness through a 2D-video image. Observations confirm the influence of injection type and of the cell wall

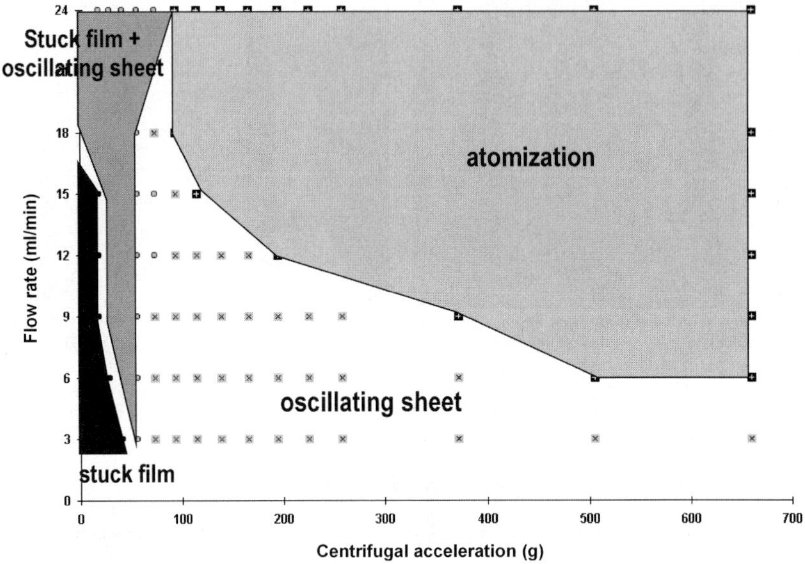

Fig. 5.16. Flow chart showing the differing flow regimes and operating conditions.

proximity. Coriolis acceleration in the rotary system has a significant effect on the mobile phase path deviation and maybe also on instabilities as its influence increases with velocity. Both centrifugal acceleration and flow rate apparently increase destabilization of film flow and mixing of the two phases in cells.

Coalescence: In case of poor phase separation or coalescence at the outlet of the cell, both phases exit the column, whereas only the mobile phase enters it. If the stationary phase volume decreases in the column (bleeding) the separation efficiency of the column will decrease too. Equilibrium is reached when coalescence is complete, i.e., when only the mobile phase leaves the cell. Coalescence depends on centrifugal acceleration, phase ratio and stirring at the cell outlet proximity. In spite of mixing, low disturbance is preferred for coalescence, and large areas can be required when low interfacial tension systems are used.

Phase ratio

Volumes of both phases in the instrument result from the hydrodynamic equilibrium in each cell, according to the flow regimes. The different volumes of the set-up are first examined. Many methods can be used for

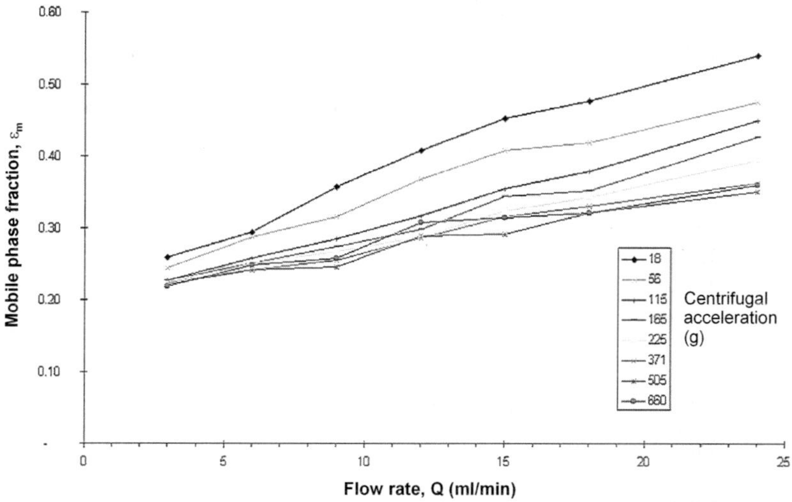

Fig. 5.17. Mobile phase fraction (1 – *Sf*) as a function of the flow rate.

volume measurements. They include pushing one phase by another immiscible one or injecting a tracer in a monophasic flow and then calculating its average residence time in the apparatus. The visual-CPC has the following volumes: extra-column volume, $V_0 = 3.11$ ml ± 5%, and column volume, $V_c = 16.1$ ml ± 5%.

The mobile phase fraction, $\varepsilon_m = 1 - Sf$, is a linear increasing function of the flow rate (Fig. 5.17), whereas it decreases with the acceleration field, reaching a limit value (Fig. 5.18). When F tends to 0, ε_m values gives the minimum of mobile phase ratio available in the instrument ε_{mo}, which corresponds to the duct overall volume [7]. $\varepsilon_{mo} = 0.20$ ± 5%. Here, 20% of the mobile phase volume does not participate in the separation step.

Evolution of mobile phase fraction with the flow rate has already been noted for this kind of two-phase system with other instruments [7], it seems to be representative of CPC working rather than related to CPC geometry. On the other hand, the observation of a decrease of ε_m with the acceleration value is quite new. At a given flow rate, the more dispersed is the mobile phase, the less important is the mobile phase fraction. Both aspects, mobile phase dispersion and low ε_m values, are beneficial to the column efficiency. The mobile phase fraction is low even for high flow rates (less than 40% at 24 ml/min). It can thus be assumed that higher

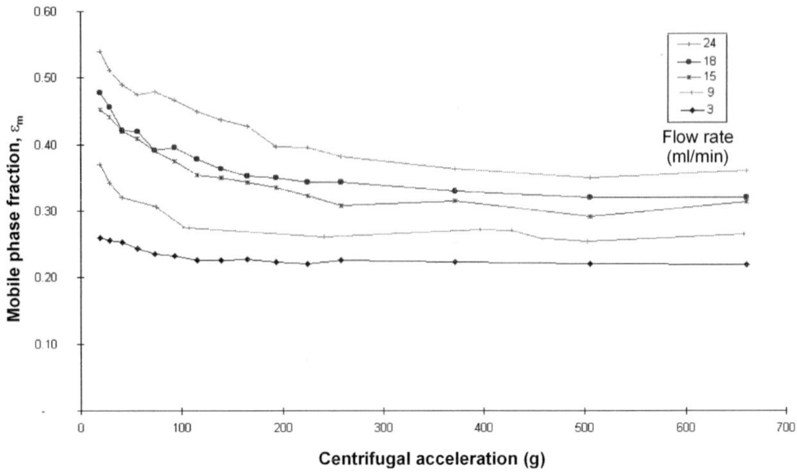

Fig. 5.18. Mobile phase fraction (1 – *Sf*) as a function of the centrifugal acceleration.

flow rates can be used, showing the potential interest of this kind of geometry for preparative applications of CPC.

Remark on the effects of physical properties of both phases

Five two-phase systems are compared here (properties and composition are given in Table 5.1). Five pictures of the flow regime of each system are presented in Fig. 5.19. Operating conditions are identical and the interface is highlighted with a fine white line as no dyes were used making the interface difficult to distinguish on a paused image.

The HMW two-phase system has an important interfacial tension according to the hydrophobic character of heptane. Formation of interface requires much more energy than for low interfacial tension systems. Minimization of interfacial area is then an important phenomenon, and the mobile phase flow still to be stuck along the cell walls.

HM and HCPMW systems have quite similar physical properties. The observed flow patterns are also similar, even if the wavy sheet seems to be more stable for the HCPMW system. The only significant difference between the two systems is mobile phase viscosity. The HCPMW upper phase viscosity is three times greater than that of HM. In the same conditions, the mobile phase velocity at the cell inlet is expected to be lower when the mobile phase viscosity is greater, minimizing instability growing.

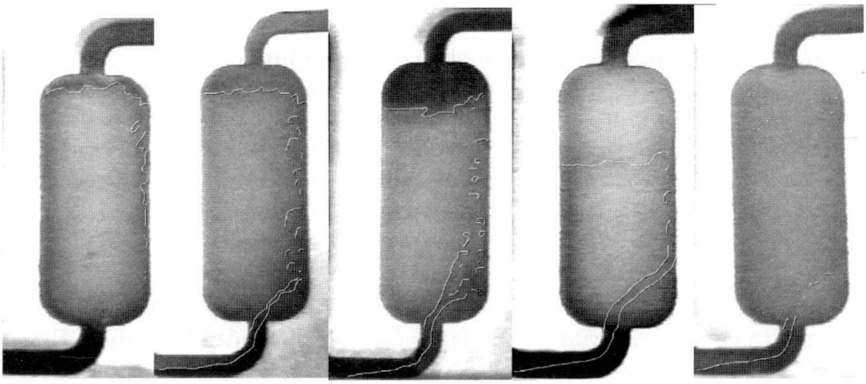

Fig. 5.19. Flow regimes for five different two-phase systems in the same operating conditions: ascending mode, 1200 rpm and 9 ml/min. Two-phase systems from the left to the right: HMW; HM; HCPMW; ATPS; BAW. The interface is highlighted with a light white line.

This is confirmed with the ATPS; mobile phase is very viscous (20 cP) and the flow is a non-oscillating sheet. ATPSs are often considered as "unstable" systems as they have a very low interfacial tension, a low density difference and a high viscosity. Visualization shows that an important mobile phase pool is formed at the cell outlet even if the flow regime is stable. The lack of stability is due to a coalescence problem.

In the case of BAW systems, very small droplets are obtained and an atomization regime is observed. However, the mobile phase viscosity is not less than that of the HM or HCPMW systems. The density differences and interfacial tensions are quite similar, but the stationary phase viscosity is more important, increasing the droplets settling time and the number of droplets present in the cell. The stationary phase viscosity effect on flow regime is not the only expected one, it is also known to have a stabilizing effect on interfacial instabilities.

Many studies are necessary for a better understanding of the physical properties' interdependent effects on the phase hydrodynamic. It is quite difficult when the two-phase system is changing to keep all properties constant except one.

5.4.4 Residence time distribution

We are interested here in the behaviour of the mobile phase in the overall set-up. For the HM system, potassium nitrate RTD is used to estimate the band broadening due to mobile phase flow in both off-column and mobile phase part of the column volume.

The off-column volume of a CPC apparatus mainly consists in a long tube (internal diameter d) and some valves and fittings that can be considered as irregularities on the tube. The flow characteristics in such a tube are well described by the dimensionless Reynolds number,

$$\mathrm{Re} = \frac{\rho U d}{\eta},$$

the ratio of inertial and viscous forces in the flow. The flow is laminar (parabolic velocity profile in the tube for established flow) when Re is lower than 1000 and turbulent when Re is greater than 2000. For example, in our set-up, with the HM liquid system in descending mode, d is equal to 1.02 mm and Re varies from 70 to 600 when the flow rate varies from 3 to 24 ml/min. The Re number is typical of a laminar flow. In Fig. 5.20, the axial dispersion coefficient is plotted versus the linear velocity, U, in the tube.

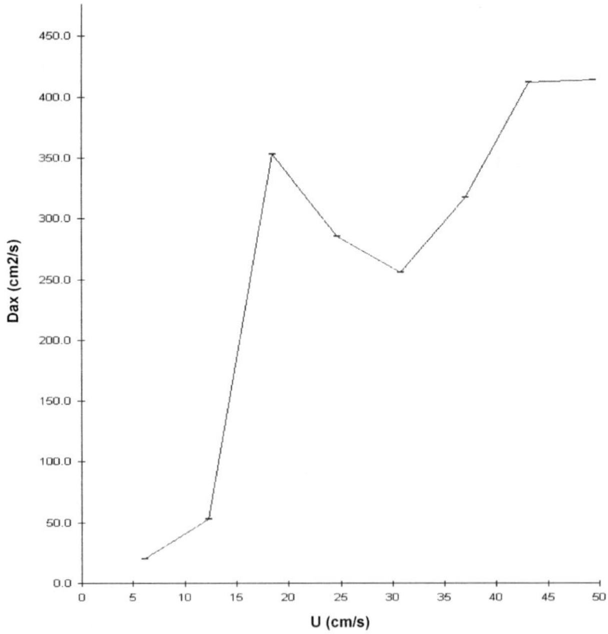

Fig. 5.20. Axial dispersion coefficient as a function of linear velocity in the off-column volume.

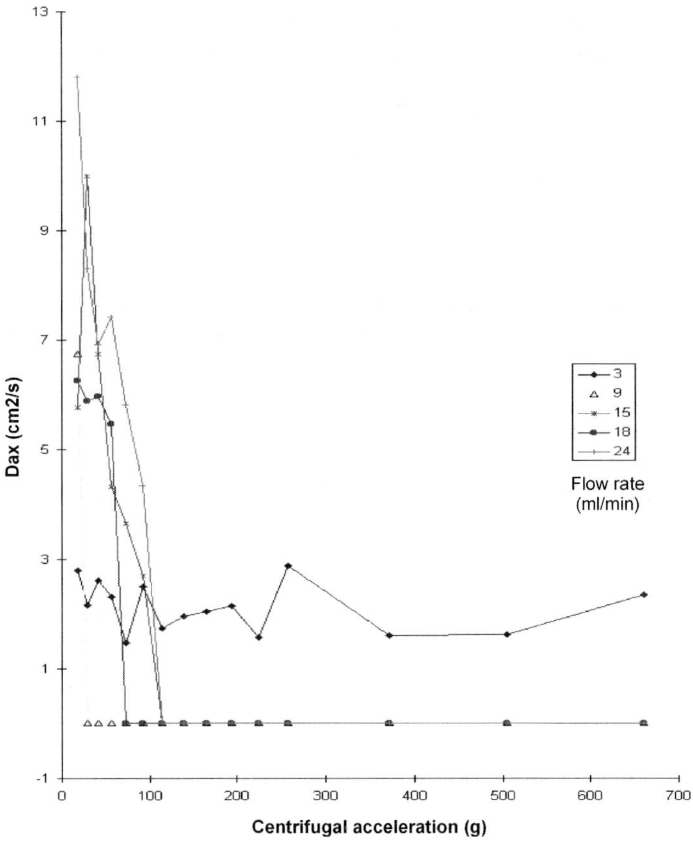

Fig. 5.21. Axial dispersion coefficient as a function of centrifugal acceleration in the column volume.

In established laminar flow in straight pipes, D_{ax} is proportional to U^2 [34]. When U is greater than 18 cm/s, the law is no longer verified and D_{ax} reaches a limit value. Bends, valves and joints create disturbances and diminish the laminar character of the flow in the off-column set-up. Off-column fittings are an important source of band broadening. Both diameter and length of tubing have to be diminished.

In the column, the mobile phase dispersion coefficients are expected to depend on the equilibrium characteristics in the cells. For the axial dispersion coefficient determination, the column length must be calculated; it is equal to the sum of duct and cell lengths. The values of the dispersion coefficients are presented in Fig. 5.21; they are hundred times lower than those obtained for the off-column volume. There is no direct influence of the phase ratio on dispersion. Inside the column, straight ducts are very short. There is insufficient length between two cells to have an established laminar flow. In all cases, the mobile phase flow behaviour is close to that of a plug flow.

The CPC column design characterized by tortuous ducts, sudden enlargements and contractions provides a plug flow mobile phase for this low viscous phases system. Those results confirm that band broadening in the CPC column itself is only due to mass transfer limitation [11] even if the ducts' geometry and flow regimes in both studies are different.

5.4.5 Mass transfer

We are interested here in the main step of the chromatographic process, the step where partitioning occurs with the transfer of the solutes between the two phases. RTD determinations with phenyl salicylate as a tracer in the HM system are then used for mass transfer determination. Previous parameters (average residence time and Péclet number in mobile phase) are kept constant in transfer function and both partition ratio and mass transfer constant time are fitted on phenyl salicylate RTD. First, T_m is fitted with experimental curves and K_D fixed at 0.7. Then, both parameters are fitted from previous values with an average quadratic error between curves of 4%. The resulting K_D is a constant whose value is equal to 0.67±3%. The overall volumetric mass transfer coefficients are calculated from T_m values and presented in Fig. 5.22 as a 2D-curve superimposed with the flow chart (Fig. 5.16). k_0a varies from 0.02 up to 2 s^{-1}. It increases with both the rotational speed and the flow rate in agreement with the visual increase of mobile phase dispersion and two-phase mixing. In fact, when part of the flow path spreads along the cell walls, k_0a is quite low (0–0.2 s^{-1}). This corresponds to the low interfacial area between the two phases when mobile phase is at the wall contact. These results are similar to those obtained by van Buel [11] with Sanki instruments. This could be explained by the fact that the flow regimes observed by van Buel correspond to jets stuck to the walls of the very thin cells (1.15 mm wide) of their apparatuses (see Fig. 5.13). Further information on the comparison between both systems can be found in Marchal et al. [21]. For wavy or oscillating sheet regimes, k_0a varies between 0.1 and 0.5 s^{-1} and within the experimental domain, the

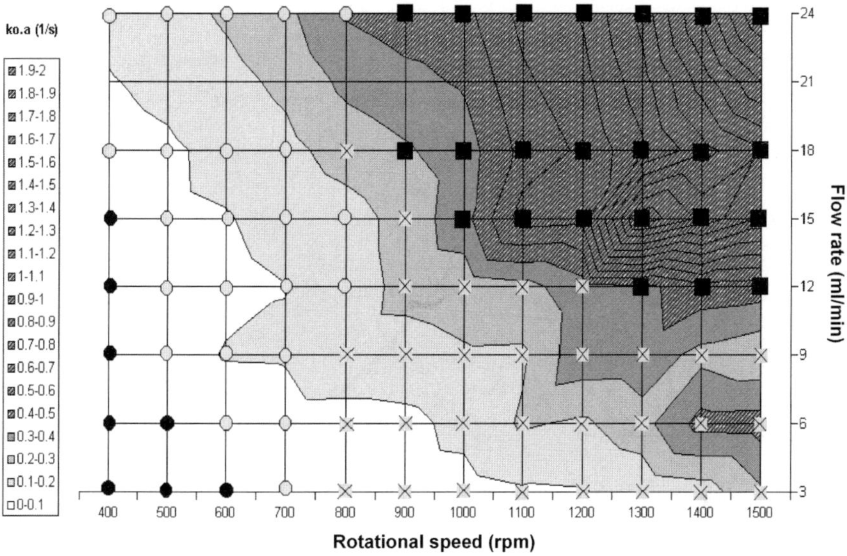

Fig. 5.22. Overall volumetric mass transfer coefficient (1/s) and flow regimes. Compare with Fig. 5.16.

rotation speed seems to have more importance than the flow rate on the k_0a variations.

Eventually, k_0a values greater than 0.5 s^{-1} correspond to atomization regimes. In this domain, there is a transition for flow rates in the range 12–15 ml/min, where k_0a increases quickly (from 0.4 up to 1.2 s^{-1}). The change in the injection conditions also occurs in this high flow rate domain. For flow rates higher than 15 ml/min, k_0a depends mainly on centrifugal acceleration, i.e., the specific interfacial area does not increase with the flow rate in this regime. This confirms the Stokes model [6] which considers that the mobile phase flows as droplets with a diameter depending only on acceleration value, while their number increases with flow rate. For droplets, a is proportional to r^{-1} (if r is their radius), and then a does not depends on flow rate.

Remark on the effects of physical properties and elution mode

Some elution curves of L-leucyl-L-tyrosine obtained at different rotational speeds but constant flow rate (15 ml/min of the BAW system) are presented by Fig. 5.23. This case is ideal for the comparison of the two elution modes since the tracer has a partition ratio of exactly one. The

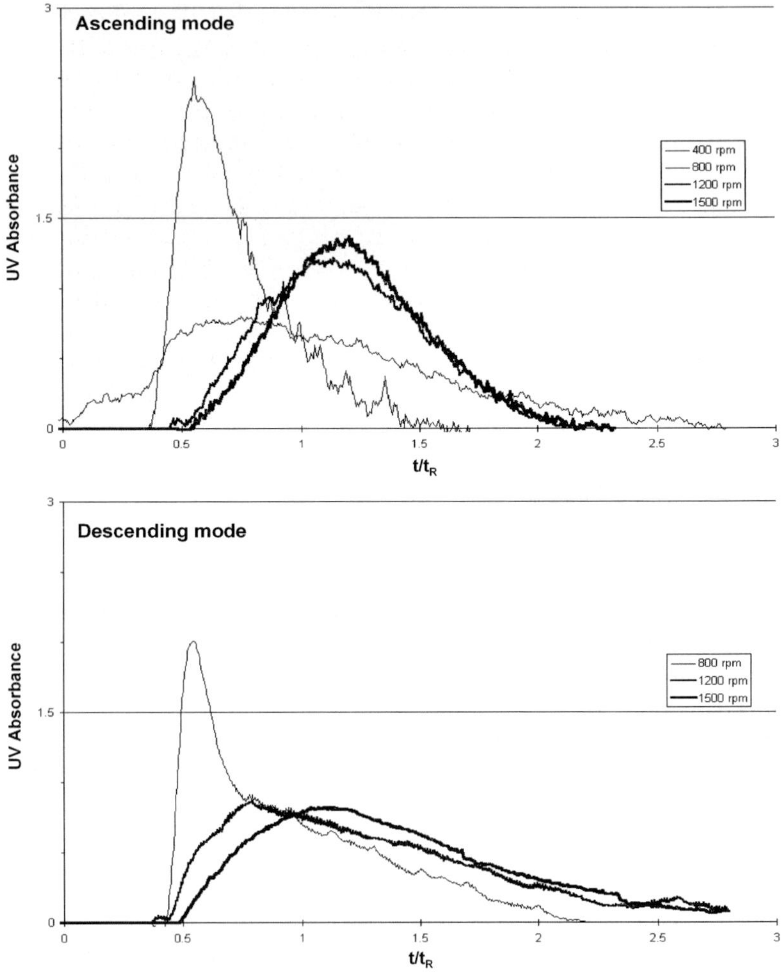

Fig. 5.23. L-Leucyl-L-Tyrosine elution curves with the BAW system in the two modes (AM and DM). F = 15 ml/min and ωε [400; 800; 1200; 1500] rpm.

model predicts that for very low overall volumetric mass transfer coefficients ($k_0a < 10^{-2}\ \mathrm{s}^{-1}$), the elution peak is asymmetric, with a long tail, as if the injected solute was coming out earlier than expected. Such conditions were observed with the BAW system, at low rotational speeds (400–800 rpm), confirming the model accuracy in this domain where no theoretical plate number calculation is possible. Increasing the rotational speed, the peak becomes Gaussian and it is eluted at the expected time. The second important result seen on Fig. 5.23 is the efficiency difference between the two modes. It appears that the mass transfer coefficient is much greater in ascending than in descending mode (0.156 s^{-1} in ascending mode, 0.076 s^{-1} in descending mode at 1500 rpm). In the two modes, as the same system is used, density difference and interfacial tension are not changed, but viscosities are interchanged. When the stationary phase is the more viscous one (in descending mode here), the interface is stabilized, and wavy sheet regime is observed rather than atomization. This confirms the remark formulated earlier on the stabilizing role of stationary phase viscosity. The lack of efficiency due to this unfavourable stabilization of flow regime is also amplified by a significant decrease of the diffusion coefficient of the solutes in the stationary phase volume with the viscosity.

5.4.6 Efficiency and resolution

T_m ($= K_D Sf/k_0 a$) is defined as the ratio of the effective stationary phase volume available for the solute and the overall mass transfer coefficient. It means that T_m is the time necessary to reach the stationary phase partition equilibrium. Let us define R, the retardation factor [35]:

$$R = \frac{t_0}{t_R} = \frac{1}{1+k'} \tag{5.40}$$

then the number of transfer units (i.e., the number of partition equilibrium achieved during elution) is:

$$\mathrm{NTU} = \frac{t_R - t_0}{T_m} = (1-R)\frac{t_R}{T_m} \tag{5.41}$$

It depends on both mass transfer kinetics and phases ratio in each cell. The height of a transfer unit (HTU) can be defined as the ratio of cells number and number of transfer units:

$$\mathrm{HTU} = N_{\mathrm{cell}}/\mathrm{NTU} \tag{5.42}$$

It gives the number of cells necessary to achieve one partition equilibrium and so it characterizes the performance of the cell geometry. Results for our system are presented in Fig. 5.24. The behaviour is the same for all flow rates within the experimental domain, and the HTU number decreases according to the centrifugal acceleration value, lead-

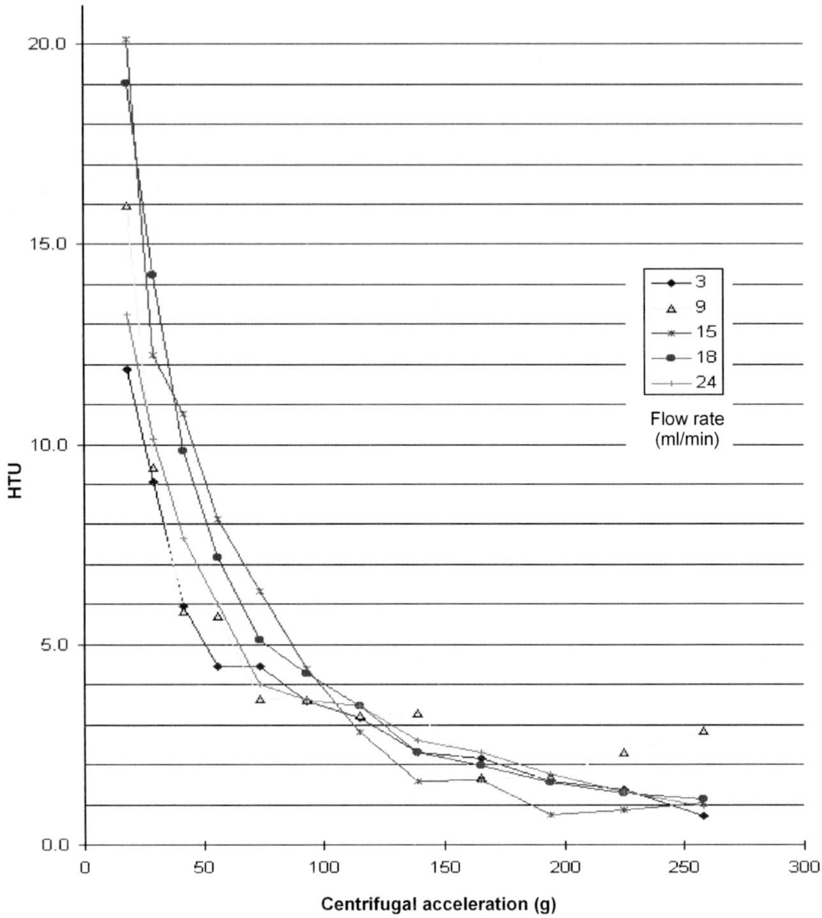

Fig. 5.24. Height of transfer unit as a function of centrifugal acceleration.

ing to a limit value around unity. Then optimization is complete since the partition equilibrium is achieved in each cell. Up to this limit, we can increase the flow rate, and so reduce the elution time, without any lack of separation efficiency due to the decrease of stationary phase volume, because of the improvement of the overall mass transfer coefficient.

By considering the transfer function, $G(s)$, of the column and the van der Laan relationship (Eq. (5.19)) between the k-order momentum (μ_k) of the impulse response of the system and its transfer function, we can calculate the mean retention time and the variance (σ^2) of a solute peak according to the column model parameters:

$$\frac{\sigma^2}{t_R^2} = \frac{2}{Pe} + \frac{2k'}{1+k'}\frac{T_m}{t_R} \tag{5.30}$$

As axial dispersion is negligible in the column volume,

$$\frac{\sigma^2}{t_R^2} = \frac{2k'}{1+k'}\frac{T_m}{t_R} \tag{5.43}$$

since the theoretical plate number is defined as:

$$N = \frac{t_R^2}{\sigma^2} \tag{5.44}$$

then, the number of theoretical plates can be calculated as:

$$N = \frac{1}{2}\left(\frac{1+k'}{k'}\right)^2 \text{NTU} \tag{5.45}$$

Effective separation of two peaks A and B, is commonly expressed by the resolution Rs defined for Gaussian peaks as:

$$Rs = \frac{t_{R_B} - t_{R_A}}{2(\sigma_A + \sigma_B)} \tag{5.46}$$

The approximate Snyder–Knox equation [36] gives an expression of resolution for two close Gaussian peaks (α close to one, where α is the separation factor equal to the ratio of the partition ratios of the two successive solutes), according to chromatographic parameters:

$$Rs = \frac{1}{4}(\alpha - 1)\sqrt{N}\,\frac{k'}{1+k'} \tag{5.47}$$

Using Eq. (5.45), Rs becomes:

$$Rs = \frac{1}{4\sqrt{2}}(\alpha - 1)\sqrt{\text{NTU}} \tag{5.48}$$

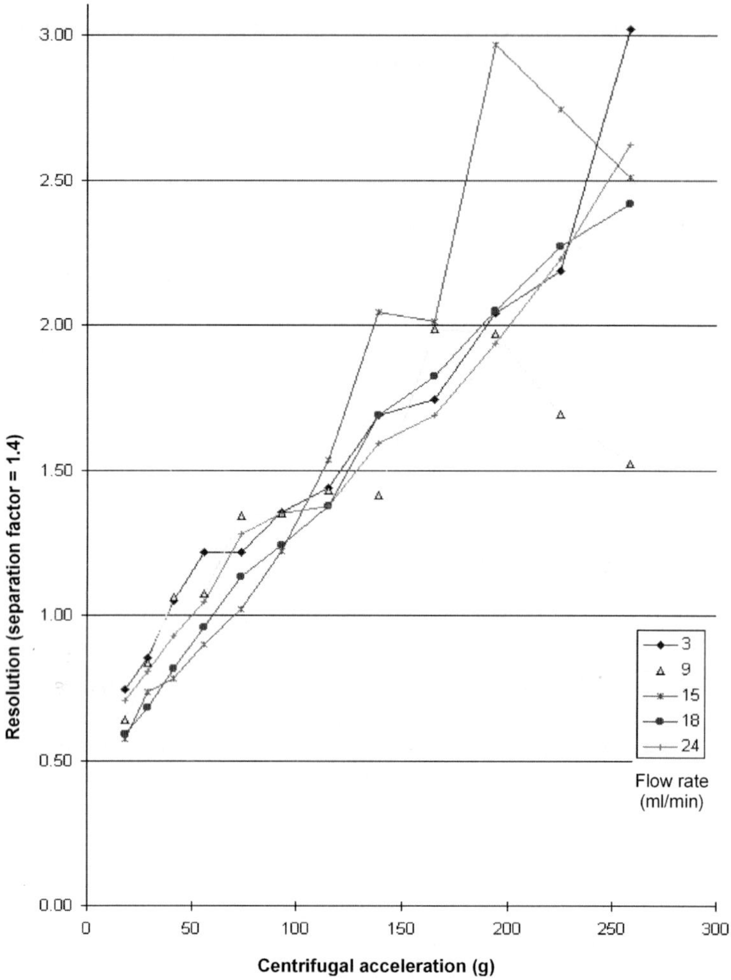

Fig. 5.25. Resolution simulation about the visual-CPC.

As an illustration of our results for a resolution simulation, we will chose α equal to 1.4 and twenty disks of 66 cells (1320 cells and a column volume of 322 ml). The results are presented in Fig. 5.25. The resolution factor is quite a linear function of the centrifugal acceleration whatever the flow rate value is. $Rs = 1.5$ is the value commonly considered as sufficient to have a baseline separation of two successive peaks. This value is reached between 100 and 150 g, i.e., between 900 and 1100 rpm.

The FCPC from Kromaton Technology [37], designed according to the preliminary results of our works, has been tested with the n-heptane–methanol–water (50–45–5) two-phase system, whose physical properties are similar to that of the HM system. Two experiments in descending mode were carried out (see Fig. 4.20): the first was done at 3 ml/min and 400 rpm, the second at 20 ml/min and 2000 rpm. These experiments correspond to two points at opposite corners of the flow chart (Fig. 5.16). Injected samples are identical mixtures of p-hydroxybenzoic acid (a marker for t_0) and the two esters diethylphthalate and dibutylphthalate. According to the observed mass transfer coefficient increase between the two points, the effective efficiency (N) is multiplied by 8 and the resolution factor increases from 1.5 to 3.4.

5.5 CONCLUSION

The importance of flow patterns on chromatographic efficiency in CPC is demonstrated. The relationship between the mass transfer coefficient and flow regimes is linked to changes of specific interfacial area. Three main flow regimes are observed. Stuck film along the cell walls, curvilinear sheet or film with oscillating instabilities and atomization. The flow path is controlled by cell design, wall proximity and both centrifugal and Coriolis accelerations. Atomization regime and re-circulation in stationary phase prevent to channelling problem often encountered in two-phase transfer processes. From an engineering point of view, the objective is to propose a suitable cell design in order to obtain easily the atomization regime of the mobile phase, followed by a well-defined coalescence of the two phases in each cell. For the heptane–methanol two-phase system, the mobile phase fraction remains low (high Sf values) within the experimental domain; the coalescence step is efficient in the cells. The two-point measurement RTD and the proposed model form an accurate method for mass transfer determination which proved to be extremely powerful for the studies to improve the CPC process. Partitioning is well described and external sources of band broadening can thus be characterized. The off-column volume is an important source of band broadening and so it has to be carefully considered for the overall process optimization. The disk (cells and ducts) width no longer favours axial dispersion apparition whatever the phase ratio and the flow regime are, owing to its tortuous geometry. Band broadening in the column itself is then only due to mass transfer limitation.

Overall volumetric mass transfer coefficient is used to characterize flow regimes whereas the number of transfer units is introduced for the separation process evaluation. NTU only depends on centrifugal acceleration; no significant flow rate effect is observed, leading to one transfer unit per cell. Lastly, relations between these parameters and the classical chromatographic parameters (theoretical plate number, retention and resolution factors) are given. The physicochemical properties of the biphasic liquid system have important effects on flow regimes and mass transfer. This is emphasised to make the reader sensitive to the importance of these parameters, in particular for the elution mode choice. Many more studies are needed for a better understanding of the effects of the liquid phase properties on efficiency and phases ratio in CPC. Work on some new cell shapes, designed in agreement with these results, is in progress.

REFERENCES

1 W. Murayama, T. Kobayaki, Y. Kosuge, H. Yano, Y. Nunogaki and K. Nunogaki, *J. Chromatogr.*, 239 (1982) 643.

2 M.J. van Buel, L.A.M. van der Wielen and K.Ch.A.M. Luyben, In: A.P. Foucault (Ed.), *Centrifugal Partition Chromatography*, Chromatographic Sciences Series, Vol. 68. Marcel Dekker, New York, 1994, p. 51, Chapter 3.

3 M. Tswett, *Proc. Warsaw Soc. Nat. Sci., Biol. Sect.*, 1903, XIV No. 6.

4 A. Berthod and D.W. Armstrong, *J. Liq. Chromatogr.*, 11 (1988) 567.

5 D.W. Armstrong, G.L. Bertrand and A. Berthod, *Anal. Chem.*, 60 (1988) 2513.

6 A.P. Foucault, E. Camacho-Frias, C.G. Bordier and F. Le Goffic, *J. Liq. Chromatogr.*, 17 (1994) 1.

7 A.P. Foucault, O. Bousquet and F. Le Goffic, *J. Liq. Chromatogr.*, 15 (1992) 2691.

8 A.P. Foucault, In: A.P. Foucault (Ed.), *Centrifugal Partition Chromatography*, Chromatographic Sciences Series, Vol. 68. Marcel Dekker, New York, 1994, Chapter 2.

9 M.J. van Buel, L.A.M. van der Wielen and K.Ch.A.M. Luyben, *J. Chromatogr. A*, 773 (1997) 1.

10 M.J. van Buel, L.A.M. van der Wielen and K.Ch.A.M. Luyben, *J. Chromatogr. A*, 773 (1997) 13.

11 M.J. van Buel, L.A.M. van der Wielen and K.Ch.A.M. Luyben, *AIChE J.*, 43 (1997) 693.

12 M.J. van Buel, Thesis, Delft University of Technology, The Netherlands, 1997.

13 M.J. van Buel, F.E.D. van Haldema, L.A.M. van der Wielen and K.Ch.A.M. Luyben, *AIChE J.*, 44 (1998) 1356.
14 J. Villermaux, D. Schweich and M. Sardin, In: G. Ganestos and P.E. Barker (Eds.), *Preparative and Production Scale Chromatography*. Marcel Dekker, New York, 1992, p. 701.
15 M.J.P. Martin and R.L.M. Synge, *Biochem. J.*, 35 (1941) 1359.
16 J.J. Van Deemter, F.J. Suiderweg and A. Klinkenberg, *Chem. Eng. Sci.*, 5 (1956) 271.
17 L. Lapidus and N.R. Amundson, *J. Phys. Chem.*, 56 (1952) 984.
18 J. Villermaux, *Génie de la Réaction Chimique*, Lavoisier, Paris, 1982.
19 N. Wakao and S. Kaguei, *Heat and Mass Transfer in Packed Beds*. Gordon and Breach Science, 1982.
20 P. Legentilhomme, L. Brujes and J. Legrand, *Chem. Eng. J.*, 67 (1997) 83.
21 L. Marchal, A. Foucault, G. Patissier, J.M. Rosant and J. Legrand, *J. Chromatogr. A*, 869 (2000) 339.
22 A.E. Dukler and Y. Taitel, In: G.F. Hewitt, J.M. Delhaye and N. Zuber (Eds.), *Multiphase Science and Technology*, Vol. 2. Hemisphere Publishing, Springer-Verlag, 1986.
23 S.B.R. Karri and V.K. Mathur, *AIChE J.*, 43 (1988) 137.
24 L. Zhao and K.S. Rezkallah, *Int. J. Multiphase Flow*, 19 (1993) 751.
25 J.C. Godfrey and C. Hanson, In: G. Hetsroni (Ed.), *Handbook of Multiphase Systems*. Hemisphere Publishing, McGraw Hill, 1982.
26 G.W. Govier and K. Aziz, *The Flow of Complex Mixtures in Pipes*. van Nostrand Reinhold, New York (1972).
27 P. Angeli and G.F. Hewitt, *Int. J. Multiphase Flow*, 24 (1998) 1183.
28 R. Clift, J.R. Grace and M.E. Weber, *Bubbles, Drops and Particles*. Academic Press, 1978.
29 E.A Ibrahim., *Chem. Eng. Sci.*, 52 (1997) 4419.
30 M.J. McCarthy and N.A Molloy, *Chem. Eng J.*, 7 (1974) 1.
31 R.P. Fraser and P. Eisenklam, *Imp. Coll. Chem. Eng. Soc. J.*, 7 (1953) 52.
32 G.G. Coriolis, *J. l'Ecole Royale Polytechnique*, 15 (1835) 142.
33 A.P. Foucault, L. Marchal, J.M. Rosant, S. Farias Neto, J. Legrand and P. Durand, *Pittsburgh Conference*, New Orleans, Abstract 878, 1998.
34 G. Taylor, *Proc. Roy. Soc. A*, 219 (1953) 186.
35 L.S. Ettre, *Pure Appl. Chem.*, 65 (1993) 819.
36 A.S. Said. *Theory and Mathematics of Chromatography*. A. Huthig, New York, 1981.
37 Kromaton Technologies, 64 rue Pasteur, 94807 Villejuif, France, http://www.kromaton.com.

Chapter 6

Liquid stationary phase retention and resolution in hydrodynamic CCC

Ian A. Sutherland

6.1 INTRODUCTION

In countercurrent chromatography (CCC) both phases are liquid. Unlike other chromatography techniques the volume of the stationary phase can be measured accurately and hence the elution behaviour of solutes of a given distribution ratio can be predicted. Furthermore, the volume of stationary phase retained can vary with such variables as mobile phase flow, rotational speed (or "g" field) and the physical properties of the solvent phase system being used.

A solvent phase is retained against the flow of mobile phase in the rotating coils of a coil planet centrifuge in a similar way that an Archimedean screw pumps water out of a river against gravity (Fig. 6.1). The mysteries of why one phase moves to one end of the coil and the other to the opposite end will not be revealed in this chapter. However, the rules will be given and the predictable nature of stationary phase retention and how sample elution can be predicted from the distribution ratio and retention of the stationary phase will be revealed.

In the 1950s Craig's countercurrent distribution (CCD) apparatus (Fig. 6.2) was extensively used for the purification of natural products based on their distribution ratio [1]. It is a beautiful example of a separation system that is completely mathematically predictable, once the solute distribution ratio and ratio of mobile to stationary phase volumes are known.

This chapter will show how mathematically predictable counter-current chromatography (CCC) is and how it can be modeled using an

Comprehensive Analytical Chemistry XXXVIII
A. Berthod (Ed.)

Fig. 6.1. A woodcut from an edition of *Virtruvius's De Architectura* by Fra Giocondo (Venice, 1511) illustrating a hand-operated Archimedean pump used for pumping water from a river against gravity in the 14th Century.

Fig. 6.2. Craig 100-400 transfer CCD apparatus (circa 1955).

eluting CCD mathematical model. Initially there will be a description of how the dead volume of the system and stationary phase retention can be measured and how this retention can be used to predict when peaks of a given distribution ratio elute. This will be illustrated by a simple spreadsheet technique for optimizing the resolution that can be achieved with the system being used. The chapter will conclude by showing examples of how the theoretical model fits with practical measurements and how the model can then be used to optimize running conditions.

6.2 HEAD/TAIL RULE

Each coil has a head and a tail. The head is the end a bubble or a bead would screw to under Archimedean action when the coil is rotated. If the coil is visible, then slowly rotating the coil by hand, whilst holding a finger against it, will move your finger towards the head. If the coil is not visible, the coil can be half filled with water and slowly rotated (as it will probably be out of balance in this condition) until water spits out of one end of the coil, which will be the head end. Note that the head and tail will switch if the direction of rotation is changed.

Basic head/tail research [2] has led to the following rules: if the lower phase is mobile, pump from head (centre) to tail (periphery) and if the upper phase is mobile, pump from tail (periphery) to head (centre). These modes of pumping will ensure best results with the Archimedean and hydrostatic forces being additive. If it is found that the head of the coil is at the periphery, then change the direction of rotation to put it at the centre.

6.3 MEASURING DEAD VOLUME (V_d) AND RETENTION OF THE STATIONARY PHASE (*Sf*) IN THE COIL PLANET CENTRIFUGE

In CCC, the coil system is initially filled with the phase intended to be the stationary phase. The coil is then rotated at a given speed and the mobile phase pumped in at a set flow rate. As the mobile phase progresses through the coil, it sets up a fluid-dynamic equilibrium between the two phases. The flow of the mobile phase is balanced against the Archimedean movement of the other phase in the opposite

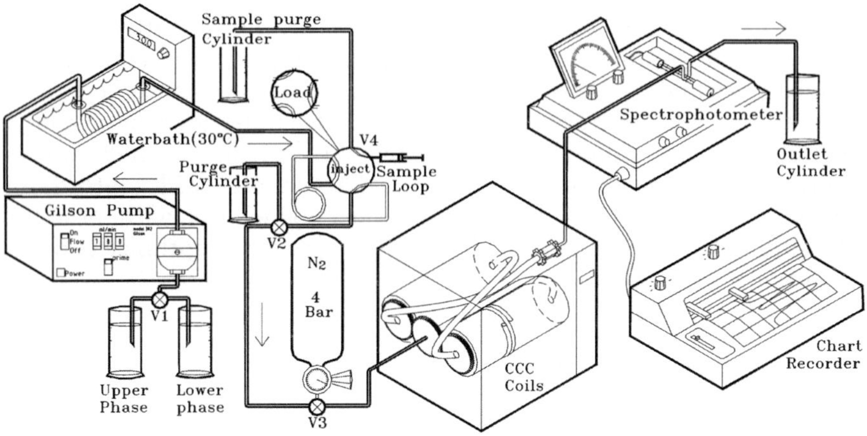

Fig. 6.3. Schematic diagram of CCC operating system.

direction—the volume occupied by the mobile phase being displaced from the system.

A typical CCC set up is shown in Fig. 6.3. After the coil system is initially filled with stationary phase, the mobile phase reservoir is then selected with valve (V1) and the pump system primed/purged with the intended mobile phase up to valve (V2). The rotor is then switched to the desired speed (typically 800 rpm for a 10 mm diameter rotor) and the temperature control system set at the desired temperature (typically 30°C). It is also advisable to temperature control the solvent supply reservoirs if repeatable results are required as phase system physical properties such as densities, viscosities and interfacial tension can change with temperature. Such changes in the physical properties will affect a solute's distribution ratio.

Once the temperature control system is in equilibrium, a fresh volumetric measuring cylinder is positioned to collect the eluent and valve (V2) is switched from "purge" to "flow through the coils". The rotor is then started and the mobile phase pumped into the coil at an initially low flow value. Typical pump rates are given in Table 6.1 for a range of different CCC units being used at Brunel. The Brunel CCC (previously known as the Quattro) is described in Ref. [3]. The Analytical and Process1 scale instruments developed as part of a BBSRC/DTI LINK funded program with industry are described in Ref. [4]. The Process2 and Process3 scale CCC instruments are currently being evaluated under an EPSRC-IMI funded research program [5].

TABLE 6.1

Typical flow rates for CCC Instruments with different tubing bore arranged in columns giving the same linear flow of the $P=1$ peak.

Instrument	Bore (mm)	Min flow (ml/min)				Maximum flow (ml/min)		
Analytical	0.8	0.25	0.5	1	–	–	–	–
Brunel CCC	1.6	1	2	4	8	–	–	–
	2.6	2.5	5	10	20	40	–	–
Process1	3.7	5	10	20	40	80	–	–
	4.5	8	16	32	64	128	256	–
Process2	5.2	10	20	40	80	160	320	–
	7.5	20	40	80	160	320	640	1280
Process3	10.4	40	80	160	320	640	1280	2560

A general rule in CCC is that a solute with a distribution ratio of $K_D = 1$ does not differentiate between the upper and lower phase and so elutes in the coil volume divided by the flow ($t_{Kd=1} = V_c/F$), as if the two phases were one (after correction for the dead volume V_d). Table 6.1 gives typical flows for a range of tubing internal diameters from analytical to process scale CCC.

Initially, stationary phase will be eluted until fluid-dynamic equilibrium is reached and the mobile phase eluted. The volume eluted (V_e) can be expressed as:

$$V_e = V_d + V_M \tag{6.1}$$

where V_d is the dead volume of mobile phase in the inlet and outlet leads and the coil feed pipes, and V_M is the active mobile phase in the coil system.

The volume of stationary phase (V_S) in the coil system can be expressed as a fraction (Sf) or percentage of the coil volume (V_C) as follows:

$$Sf = V_S/V_C \text{ or } V_S/V_C{*}100\% \tag{6.2}$$

where

$$V_M + V_S = V_C \tag{6.3}$$

The manufacturers of CCC instruments normally give the coil volumes with their operating instructions. If not, it is best to empty the coils while pressurizing them from tail to head with nitrogen and slowly rotating the coils at about 300–500 rpm. Weigh the coils off the machine while they are empty and then weigh them again once they are filled with water. Care has to be taken with large-bore tubing to ensure that there is no air entrapment. The difference in weight will give the weight of water in the coil and hence the volume.

V_d is more difficult to measure, as it will not be known whether there is any non-active retention of stationary phase in the inlet outlet leads. It is important because the dead volume makes no contribution to the separation process and has to be eliminated if accurate distribution ratios are required from chromatograms. One method of overcoming this is to construct a graph that will plot the volume of eluted stationary phase versus the square root of the mobile phase flow rate as described by Wood [6]. A linear relationship is observed:

$$V_e = V_d + AF^{1/2} \tag{6.4}$$

Such a typical V_e versus $F^{1/2}$ plot is shown in Fig. 6.4. It is obtained by incrementally increasing the mobile phase flow until a new equilibrium is reached and plotting the new eluted volume of stationary phase against the square root of the flow. The intercept at $F = 0$ gives the value of V_d and any value above this gives the volume of mobile phase in the coil at any given flow in excess of this value. It follows by eliminating V_e and V_d in Eqs. (6.1) and (6.4) that:

$$V_M = AF^{1/2} \tag{6.5}$$

Substituting for V_M from Eq. (6.3) and rearranging gives:

$$Sf = V_S/V_C = 1 - BF^{1/2} \tag{6.6}$$

where $B = A/V_C$.

Plotting the stationary phase retention (Sf) versus the square root of flow ($F^{1/2}$) as in Eq. (6.6) produces also a straight line [7]. Figure 6.5 shows typical Sf versus $F^{1/2}$ plots for different heptane–ethyl acetate–methanol–water phase systems from data given in [7].

Such plots, if corrected for dead volume (V_d), should pass through the 100% retention point at $F = 0$. This means that it is only necessary to

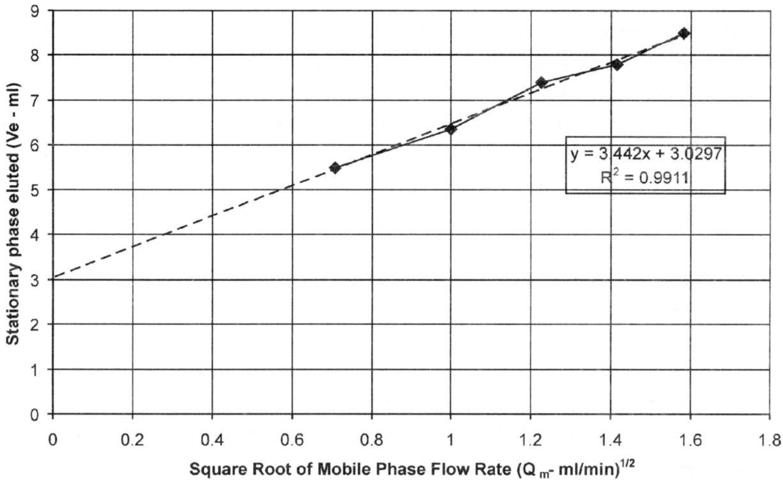

Fig. 6.4. Plot of stationary phase eluted versus the square root of volume flow giving the dead volume (V_d) as the intercept of 3.0297 ml (Brunel high speed analytical CCC, R = 110 mm. Phase system: heptane–ethyl acetate–methanol–water (1.4–0.6–1.0–1.0), N = 1400 rpm, V_c = 25.06 ml, β = 0.884, d_c = 0.762 mm).

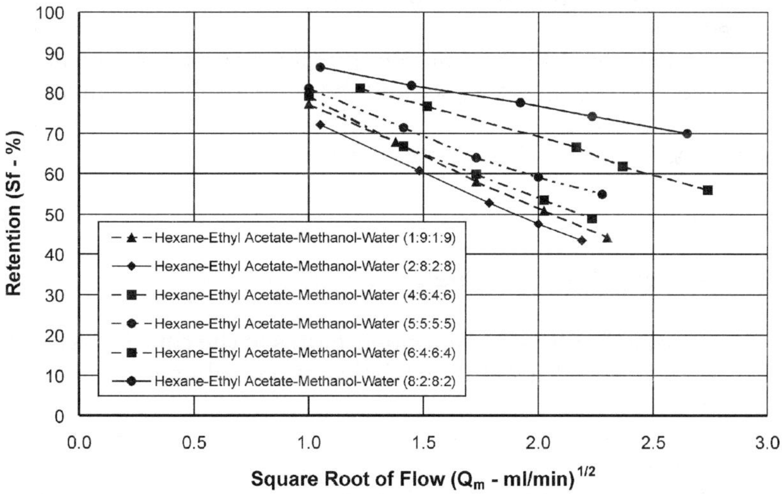

Fig. 6.5. Typical stationary phase retention versus square root of the mobile phase flow rate plots for hexane–ethyl acetate–methanol–water phase systems from Ref. [4] (J Type CCC Apparatus constructed by the Institute of New Technology Application, Beijing; R = 100 mm, N = 750 rpm, V_c = 230 m;, β = 0.4–0.78, d_c = 1.6 mm).

measure retention accurately at one flow value to calculate the gradient (B) and be able to predict the stationary phase retention, Sf, for other flows. In practice, however, it is more reliable to construct V_e versus $F^{1/2}$ and/or Sf versus $F^{1/2}$ plots for a range of flows to minimize errors. Note that the gradient from the V_e plot can be used to derive the gradient of the corresponding Sf plot by simply dividing by the system volume (V_C) as in Eq. (6.6).

6.4 PREDICTING PEAK ELUTION

Once fluid-dynamic equilibrium has been established, a sample can be injected into the mobile phase at any time, either in the mobile phase, stationary phase, or a combination of the two. It will be assumed below that the sample is injected with the mobile phase with the solvent front (i.e., the point at which the change is made from filling with stationary phase to pumping the mobile phase). The dead volume (V_d) of the inlet and outlet tubes will, in practice, need to be taken into account if it is significant. It has been neglected in the analysis below.

6.4.1 Elution of solvent front ($K_D = 0$)

The solvent front will elute in the time it takes the mobile phase to pass through the inlet and outlet tubes and the volume of mobile phase in the coil. This can be expressed mathematically as follows, with reference to the schematic elution profile given in Fig. 6.6.

$$t_{Kd=0} = V_M/F \tag{6.7}$$

6.4.2 Elution of $K_D = 1$ Peak

The $K_D = 1$ peak, as already mentioned, elutes in the time it takes for the mobile phase to pass through the whole system volume.

$$t_{Kd=1} = V_c/F \tag{6.8}$$

The volume position of the $K_D = 1$ peak in CCC will always remain the same whichever phase system or flow is used, whereas the $K_D = 0$ position will move toward the $K_D = 1$ elution point as the retention decreases.

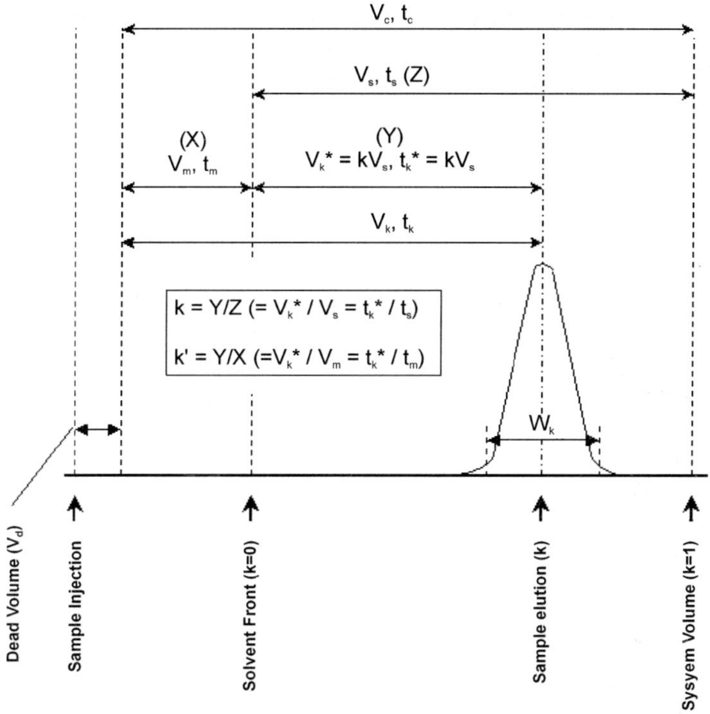

Fig. 6.6. Schematic diagram of CCC elution profile for a solute with partition coefficient (K_D).

6.4.3 Elution of a solute with distribution ratio (K_D)

The elution time, t_R, for a solute with distribution ratio (K_D) is as follows:

$$t_R = (V_M + K_D V_S)/F = t_{Kd=0} + K_D V_S/F \quad (6.9)$$

It can be seen from Fig. 6.6 that the distribution ratio of a given peak can easily be worked out by dividing the elution volume or time from the solvent front (V_P^*, t_P^*) by the stationary phase volume (V_S) or simply, with reference to Fig. 6.6, dividing Y by Z. This is a good reason to know the retention of the stationary phase (Sf) accurately.

$$K_D = V_P^*/V_S = t_P^*/t_s = Y/Z \quad (6.10)$$

The solute retention factor k' (k-prime), which is the ratio of sample mass in the respective phases regardless of volume ratio, can also be obtained from the elution profile as follows:

$$k' = K_D V_S / V_M = V_P^* / V_M = t_P^* / t_m = Y/X \tag{6.11}$$

6.5 MODELLING CCC USING AN ELUTING CCD MODEL

The theory of Countercurrent Distribution was established by Martin and Synge in the 1940s [8] and re-used in some depth by Ruth and Mandava [9] in their book on CCC. The distribution of a sample, with distribution ratio (K_D) and with "τ" test tube steps, can be described as follows:

$$(x+y)^{\tau} = x^{\tau} + {}_{\tau}C_1 x^{\tau-1} y + {}_{\tau}C_2 x^{\tau-} y^2 + \ldots + y^{\tau} \tag{6.12}$$

where

$${}_{\tau}C_r = \frac{\tau!}{(\tau - r)!\,r!}$$

$$x = \frac{1}{(1+k')} \text{ and } y = \frac{k'}{(1+k')}$$

Now in CCD the distribution factor is defined as the mass in the upper transferred phase divided by the mass in the lower stationary phase as follows:

$$k'_{CCD} = \frac{m_U}{m_L} = \frac{C_U V_U}{C_L V_L} = \frac{K_D V_M}{V_S} = \frac{K_D (1 - Sf)}{Sf} \tag{6.13}$$

In countercurrent chromatography the separation is not restricted to the chain of "τ" test tubes. Elution takes place in such a way that the upper phase volume is passed on to a fraction collector with fraction 1 being the "τ+1" transfer. The quantities in each eluted fraction from the CCD chain can be described [10] as follows:

$$[\text{Fraction}]_{(r-\tau)} = \sum_{r=\tau}^{n} \frac{r!\,x^{(r-\tau)} y^{\tau}}{(r-\tau)!\,\tau!} \tag{6.14}$$

where τ is the number of transfers to elution (i.e., the size of the CCD chain), n is the total number of transfers and $(r - \tau)$ is the fraction number at the rth transfer.

A computer CCD model was developed based on Eqs. (6.12)–(6.14) but defining distribution ratio from the perspective of countercurrent chromatography:

$$k'_{\mathrm{CCC}} = \frac{m_{\mathrm{S}}}{m_{\mathrm{M}}} = \frac{C_{\mathrm{S}}V_{\mathrm{S}}}{C_{\mathrm{M}}V_{\mathrm{M}}} = \frac{K_{\mathrm{D}}V_{\mathrm{S}}}{V_{\mathrm{M}}} = \frac{K_{\mathrm{D}}Sf}{(1-Sf)} \tag{6.15}$$

where

$$k'_{\mathrm{CCC}} = \frac{1}{k'_{\mathrm{CCD}}}$$

Input parameters for the model are the distribution ratios of sample 1 and sample 2, the percentage retention of stationary phase (Sf – %), the number of transfers to elution (τ) and the total number of transfers (n). The model assumes a sample mass of 100% and calculates the mass distribution as transfers progress from $r = 0$ to n. A typical trace is shown in Fig. 6.7 for $K_{\mathrm{D1}} = 0.35$, $K_{\mathrm{D2}} = 0.7$, $Sf = 85\%$, $\tau = 400$ and $n = 2500$.

It is well known in chromatography that resolution can be increased by increasing the length of the column [11] and this has been confirmed recently for CCC [12]. Doubling the column length increases the resolution by a factor $\sqrt{2}$ (i.e. $R_{\mathrm{s}} \approx \sqrt{L}$). The resolution between two samples is commonly expressed [11] as:

$$R_{\mathrm{s}} = \frac{2(t_2 - t_1)}{w_1 + w_2} \tag{6.16}$$

where t_1 and t_2 are the retention times of two solutes and w_1 and w_2 their respective baseline widths.

It follows, therefore, that as the time for peak elution increases in proportion to the length of the column ($t_{\mathrm{R}} \approx L$), the peak width increases with the square root of the length of the column ($w_{\mathrm{R}} \approx \sqrt{L}$). This was verified using the model [10] and the following relationship established:

$$w_{\mathrm{R}} = 4\sqrt{\frac{t_{\mathrm{R}} \cdot k'}{\chi}} \tag{6.17}$$

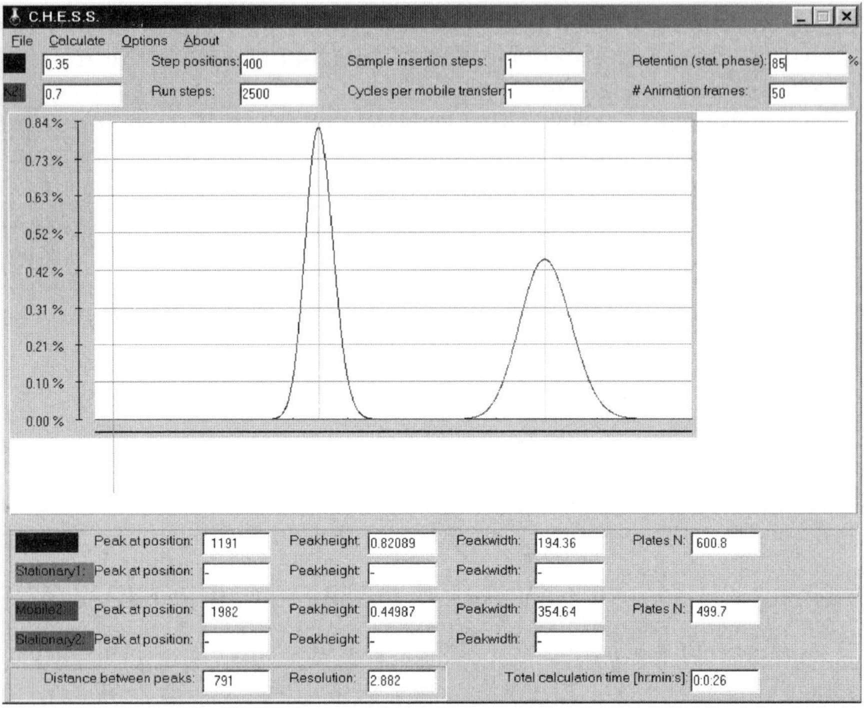

Fig. 6.7. Model chromatogram for solutes with partition coefficients 0.35 and 0.7, a stationary retention of 85% and a coil giving the equivalent of 400 CCD transfers to elution and 2500 CCD transfers in total.

where t_R is the retention time (time of elution) of a peak with distribution ratio (K_D) and retention factor (k') and χ is a scaling constant with units of steps/unit time. For CCD where the unit of time is a step then $\chi = 1$. It is also well known in chromatography that the efficiency (N) or number of theoretical plates is commonly expressed [11] as:

$$N = 16\left(\frac{t_R}{w_R}\right)^2 \tag{6.18}$$

Substituting for w_R, the peak width at peak base, in Eqs. (6.18) from (6.17) gives:

$$\chi = \frac{N\,k'}{t_R} \tag{6.19}$$

If peak elution times are being measured in minutes for example then χ would have units of steps/min. If multiplied by the elution time of the solvent front ($t_m = V_M/F$), this would give the equivalent number of transfers for a given countercurrent chromatography chromatogram:

$$n_t = \frac{\chi V_M}{F} \tag{6.20}$$

6.6 PRACTICAL APPLICATION OF THE MODEL

The example below utilizes a sample set of results from the Brunel Process CCC unit [4] with $R = 110$ mm, $N = 1200$ rpm, $V_C = 928$ ml, $\beta = 0.63$–0.87 and $d_c = 3.7$ mm. In practice the flow range used was 10, 20, 40 and 80 ml/min. The model allows elution predictions outside this range.

The procedure for developing a model for a given set of running conditions is:

(1) Prepare a Sf or V_e versus $F^{1/2}$ plot as shown in Section 6.3 (Eq. 6.6) to establish the relationship between Stationary Phase Retention and Flow (a gradient of $B = -2.3847$ was obtained from the Sf plot). Knowing the volume of the stationary phase allows the mobile phase volume to be calculated. This will establish the point of elution of the $K_D = 0$ peak (Eq. 6.7).

(2) Run a chromatogram with the phase system intended to be used and a solute of known distribution ratio. Measure t_R and w_R from the chromatogram and calculate N and k' from Eqs. (6.18) and (6.11), respectively, and hence calculate χ the number of CCD steps per minute from Eq. (6.19) (this was found to be 67.6 CCD steps/min).

(3) Once χ is known, a spread-sheet like the one in Table 6.2 can be prepared to give resolution (R_s) and any other separation parameters, like total run time, that are required.

Figure 6.8 shows theoretical and measured resolutions versus retention for the three different CCC units referenced above and running at three different conditions. Here the scaling factor (χ) was calculated from one of the eluted peaks by measuring the time of elution (t_R), the base width of the peak (w_R), calculating the distribution ratio from Eq. (6.10) and the number of CCD steps per minute from Eq. (6.18). All other predictions were made from Eqs. (6.9) and (6.19) assuming distribution ratios remain the same and that CCC follows Du's flow relationship (6.6) [7].

TABLE 6.2

Spreadsheet for deriving retention and resolution from the B gradient and the number of CCD steps per minute

Retention Relationship with Flow-Process CCC

Vc =	928 ml	$\alpha = P_2/P_1 =$	1.8182	
B =	2.3847	slope of the S_f versus F1.2 plot		
P1 =	0.22	$\chi =$	67.64	steps/min
P2 =	0.4		0.0148	min/step

F (ml/min)	S_f (%)	V_m (ml)	V_s (ml)	$t_{P=0}$ (min)	t_{P1} (min	t_{P2} (min	Δt (min)	W_{P1} (min)	W_{P2} (min)	N_{P1}	N_{P2}	R_s
4	95.23	44.26	883.74	11.07	59.67	99.44	39.77	7.87	13.71	918.82	842.15	3.69
8	93.26	62.59	865.41	7.82	31.62	51.09	19.47	4.77	8.18	703.21	624.92	3.01
16	90.46	88.52	839.48	5.53	17.08	26.52	9.44	2.90	4.88	553.58	472.87	2.43
32	86.51	125.19	802.81	3.91	9.43	13.95	4.52	1.77	2.91	452.17	367.77	1.93
64	80.92	177.04	750.96	2.77	5.35	7.46	2.11	1.09	1.73	387.62	297.39	1.50
128	73.02	250.37	677.63	1.96	3.12	4.07	0.95	0.66	1.02	354.51	254.52	1.13

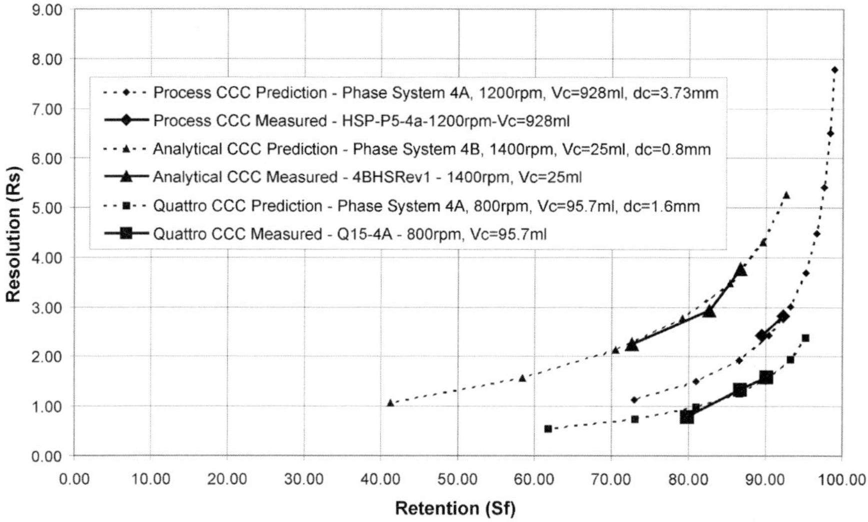

Fig. 6.8. Theoretical and measured resolutions verses retention for three different CCC instruments running at three different conditions.

Figure 6.9 shows the theoretical and measured resolutions versus mobile phase flow for the same three instruments running at the same conditions as for Fig. 6.8. The results show that the mixing efficiency remain substantially constant for the Brunel CCC (Quattro), decreases slightly with increased flow for the analytical CCC and increases slightly with increased flow for the Process CCC. These examples show the potential of using this CCD model for evaluating CCC variables and can be used also to calculate the mixing efficiency of a CCC run.

The number of CCD steps per minute for each instrument is given in Table 6.3 where it can be compared to the rotational speed in rpm, which is equivalent to the number of mixing and settling steps per minute. An operational efficiency of each instrument operating with a particular phase system can then be calculated as follows:

$$E = \frac{100\chi}{\omega} \tag{6.21}$$

where E is the percentage efficiency, χ the number of equivalent CCD steps per minute and ω the rotational speed in revs/min.

The results of Table 6.3 show that mixing efficiency per mixing and settling cycle can vary between 5% and 17% as the process is scaled

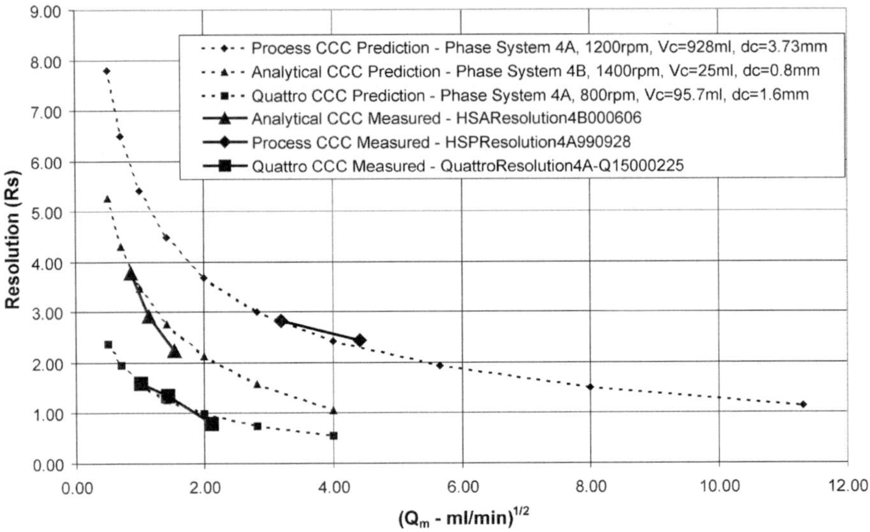

Fig. 6.9. Theoretical and measured resolutions versus the square root of mobile phase flow for three different CCC instruments running at three different conditions.

TABLE 6.3

Comparison of the efficiency of various CCC instruments running under different conditions

Instrument	V_i (%V_c)	χ (CCD steps/min)	N (rpm)	E (%)
Analytical CCC	0.2	240	1400	17.1
Quattro CCC	1.0	14.1	800	1.8
Process CCC	0.1	67.6	1200	5.6

down. It also shows how the effective efficiency increases with the speed of rotation and can be reduced as the sample volume (V_i) increases. The Brunel CCC (Quattro) would normally be running at an efficiency of 10% for sample volumes of equivalent size.

At the time of writing, the scientific understanding of the parameters affecting stationary phase retention is only just evolving. It appears [6] that the "J" type coil planet centrifuge behaves like a constant pressure pump with a pressure generated by the sum of the pumping pressure and the internally generated Archimedean screw action of the phases in the coils towards the head end. The stability of a given phase system in a coil planet centrifuge will be influenced by the density difference of the phase system and the "g" field, which have been found to directly affect this pumping pressure and the retention achieved [13]. It would be premature in this chapter to summarize how each operating parameter affects retention, but it is likely that in the next few years this will become well documented and understood.

6.7 CONCLUSIONS

This chapter has set out to show the importance of stationary phase retention in being able to predict the elution time of the solvent front and solutes of a given distribution ratio. It also shows how the equivalent number of CCD steps per minute can be calculated from a sample chromatogram and how this can then be used to make predictions for a whole range of separations at different flow, retention and distribution ratios.

Finally, a way of measuring the efficiency is given which can be used to compare different CCC Instruments if the same phase system and percentage sample volume is used.

REFERENCES

1 L.C. Craig and O. Post, Apparatus for countercurrent distribution. *Anal. Chem.*, 21 (1949) 500.

2 I.A. Sutherland, J. Muytjens, M. Prins and P. Wood, A new hypothesis on phase distribution in countercurrent chromatography. *J. Liquid Chromatogr. Relat. Technol.*, 23 (2000) 2259.

3 I.A. Sutherland, L. Brown, S. Forbes, D. Games, D. Hawes, K. Hostettmann, E.H. McKerrell, A. Marston, D. Wheatley and P. Wood, Countercurrent Chromatography (CCC) and its versatile application as an industrial purification and production process. *J. Liquid Chromatogr. Relat. Technol.*, 21 (1998) 279.

4 I.A. Sutherland, A. Booth, L. Brown, B. Kemp, H. Kidwell, D. Games, A.S. Graham, G.G. Guillon, D. Hawes, M. Hayes, L. Janaway, G. Lye, P. Massey, C. Preston, P. Shering, T. Shoulder, C. Strawson and P. Wood, Industrial scale-up of countercurrent chromatography, *J. Liq. Chromatogr. Relat. Technol.* Special Issue on CCC, 24 (11–12) (2001) 1533.

5 EPSRC IMI grant No. GR/RO3143/01 on Responsive Processing, "Realisation of Process Scale Countercurrent Chromatography", Dec 2000–Nov 2002.
6 P. Wood, Hydrodynamics of CCC, PhD Thesis, Brunel University (2001).
7 Q. Du, C. Wu, G. Qian, P. Wu and Y. Ito, Relationship between the flow-rate of the mobile phase and retention of the stationary phase in CCC. *J. Chromatogr. A*, 835 (1999) 231.
8 A.J.P. Martin and R.L.M. Synge, *Biochem. J.*, 35 (1941) 1358.
9 N.B. Mandava and J.M. Ruth, The origins of CCC. In: Y. Ito and N.B. Mandava (Eds.), *Countercurrent Chromatography: Theory and Practice*, Chromatographic Science Series, Vol. 44. Marcel Dekker, New York, 1988, pp. 27–77.
10 I.A. Sutherland, J. de Folter and P. Wood, Practical and theoretical models for CCC. *J. Liq. Chromatogr. Relat. Technol.* (in press).
11 J.C. Giddings, *Dynamics of Chromatography: Principles and Theory*, Chromatographic Science Series, Vol. 1. Marcel Dekker, New York, 1965.
12 Q. Du, C.Q. Ke and Y. Ito, Separation of epigallocatechin gallate and gallocatechin gallate using multiple instruments connected in series. *J. Liq. Chromatogr. Relat. Technol.*, 21 (1998) 203.
13 I.A. Sutherland, Q. Du and P.L. Woods, The relationship between retention, linear velocity, flow and density difference for CCC. *J. Liq. Chromatogr. Relat. Technol.* 24 (2001) 1669.

Chapter 7

Special uses of a support-free liquid stationary phase

Béatrice Billardello and Alain Berthod

7.1 INTRODUCTION

The unique feature of CCC is its support-free liquid stationary phase; most other chromatographic techniques use a solid as stationary phase. Solute interactions between a liquid phase and a solid phase occur at the surface of the liquid–solid interface. Solutes can access the volume of the two liquid phases in CCC. This observation explains the preparative character of the CCC technique [1], a point which is extensively addressed in several chapters of this book. A liquid stationary phase can be used in many other ways.

The unique capabilities of the support-free liquid stationary phase that will be developed in this chapter are:

- preparative displacement chromatography;
- a particular interest in ionizable compounds' displacement, a technique known as "pH-zone refining" by its inventor, Yoichiro Ito [2];
- liquid–liquid partition coefficient determination; and
- other special uses such as continuous liquid extraction, partition or distribution isotherm determination, and as a chemical reactor.

7.2 DISPLACEMENT CHROMATOGRAPHY

7.2.1 Principle

Displacement chromatography defines any chromatographic technique in which a molecule, called the displacer, is used to displace other

Comprehensive Analytical Chemistry XXXVIII
A. Berthod (Ed.)

molecules adsorbed into the stationary phase. Most CCC separations are made using liquid–liquid partitioning. In this mode, displacement rarely occurs. Two molecules can be distributed at the same time in the liquid stationary phase. However, the partition of a second molecule does change somewhat the distribution isotherm of the first partitioned molecules [3].

7.2.2 A practical example: separating cations

Principle

Lipophilic ligands, such as diethylhexylphosphoric acid (DEHPA), were added to a heptane liquid phase that was used as a stationary phase to separate metallic ions contained in the mobile aqueous phase [4] (see also Chapter 9). The M^{n+} ions in the aqueous phase are extracted into the heptane phase according to:

$$M^{n+}{}_{aq} + n\,HA_{org} \leftrightarrow M(A)_{n,\,org} + n\,H^{+}{}_{aq} \tag{7.1}$$

The ligand is denoted HA in its protonated form and A^- in its anionic complexing form. The subscripts aq and org refer to the aqueous and organic phase, respectively. The equilibrium constant is:

$$K_D = \frac{[MA_n]_{org}[H^+]^n_{aq}}{[M^{n+}]_{aq}[HA]^n_{org}} \tag{7.2}$$

This constant, K_D, is related to the apparent distribution coefficient of the metal cation, P_{ex}, that is obtained experimentally using the retention volume, V_R, of the cation:

$$P_{ex} = K\left(\frac{[HA]_{org}}{[H^+]_{aq}}\right)^n \tag{7.3}$$

Rare-earth cations were separated by CCC using DEHPA and a heptane/water biphasic liquid system [4,5]. The cation retention volumes depended critically on the mobile phase pH. Equation (7.3) can be rewritten as:

$$\log P_{ex} = \log K_D + n \log [HA]_{org} + n\,pH \tag{7.4}$$

clearly showing that P_{ex} is dramatically dependent on the mobile phase pH.

A practical example

The ligand cannot complex two cations at the same time, so the cation with the highest K_D constant will displace cations with lower K_D value. Displacement chromatography can be used to separate large amounts of cations. Figure 7.1 illustrates the principle of the method for transition metal cations. In Step 1, the CCC machine is filled by the heptane stationary phase containing 0.2 *M* of DEHPA ligand and equilibrated with water. In Step 2, displacement chromatography starts. A mixture of 0.01 *M* nickel and 0.01 *M* cobalt acetate is continuously fed into the CCC machine (no plug injection). The two cations are complexed by the ligand in the heptane stationary phase. The mobile phase is deionized, and since the cations are replaced by protons, its pH drops to ~3. Inside the CCC machine, bands of pure cations form. This is because the cobalt ions displace the nickel ions. The latter cations are pushed further into the stationary phase, forming a band of nickel complex in the stationary phase in front of the band of cobalt complex. This process occurs as long

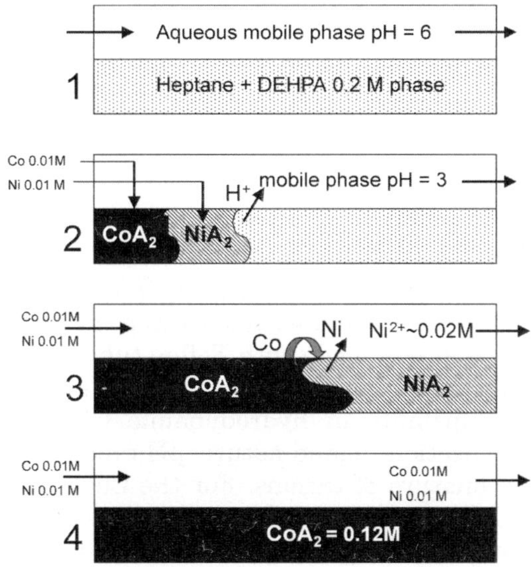

Fig. 7.1. Displacement CCC. Step 1: The CCC machine is equilibrated with a heptane/water liquid system. Step 2: An ionic solution is continuously fed into the machine. All ions are retained, an acidic solution exits the machine. Step 3: Bands of pure cations form. The cobalt ions displace the nickel ions that leave the machine with a higher concentration. Step 4: the stationary phase is saturated in cobalt ion.

as the cations can find free ligands in the stationary phase. In Step 3, the nickel ions cannot find free ligands, so they cannot form a complex and are eluted without cobalt ions, at a concentration of 0.02 *M*, twice as high as the entering concentration. The entering 0.01 *M* of nickel cations cannot find free ligands so they elute out of the column. Their accompanying 0.01 *M* cobalt cations displace complexed nickel in the stationary phase, adding 0.01 *M* of nickel cations in the mobile phase. In Step 4, this process lasts as long as the cobalt cations find some nickel to displace. When the stationary phase is saturated in cobalt cations, there is no more separation. The liquid nature of the stationary phase is used: the rotor is stopped, and the dark blue heptane phase is collected and exposed to a pH 1 aqueous phase to extract the cobalt cations and regenerate the ligand (Eq. 7.1). The process can start again.

In the experiment carried out with a 53 ml CCC hydrodynamic machine, 240 ml of the 0.01 *M* nickel and 0.01 *M* cobalt acetate were separated in 120 ml of 0.02 *M* nickel acetate solution and 20 ml of a 0.12 *M* cobalt–DEHPA heptane phase, a 12-fold concentration increase [5]. The respective purities of the nickel and cobalt salts were 97.4% for the aqueous phase and 99.3% in the heptane phase. It was demonstrated that the copper cations displaced the cobalt cations, forming three bands—the copper band pushing the cobalt band which itself pushed the nickel band [5].

Other applications

The method was tried with rare-earth elements, and it was found that the trivalent cations (La^{3+}, Ce^{3+} and Nd^{3+}) formed a gel upon complexation with the DEHPA ligand in the heptane phase. The gel formation increased the back-pressure and we had to stop the experiment to save the CCC machine, that was made with Teflon tubes. It was not possible to obtain the gel formation in shaken test tubes. This demonstrates the powerful mixing capability of hydrodynamic CCC machines. Upon experimental optimization (ligand nature, pH range), the method could be used with any mixture of cations. For the Cu^{2+}, Co^{2+}, Ni^{2+} cations with DEHPA as the ligand, the working pH range was $3 < pH < 6$ [5]. For rare-earth elements, the pH should be much lower.

Chiral selectors were used as ligand in the stationary phase [6]. If the binding constants between the enantiomers and the chiral ligand in the stationary phase are sufficiently different, displacement chromatography can be used to produce significant amount of pure enantiomers [7] (see also Chapter 11).

7.3 pH-ZONE REFINING

pH-Zone refining could be defined as displacement chromatography of ionizable compounds. The main difference is that, most often, the solutes to be separated are not fed continuously into the CCC machine. A plug of solution containing the compounds to be separated is injected in front of the displacer agent. The latter is then introduced continuously in the CCC machine. The injected compounds should be ionizable. Ionizable compounds are characterized by a sharp change in hydrophobicity occurring upon ionization of a functional group of their molecule [8]. For example, the octanol/water distribution coefficient of piretanide, a diuretic with a carboxylic acid group, is 160. It drops to 0.02 for the ionized piretanide carboxylate [8]. When a biphasic liquid system is used, such huge changes in hydrophobicity produce a change in the solute location. When the solute has its molecular form, it is located in the apolar organic phase, and its polar ionized form will be located in the aqueous phase. In CCC, one phase is mobile, the other is stationary. Upon pH change, the location—and consequently the motion—of the solutes will change. If the dissociation constants, K_A, of the solutes are different, they will ionize for different pH values, and bands of solutes will form.

7.3.1 Principle

Yoichiro Ito invented and named the pH-zone-refining technique. He described it in several excellent review articles [9,10]. Only the principle as illustrated by Fig 7.2 will be recalled. The stationary organic phase contains trifluoroacetic acid, and is loaded into the CCC machine. The mobile phase is an aqueous solution of ammonia. The machine is *not* equilibrated—the basic mobile phase would neutralize the acidic stationary phase and cancel any pH gradient. The solutes, e.g., an aqueous mixture of three acids, are injected into the stationary phase, and the basic mobile phase is used to push them in the CCC machine.

Inside the CCC machine

Figure 7.2 (top) shows that the three acids arrange themselves in zones. The basic mobile phase ionizes them. As ions, they are hydrophilic, and move with the mobile phase to be exposed to a strongly acidic stationary phase that protonates them. The molecular, protonated forms of the solutes are lipophilic. They solubilize in the stationary phase and stop.

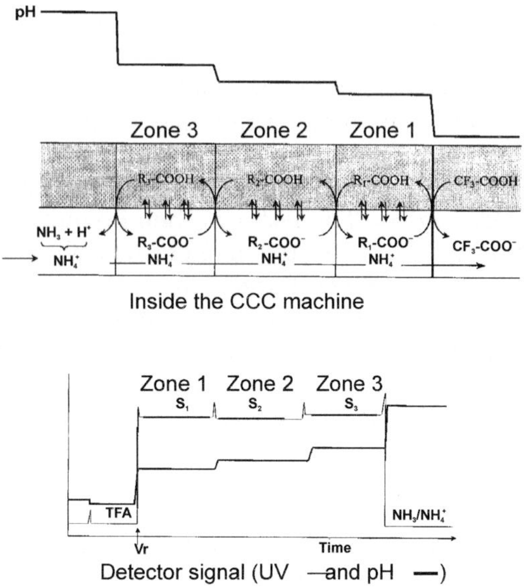

Fig. 7.2. pH-Zone-refining CCC. The aqueous mobile phase contains a basic displacer (NH_3). The organic liquid stationary phase contains an acid retainer (CF_3COOH). Three acids were injected; they form the zones, the stronger acid being first. The pH of a zone is constant.

More mobile phase enter the machine to again ionize the solutes, and so forth. Thus, the molecular forms of the acids are retained and the ionized, basic forms are eluted. Furthermore, the solute with the higher dissociation constant (the lower pK_A) can protonate the ionized form of a solute with a lower dissociation constant (a higher pK_A). This process forms the zones as more mobile phase re-ionizes the solutes. Zone 1 corresponds to the stronger injected acid, and can only be protonated by the stationary phase trifluoroacetic acid; it is the fastest-moving zone. Zone 3 contains the weakest acid, that is protonated in its retained molecular form by all other acids. It is the slowest-moving zone. The zones form a pH gradient inside the machine. The basic mobile phase has a high pH, then inside the machine the pH decreases in steps corresponding to the three zones, down to the low pH value of the acid stationary phase that is located at the column exit (bold line).

Recorded signal

Figure 7.2 (bottom) shows the signal obtained at the machine outlet. From the injection point to the beginning of the Zone 1 band, the acidic organic phase exits the CCC machine: the pH is low. At the V_M volume, the nature of the liquid phase changes; the aqueous phase containing the stronger acid of the sample mixture produces a sharp detector signal and the pH changes. Then the two other acid zones are eluted, showing step changes in the pH and detector signal. After elution of Zone 3, the organic stationary phase no longer contains trifluoroacetic acid; this has been completely stripped out, in the form of ammonium trifluoroacetate, by the basic mobile phase. The observed pH is that of the mobile phase.

pH-Zone refining was described with an acid stationary phase and a basic mobile phase. There are no objections to working in the opposite way if needed. In that case, the retaining agent in the stationary phase would be a base and the displacer in the mobile phase would be an acid. The zone order and pH changes would be reversed [2,9,10].

7.3.2 Simple modelling of the phenomenon

pH-Zone refining was mathematically well modelled by Ito, who assumed than ions do not go in the organic phase [2,9]. He showed that the retainer concentration in the stationary phase, C_R, the displacer concentration in the mobile phase, C_D, and the retention-phase ratio, $\Phi = V_S/V_T$, are the three essential parameters in pH-zone-refining separation.

$$C_D V_S = C_R (V_r - V_M) \tag{7.5}$$

with

$$V_r = V_M + K_{R,app} V_S \tag{7.6}$$

Here, V_r is the mobile-phase volume needed to see the sharp pH change (beginning of Zone 1, Fig. 7.2, bottom), and $K_{R,\ app}$ is the apparent distribution coefficient of the retainer between the two phases used. It depends on the true distribution coefficient, K_R, on the dissociation constant, K_A (with $pK_A = -\log K_A$), and on the displacer and retainer concentrations, C_D and C_R [9].

$$K_{R,app} = \frac{1}{\dfrac{C_D}{2C_{FR}} + \dfrac{1}{K_R} + \sqrt{\left(\dfrac{C_D}{2C_R}\right)^2 + \dfrac{K_A}{C_{R_R}}}} \tag{7.7}$$

Furthermore, it was demonstrated that the apparent distribution coefficients of all injected ionizable compounds, $K_{i,\,app}$, was equal to $K_{R,\,app}$ [9]. This means that the chosen experimental concentrations in the retainer and displacer, C_R and C_D, respectively, are the main factors affecting the solute zone formation.

The pH of the zone, pH_Z, is calculated [9] using:

$$pH_{Zi} = pK_{Ai} + \log\{(K_i/K_{i,\,app}) - 1\} \tag{7.8}$$

The pH-zone "selectivity" can be defined by the difference between the pH of two successive zones:

$$\Delta pH = pH_{Zi+1} - pH_{Zi} = (pK_{Ai+1} - pK_{Ai}) + \log\{(K_{i+1} - K_{R,\,app})/(K_i - K_{R,\,app})\} \tag{7.9}$$

Equation (7.8) shows that the pH of the zone is directly proportional to the dissociation constant of the contained compound. Equation (7.9) shows that this is true as long as the hydrophobicities of the compounds are not very different. In the latter case, the second terms of Eq. (7.8) and Eq. (7.9), that represent the hydrophobicity contribution, may compensate the first term (acidity term) contribution. It should be noted that this second term also depends on the experimental conditions, since $K_{i,\,app} = K_{R,\,app}$ (Eq. 7.7). We chose to study this problem, which is responsible for the zone overlaps observed experimentally [11].

7.3.3 Modelling zone formation

Computer program

A computer program was developed by Scher and Ito [12]; it assumes that the distribution ratios, K_R, and the dissociation constants, K_A, do not vary when the solute concentrations change. It also assumes that all ionized forms do not partition with the organic phase. Then, using the Craig model that divides the machine volume into N equal chambers in which equilibria occur, they propagated constant volumes of mobile phase from the machine inlet to the outlet, calculating the acid–base + partitioning equilibria in every chamber [12].

We chose to prepare a similar computer program that, for every chamber corresponding to a theoretical plate, calculates all species' concentrations using the conservation of mass- and electro-neutrality in the chamber. The program allows one to input an adjustable value for:

- the number of plates (Craig chambers) of the CCC machine;
- its volume;
- the *Sf* phase retention factor ($Sf = V_S/V_M$);
- the concentration and hydrophobicity of the retainer;
- the concentration and hydrophobicity of the displacer;
- the number of ionizable compounds injected;
- their respective concentration and hydrophobicity;
- the volume injected.

The program produces a set of points that can be treated by any modern spreadsheet software.

Validation of the computer program

Figure 7.3 shows the actual experimental profile obtained with injections of large amounts (930 mg) of a mixture of two hydroxybenzoic acid isomers. With these isomers, there is a large overlap of the two zones. The computer program was used to calculate the zone profile with the same conditions. Figure 7.3 shows that the calculated zone profiles correctly match the experiments, thus validating the program. All the experimental conditions will be described in the next Section. Having been validated, the computer program will allow us to study theoretically all possible cases. We will focus on the zone-overlap problem, but other cases can be studied as well, as has been done before [12].

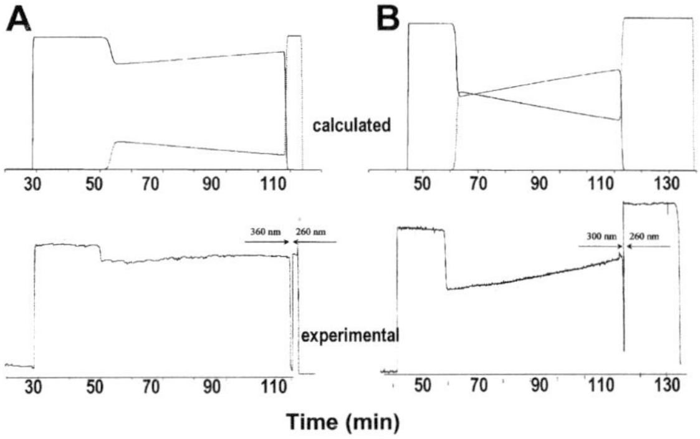

Fig. 7.3. Validation of the computer program. (A) Retainer concentration, 10 m*M*, displacer concentration (NH_3), 25 m*M*, injection of 8 ml of salycilic acid (0.72 *M*) and *p*-hydroxybenzoic acid (0.15 *M*). (B) Retainer concentration, 20 m*M*, displacer concentration (NH_3), 20 m*M*, injection of 8 ml of salicylic acid (0.43 *M*) and *p*-hydroxybenzoic acid (0.43 *M*). Efficiency: 90 plates; *Sf*: 0.65. Machine volume: 152 ml; rotor rotation: 700 rpm; flow rate: 3 ml/min, aqueous mobile phase in the head-to-tail direction. UV detection at 360 nm (only salicylic acid is detected) and switched to 260 nm after its elution.

7.3.4 Hydrophobicity *versus* acid strength: case study of zone overlap

Experimental section

Chemicals: The selected ionizable solutes were the three hydroxybenzoic acid isomers (Merck Prolabo, Paris, France). Table 7.1 lists their relevant physico-chemical parameters [13]. The solvents, acetonitrile (ACN) and methyl-*tert*-butyl ether (MTBE) were obtained from Sigma-Aldrich (L'Isles d'Abeau Chesnes, France). Trifluoroacetic acid and ammonia were from Fluka (Milwaukee, Wisconsin, USA). Water was deionized, distilled and filtered on a Barnstead-Thermolyne cartridge (Dubuque, Iowa, USA).

The pH-zone refining system: A classical biphasic liquid system was used: MTBE:ACN:water, 4:1:5 v/v/v. The organic upper phase was the stationary phase containing the trifluoroacetic acid retainer. The denser aqueous phase with ammonia as the displacer base was the mobile phase entering the CCC machine in the head-to-tail direction. The CCC machine was a single coil 52 ml hydrodynamic machine, Ito type IV. The spool contained 133 turns of PTFE tubing of 1.6 mm i.d. (26 metres in 7 layers of 19 turns). The spool radii ranged from 2.2 cm for the first inner tube layer to 4.5 cm for the outermost layer. The rotor radius was 6 cm, giving an Ito β value ranging from 0.37 to 0.75. The temperature was air-regulated at 22±2°C. The CCC chromatograph was connected to a Shimadzu LC6-A HPLC pump, a Shimadzu SPD-6A UV detector and a

TABLE 7.1

Physicochemical properties of the hydroxybenzoic acid isomers (f.w. 138.12)

Compound	pK_{A1}	pK_{A2}	$K_{oct/w}$	$K_{MTBE/w}$	CAS
Ortho-Hydroxybenzoic (salicylic) acid	2.97	13.4	182	75	69-72-7
Meta-Hydroxybenzoic acid	4.06	9.92	32	18	99-06-9
Para-Hydroxybenzoic acid	4.48	9.32	38	16	99-96-7
Trifluoroacetic acid	0.30	–	3*	90*	76-05-1
Ammonia	9.25	–	0.04	<0.01	7664-41-7

$K_{MTBE/w}$ are the distribution coefficients of the compounds at pH 1.4 in the biphasic liquid system obtained by mixing MTBE: ACN: water, 4:1:5, v/v/v (trifluoroacetic acid added, 0.1 *M* or 7.5 ml/l).

*Value for the molecular form CF_3COOH (TFA) see Ref. [8]. The apparent value at pH 1 is $K_{MTBE/w}$ = 1.8 owing to the CF_3COO^- anion. The listed $K_{oct/w}$ is the uncorrected value calculated using the LogKow (KowWin) software [14].

Shimadzu CR501 Chromatopac integrator (Touzard & Matignon, Les Ulis, France).

Protocol: Different volumes of a 0.43 *M* solution (60 g/l) of each solute were injected. The 8 ml injection loop allowed the injection of a maximum of 480 mg of each solute. The fact that the salicylic acid has a significant UV absorbance at 300 nm, and *p*-hydroxybenzoic acid shows low absorbance at this wavelength, allowed the use of on-line UV detection to identify the solute zones. Both solutes absorb UV light at 260 nm. The mobile-phase pH was also used to delineate the zones.

The organic upper phase is selected as the stationary phase containing the retainer acid (TFA). The machine is completely filled with the acid stationary phase. The denser aqueous phase received the displacer base (NH_3). The desired volume of sample is prepared in the mobile phase and loaded in the injection loop. The injection valve is then turned to the "inject" position and the aqueous mobile phase is pushed into the CCC machine in the head-to-tail direction at 2 ml/min with the rotor set at 700 rpm. The chromatogram is developed, monitoring continuously the 300 nm UV signal and the pH. At the end of the salicylic acid zone, the UV detector was switched to 260 nm, to monitor the *p*-hydroxybenzoic acid zone. Figure 7.3 shows two examples of experiments done.

pH Difference between adjacent zones

Equation 7.9 shows that the pH difference between two adjacent zones depends on the acidity and hydrophobicity difference between the compounds forming the zones. Three cases can be distinguished:

1. The stronger acid has the lower hydrophobicity. It is always eluted first. This case is modelled as $pK_{A1} < pK_{A2}$ and $K_1 < K_2$, where the pK_A values are $-\log K_A$ the co-logarithm of the dissociation constant, and K_i is the solute distribution coefficient.
2. The stronger acid also has the higher hydrophobicity, and it is still eluted first. This case is modelled as $pK_{A1} < pK_{A2}$ and $K_1 > K_2$.
3. As for Case 2, the last case is when the stronger acid also has the higher hydrophobicity, but it is eluted last, so it is referred to by $i = 2$. This case is modelled as $pK_{A1} > pK_{A2}$ and $K_1 < K_2$.

The first case is the most common since strong acids are often polar molecules. However, it is possible to find compounds falling in Case 2 or 3. In the case of hydroxybenzoic acids, the *meta-* and *para* isomers are

Case 1 examples. The *ortho* isomer and either of the other two isomers correspond to Case 2: due to chelation between the carboxylic acid and the adjacent phenol group, salicylic acid—the *ortho* isomer—is the strongest acid, it also has the highest $K_{oct/w}$ value (Table 7.1).

From a practical point of view, small changes in the liquid system composition may greatly affect the distribution constants. However, the acid dissociation constants of the solutes are much less affected by these changes. The effect of hydrophobicity and strength of acidity will be studied successively.

Effect of solute hydrophobicity on zone profiles

Case 1: Since both the acid strength and hydrophobicity act in the same way, the stronger and more polar compound is always eluted first. There is no zone overlap. The calculations show that, when the distribution coefficients become very low (close to or below unity), the zones lose their rectangular shapes and start to form peaks that may overlap when the K_i values are low and close (polar solutes). These results make sense: the zones form because the molecular forms of the ionizable compounds interact with the stationary phase, and the ionic form does not (Fig. 7.2). If the molecular forms interact weakly (low K_i values), or not at all with the stationary phase, the zones cannot form.

Case 2: In this case the acid strength and hydrophobicity act in opposite ways. Solute 1 is a stronger acid than Solute 2, so it should elute first. At the same time, it has a higher affinity for the organic phase than Solute 2, so without a pH gradient it would be more retained than Solute 2. However, the pK_A difference between the two solutes is high enough so that the acidity term prevails and Solute 1 elutes first.

Figure 7.4 shows with a bold black line the experimental pH-zone-refining profile obtained with salicylic acid (Solute 1) and *p*-hydroxybenzoic acid (Solute 2). With the experimental conditions selected (Fig. 7.4 caption), the two zones overlap, forming a third zone containing about 2/3 of Solute 1 and 1/3 of Solute 2. Only 65% of Solute 1 and 80% of Solute 2 can be recovered with a high degree of purity. The computer program was used to draw the other zone profiles shown by Fig. 7.4. For the calculations, the pK_A values were not changed. The hydrophobicity was changed in such a way that the "selectivity" factor, ΔpH (Eq. 7.9), remains constant. The K_i distribution coefficients are indicated on the left of Fig. 7.4 for Solute 1 and on the right for Solute 2.

It is clear that when the hydrophobicity of the solutes increases, the zone overlap decreases, all other parameters being constant. For K_1 and

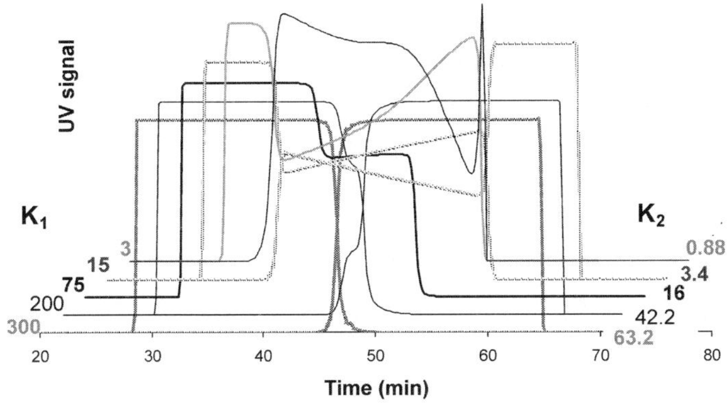

Fig. 7.4. Hydrophobic effect on zone shape with Case 2 solutes. The distribution coefficients K_1 and K_2 were calculated so that the ΔpH value is constant (Eq. 7.9, ΔpH = 0.832, the experimental value). The pK_A values are 2.97 and 4.48. [NH_3] = 60 mM, [TFA] = 20 mM. All other experimental values are listed in the caption of Fig. 7.3B.

K_2 values higher than 300 and 63.2, respectively, there is no overlap, and two clean rectangular zones form. This means that the molecular forms of the solutes interact well with the stationary phase for both compounds. The difference in acid strength makes the separation occur in the aqueous mobile phase from which the weaker acid is expelled. As the distribution coefficients decrease, the solutes have less affinity for the stationary phase. The zone corresponding to Solute 2 tends to elute faster and seems to crawl underneath the Solute 1 zone, producing overlapping. With K_1 = 3 and K_2 = 0.88, it is possible to recover only ~20% of pure Solute 1 and no pure Solute 2 at all. The overlap is so severe than the Solute 2 pure zone is reduced to a sharp trailing peak (Fig. 7.4). With lower distribution coefficients (e.g., K_1 = 0.5 and K_2 = 0.36), the rectangular shape of the zones is completely lost and the solutes elute in the form of two mixed non-Gaussian peaks

Case 3: With pK_{A1} > pK_{A2} and $K_1 < K_2$ Case 3 is similar to Case 2, but the stronger acid elutes last. Hydrophobicity is the dominant factor in the retention process. Several zone profiles were calculated with the same ΔpH value that was obtained with the hydroxybenzoic acids and with pK_{A1} = 3.2 and pK_{A2} = 3.0. For any K_i couple giving a ΔpH value of 0.832, two adjacent rectangular zones were obtained with a minimum overlap. With K_1 = 0.5 and K_2 = 2.23, the Solute 1 zone was distorted, forming a broad peak. This is again due to the poor interaction of Solute 1 with the stationary phase. The Solute 2 zone was rectangular and did

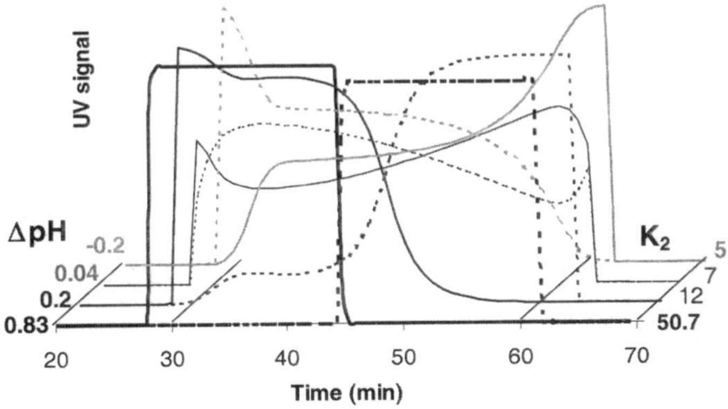

Fig. 7.5. Hydrophobic effect on zone shape with Case-3 solutes. The distribution coefficient K_1 is constant (= 5) and K_2 varies from 5 to 50.7. The corresponding ΔpH values are indicated (Eq. 7.9). The pK_A values are 3.2 and 3.0. [NH_3] = 60 m*M*, [TFA] = 20 m*M*. All other experimental values are listed in the caption of Fig. 7.3B.

not overlap with Solute 1. Clearly, in Case 3, a ΔpH difference of 0.832 units is enough to obtain a good zone separation with the 0.2-pK_A weaker acid eluting first.

Maintaining the ΔpK_A of 0.2 units, the ΔpH "selectivity" was decreased by reducing the hydrophobicity difference between Solute 1 and Solute 2. The K_1 coefficient was 5 and the K_2 coefficient was reduced from 50.7, giving a ΔpH of 0.832, to 5 (identical hydrophobicity) giving a theoretical ΔpH of –0.2 that suggests that Solute 2 should actually be Solute 1 (Case 1). Figure 7.5 shows the zone profiles obtained. The Solute 1 zones are drawn in solid lines and the Solute 2 zones have dotted lines. For a ΔpH value of 0.2, the two zones overlap. Only 10% and 40% of Solutes 1 and 2, respectively, could be recovered with 99% purity. Logically, when there is no pH difference between the zones, they overlap completely. This is shown in Fig. 7.5 by the zone profiles obtained with K_1 = 5 and K_2 = 7, producing a ΔpH = 0.04. The last zone profiles were calculated with K_1 = K_2 = 5 to confirm that this configuration belongs to Case 1. Solute 2 elutes first with a severe overlap with Solute 2.

Effect of solute acid strength on zone profiles

The effect of the acid strength on zone profiles is much easier to explain than the effect of hydrophobicity. One sentence may suffice: "pH-zone refining will not work well with a very weak acid". The pH-zone refining is based on acid–base reactions in the aqueous phase and hydrophobic partitioning with the organic phase. If any one of these two interactions cannot take place, the zones cannot form properly. For the three cases defined, any zone overlap decreased and disappeared when the solute pK_A difference increased. The ΔpH difference between zones increased proportionally (Eq. 7.5). Keeping the pK_A difference constant, all zone overlaps worsened when the pK_i values were higher than 6. Very weak acids ionize poorly in the aqueous phase, making the zone formation more difficult.

Experimental conditions and zone overlap

Nature of the retainer and displacer: Figure 7.2 suggests that the best retainer should be the strongest acid possible with a high affinity for the organic phase in its molecular form. The best displacer should be the strongest base with a low affinity for the organic phase. Trifluoroacetic acid is rarely challenged as an excellent retainer acid [2,7,11,12]. Ammonia is most often used, but sodium hydroxide was also used successfully. It should be noted that pH-zone refining could be also performed in the opposite way, with a retainer base such as triethylamine in the organic stationary phase and a displacer acid such as hydrochloric acid in the aqueous mobile phase [2].

Concentrations of the displacer: The pH-zone theory shows that, when zone formation occurs correctly, the solute concentrations in the zones are equal to the displacer concentration [2,11]. If the displacer concentration increases, with the retainer concentration remaining constant, the zone height will increase and the zone width will narrow. The duration of the experiment will shorten. The extra benefit is that the severe peak overlap observed for Case-2 solutes (Fig. 7.4), decreases when the displacer concentration, $[NH_3]$, increases in the aqueous mobile phase.

Concentration of the retainer: The zone heights do not change when the retainer concentration changes in the stationary phase. The zone positions do change, however. At constant displacer concentrations, an increase in retainer concentration logically increases the zone retention. The good point is that the zone overlap observed for hydroxybenzoic acid isomers (Case 2) decreased significantly when the TFA concentration in the organic phase increased.

Amount of material to separate: The concentration and injected volume of the sample are critical parameters. The goal is always to separate a maximum amount of material, in the minimum time, with the highest possible purity. Increasing the amount of injected material significantly increases the chances of zone overlaps. So higher concentrations of retainer and displacer should be used. Solubility problems will be the limit.

Conclusion

Our goal was to point out that zone overlap may occur in pH-zone refining. This way of using CCC machines was found to be the most productive in preparative separation [1]. The chances of obtaining zone overlaps increase when the amount of material injected in the CCC machine increases. It increases when the difference between the compounds to be separated is low (similar pK_As and similar K_is distribution coefficients). The rule-of-thumb for solving the problem is to increase both the retainer and displacer concentrations. Then, if needed, dilution of the injected sample should allow one to obtain clean rectangular zone profiles.

7.4 LIQUID–LIQUID PARTITION COEFFICIENT DETERMINATION

The fundamental solute retention equation in CCC is:

$$V_R = V_M + K_D V_S \tag{7.10}$$

The solute retention volume, V_R, depends on the liquid phase ratio in the CCC machine of volume V_T, i.e., the mobile phase volume, V_M, and the stationary phase volume, V_S, with $V_M + V_S = V_T$. The only solute-dependent parameter is K_D, (= $[\text{solute}]_{\text{stationary phase}}/[\text{solute}]_{\text{mobile phase}}$), the distribution constant. This constant is an intrinsic physico-chemical parameter of the solute; K_D does not depend on the CCC machine used or on the technique itself. However, the CCC technique can be used to measure K_D. This is another unique advantage of the liquid nature of the stationary phase in CCC.

7.4.1 Direct measurement of liquid–liquid distribution constants

From Eq. (7.10), the distribution constant is easily derived:

$$K_D = (V_R - V_M)/V_S \tag{7.11}$$

Tsai pointed out that the K_D determination by CCC depends on the mobile phase flow rate, F, [15]:

$$K_D = (t_R - t_o)\,F/(V_T - Ft_o) \quad (7.12)$$

where t_R and t_o are respectively the solute retention time and dead time, depending on the liquid phase volume in the CCC machine ($t_o = V_M/F$): V_T is the machine volume. Since it is crucial to know exactly the volume V_M or the time t_o, it is recommended that a compound with known distribution coefficient be used as a mobile phase (or dead-time) marker.

The range of K_D values that can be measured directly is relatively narrow. High retention time (or volume) and detection problems limit the range on the high K_D value side. It is difficult to obtain directly K_D values higher than 200 or lower than 0.05. This corresponds to a log K_D range of almost 4 units, from –1.3 to +2.3 [16]. Very low flow rates were recommended to measure low K_D values [15]. For $K_{oct/w}$ distribution coefficient measurement, the aqueous phase is the mobile phase and octanol is the stationary phase. The phase roles can be reversed. With octanol as the mobile phase, the $K_{w/oct}$ coefficients obtained are the inverse of the usual $K_{oct/w}$.

7.4.2 Dual-mode CCC

It is important to measure $K_{oct/w}$ values for very hydrophobic solutes. The upper limit of 200 in the directly measured K_D value is not high enough. Dual-mode CCC was developed to increase the range of K_D values measurable by CCC. The dual-mode method was explained in Chapter 2. It was demonstrated that the solute partition coefficient was simply [17,18]

$$K_{oct/w} = V_w/V_{oct} \quad (7.13)$$

in which V_w is the volume of aqueous phase pumped after the solute injection in the head-to-tail (descending) mode up to the phase inversion point, and V_{oct} is the volume of octanol pumped in the tail-to-head (ascending) mode up to the maximum of the peak corresponding to the injected solute [15,17,18]. The dual-mode method allowed the measurement of K_D values as high as 10,000 (log K_D = 4) [18].

7.4.3 Other liquid–liquid distribution coefficient measurements

Cocurrent CCC: the moving stationary phase

A unique method for measuring very high K_D values used the liquid character of the stationary phase. The liquid stationary phase is slowly pushed in the same direction as the mobile phase. In the case of octanol/water distribution coefficients, the octanol phase is pumped at a flow rate, F_{oct}, in the same direction as the aqueous phase flow rate, F_{aq}. The $K_{oct/w}$ value is computed using [19]:

$$K_{oct/w} = (t_R F_{aq} - V_{M,aq})/(V_{S,oct} - t_R F_{oct}) \qquad (7.14)$$

In Eq. 7.14, $V_{M,aq}$ is the volume of aqueous phase contained in the CCC machine and $V_{S,oct}$ is the volume of "stationary" phase, although octanol is not strictly stationary since it is pumped at a slow flow rate. In all practical cases, F_{oct} was equal to, or lower than, $F_{aq}/100$. The method is not commonly used to measure $K_{oct/w}$ coefficients because it cannot extend the measurable range significantly compared to the dual-mode method. Also, since a mixture of the two phases exits the CCC machine, a clarifying agent, e.g., 2-propanol, should be added to allow continuous detection. Such an addition precludes solvent recycling.

Ionizable compounds

Ionizable compounds evolve from a molecular form to a charged form as the pH changes. Since the affinities for the aqueous phase of the molecular and ionic forms of the compound are very different, the amount of compound distributed in the aqueous and organic phases is extremely dependent on the pH. Only an apparent distribution, K_{app}, can be measured:

$$K_{app} = ([AH]_{org} + [A^-]_{org})/([AH]_{aq} + [A^-]_{aq}) \qquad (7.15)$$

assuming the ionizable compound is an acid, AH, that can give the ionized form, A^-. The K_{app} distribution coefficient depends on the aqueous pH according to [8,20]:

$$K_{app} = \frac{K^0 + K^-(K_A / H^+)}{1 + (K_A / H^+)} \qquad (7.16)$$

in which K^0 and K^- are the true distribution coefficients of the AH molecular form and A^- ionic form, respectively. In the case of a cationic acid/molecular base couple (e.g., amines), biprotic systems with two acidic or basic groups (e.g., phthalic acid) and biprotic systems with ionizable groups showing different charges (e.g., amino acids), the corresponding equations were also derived [8]. We demonstrated that the measurement of three K_{app} coefficients for the same compound at three different pH values permitted the determination of both the molecular K^0 and the ionic K^- values of the compound.

The results found for a large set of ionizable compounds and the octanol–water biphasic system showed that, in most cases, the assumptions made to model the pH-zone refining CCC separation method were correct. The $K^-_{oct/w}$ of the ionized forms of the compounds were found to be zero, or several orders of magnitude lower than the corresponding $K_{oct/w}$ for the molecule for many compounds [8,20]. However, this is not always true: the $K_{oct/w}$ coefficients of several ions were not very small. The ionized form of trichloromethiazide, a diuretic, has a $K^0_{oct/w}$ value of 10 for the molecule and 0.7 for the ionized form [8].

7.5 OTHER USES OF THE LIQUID NATURE OF THE STATIONARY PHASE

7.5.1 Continuous extraction

A CCC machine can be used to concentrate solutes in the liquid stationary phase. This was already shown in the displacement chromatography description (Section 7.2.2). The example presented demonstrated that the cobalt ions were concentrated by twelve times in the organic stationary phase being extracted from the aqueous phase. It was shown that a CCC machine could be used as a continuous extractor to remove trace amounts of a surfactant molecule from waste water [21]. With a hydrostatic Sanki LLN machine, it was possible to concentrate in 24 ml of ethyl acetate stationary phase 99.6% of the nonylphenol ethoxylate (NPEO) contained in 10 litres of aqueous phase. The concentration factor was 420. The NPEO concentration in the wastewater was 15 mg/l. The CCC machine concentrated it up to 6.25 g/l in the ethyl acetate liquid stationary phase [21]. There are no theoretical objections to using a CCC machine as a continuous extractor for different liquid systems and different solutes.

7.5.2 Rapid determination of liquid–liquid isotherms

Most often, the solute distribution coefficient, K_D, is said to be a constant. This is only true for diluted solutions. For concentrated solutions, Henry's Law is no longer followed. Langmuir-type equations can represent the shape of the liquid–liquid distribution isotherm. Using $[A]_{aq}$ and $[A]_{org}$ to define the concentration of a solute A in the aqueous and organic phases, respectively, the distribution coefficient is expressed by:

$$K_D = [A]_{aq}/[A]_{org} = K_D{}^0/(1 + p\,[A]_{org}) \qquad (7.17)$$

In Eq. (7.17), $K_D{}^0$ is the distribution coefficient of Solute A at infinite dilution and p is a constant. Figure 7.6A shows partition isotherms with the same $K_D{}^0$ coefficient and different p value. The arrows show the concentrations at which the actual distribution coefficients, K_D, differ from $K_D{}^0$ by more than 2%.

Since in CCC the K_D coefficient is always the ratio of the solute concentration in the stationary phase to the concentration in the mobile phase, K_D in the tail-to-head direction, with the light organic phase being mobile, becomes $K'_D = 1/K_D$ in the head-to-tail mode when the mobile phase is the denser aqueous phase. Trivially, K'_D is expressed by:

$$K'_D\ [A]_{org}/[A]_{aq} = K'_D{}^0\,(1 + p\,[A]_{org}) \qquad (7.18)$$

It is important to point out that the convex isotherm represented by Eq. (7.17) and Fig. 7.6A becomes a concave isotherm (Eq. 7.18 and Fig. 7.6B) for the same solute and the same liquid system when the phase role is switched. This situation can be obtained because of the liquid nature of the stationary phase. So CCC can be used to determine concave as well as convex liquid–liquid partition isotherms [22]. Two steps are performed after the CCC machine is equilibrated with the biphasic liquid system selected. In step 1, the pure mobile phase is switched to the mobile phase containing a concentrated amount of the solute. This phase is fed long enough so that the liquid stationary phase is equilibrated with the solute. In step 2, the CCC machine inlet is switched again to pure mobile phase. Step 1 produces a signal rising from the baseline to a constant value. Step 2 produces a decreasing signal, back to the baseline. The analysis of these two transient signals allows one to obtain the solute isotherm in the biphasic liquid system used [22].

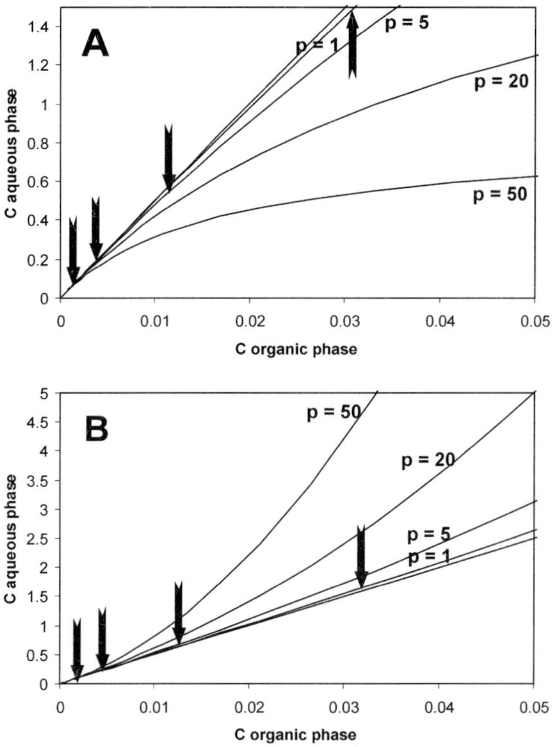

Fig. 7.6. Adsorption isotherms. (A) Concave isotherms with $K_D{}^0 = 50$ and different p values (Eq. 7.17). (B) Convex isotherms also with $K_D{}^0 = 50$ and different p values (Eq. 7.18). The arrows point to the concentrations above which K_D differs from $K_D{}^0$ by more than 2%.

7.5.3 A liquid–liquid reactor for catalytic reactions

In recent work, we demonstrated that chemical reactions could be performed continuously in a CCC machine [23]. The reduction of benzaldehyde by sodium formate in benzyl alcohol can be performed at room temperature in the biphasic liquid system cyclohexane/water when a ruthenium complex is used:

$$C_6H_5\text{–}CHO + HCOO^- + H_2O \rightarrow C_6H_5\text{–}CH_2OH + HCO_3^- \quad (7.19)$$

A hydrodynamic CCC machine was loaded with an aqueous phase containing 1 M sodium formate and 3.4×10^{-3} M of the ruthenium complex

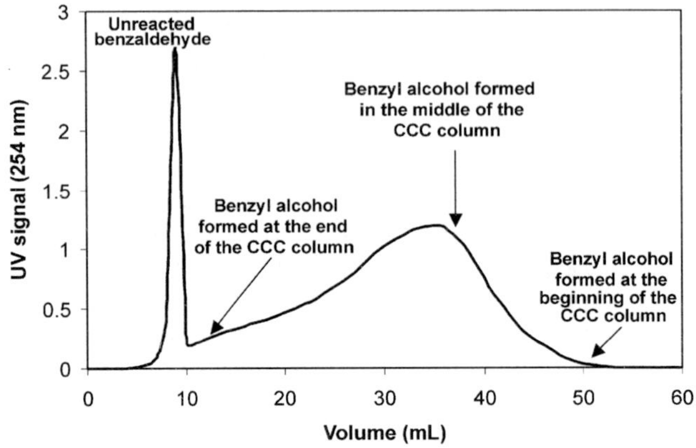

Fig. 7.7. Benzaldehyde and benzyl alcohol peak profiles. Injection volume = 200 μl of 10 mM solution. CCC machine volume = 59 ml, 5 M sodium formate aqueous stationary phase, Sf = 0.88, mobile phase cyclohexane 1.5 ml/min in the tail-to-head direction. Adapted from Ref. [23].

catalyst. The mobile phase was cyclohexane containing 0.117 M benzaldehyde pumped in the tail-to-head direction at 3 ml/min. A conversion factor of 30% was obtained. This factor was higher than the one obtained in a batch reactor because of the benzyl alcohol retention in the aqueous phase that displaced the reaction toward benzaldehyde reduction [23].

The CCC machine could also be used to study the properties of a new catalyst for biphasic liquid reactions. The catalyst is loaded in the liquid stationary phase. The CCC equilibrium is achieved with the mobile phase but without reagent. Then, the reagent is injected as a short plug. When it travels through the stationary phase, the reaction (Eq. 7.19) occurs and some product is created. Since the distribution coefficients of reagents and products are different, a profile as shown by Fig. 7.7 can be obtained. This profile is related to the properties of the liquid catalyst studied [22,23].

7.6 CONCLUSION

In conclusion, the main feature of CCC is its support-free liquid stationary phase. The large amount of solutes that can be dissolved in a liquid stationary phase explains the preparative capabilities of the technique. Because chemical reactions can take place in a liquid, ion-exchange displacement chromatography and pH-zone refining separations could be performed with a CCC machine. The measurement of liquid–liquid distribution coefficients and partition isotherms are other possibilities of the technique made possible by the liquid nature of its stationary phase.

A new technique which is showing great potential in preparative HPLC is called simulated moving bed chromatography (SMB) [24]. By using several HPLC columns and changing the injection and collection ports, it is possible to simulate motion of the stationary phase and continuously produce an extract and a refinate from a continuously fed mixture. Since CCC uses a liquid stationary phase, there would be no need to "simulate" the stationary phase motion—it is actually possible to make it move. The CCC machine should be adapted so that it has two exit ports, for the extract and the refinate, and also a stationary-phase inlet port, but there are no theoretical objections to "moving bed" CCC, the "bed" being a liquid.

REFERENCES

1 A. Berthod and B. Billardello, In: P. Brown and E. Grushka (Eds.), *Advances in Chromatography*, Vol. 40. Marcel Dekker, New York, 2000, Ch. 10, pp. 503–538.
2 Y. Ito, In: Y. Ito and W.D. Conway (Eds.), *High Speed Countercurrent Chromatography*, *Chemical Analysis*, Vol. 132. Wiley, New York, 1996, Ch. 6, pp. 121–175.
3 G. Guiochon, S. Golshan Shirazi and A.M. Katti, *Fundamentals of Preparative and Non-linear Chromatography*. Academic Press, Boston, MA, 1994.
4 T. Araki, In: A.P. Foucault (Ed.), *Centrifugal Partition Chromatography*, Chromatographic Science Series, Vol. 68. Marcel Dekker, New York, 1995, Ch. 10, pp. 241–316.
5 A. Berthod, J. Xiang, S. Alex and C. Gonnet-Collet, *Can. J. Chem.*, 74 (1996) 277.
6 Y. Ma, Y. Ito and A. Foucault, *J. Chromatogr. A*, 704 (1995) 75.
7 Y. Ma, Y. Ito and A. Berthod, In: J.M. Menet and D. Thiebaut (Eds.), *Countercurrent Chromatography*, Chromatographic Science Series, Vol. 82. Marcel Dekker, New York, 2000, Ch. 7, pp. 223–248.
8 A. Berthod, S. Carda-Broch and M.C. Garcia-Alvarez-Coque, *Anal. Chem.*, 71 (1999) 879.
9 Y. Ito, K. Shinomiya, H.M. Fales, A. Weisz and A.L. Scher, In: W.D. Conway and R.J. Petrowski (Eds.), *Modern Countercurrent Chromatography*, ACS Symposium Series, Vol. 593. ACS, Washington, DC, 1995, pp. 156–183.

10 Y. Ito and Y. Ma, *J. Chromatogr. A*, 753 (1996) 1.
11 W.D. Conway and R.J. Petrowski (Eds.), *Modern Countercurrent Chromatography*, ACS Symposium Series, Vol. 593. ACS, Washington, DC, Part III, 1995, pp. 153–230.
12 A.L. Scher and Y. Ito, In: W.D. Conway and R.J. Petrowski (Eds.), *Modern Countercurrent Chromatography*, ACS Symposium Series, Vol. 593. ACS, Washington, DC, 1995, pp. 184–202.
13 M.J. O'Neil, *The Merck Index*. 13th Edn. Merck & Co., Whitehouse Station, N.J., 2001.
14 http://esc-plaza.syrres.com/interkow/kowdemo.htm
15 R.S. Tsai, P.A. Carrupt and B. Testa, In: W.D. Conway and R.J. Petrowski (Eds.), *Modern Countercurrent Chromatography*, ACS Symposium Series, Vol. 593. ACS, Washington, DC, 1995, pp. 143–154.
16 A. Berthod, In: A.P. Foucault (Ed.), *Centrifugal Partition Chromatography*, Chromatographic Science Series, Vol. 68. Marcel Dekker, NewYork, 1995, Ch. 7, pp. 167–197.
17 S.J. Gluck and E.J. Martin, *J. Liq. Chromatogr.*, 13 (1990) 3559.
18 R.A. Menges, G.L. Bertrand and D.W. Armstrong, *J. Liq. Chromatogr.*, 13 (1990) 3061.
19 A. Berthod, *Analusis*, 18 (1990) 352.
20 A. Berthod and M. De Carvalho, *Anal. Chem.*, 64 (1992) 2267.
21 A. Berthod, In: R.A. Meyers (Ed.), *Encyclopedia of Environmental Analysis and Remediation*. Wiley, New York, 1998, pp. 1312–1327.
22 S. Caravieilhes, C. DeBellefon and N. Tanchoux, *Catalysis Today*, 66 (2001) 145–155.
23 A. Berthod, K. Talabardon, S. Caravieilhes and C. DeBellefon, *J. Chromatogr. A*, 828 (1998) 523.
24 R.M. Nicoud and R.E. Majors, *LC-GC (Europe)*, 13 (2000) 887.

Chapter 8

Separation and purification of natural products (medicinal herbs) by high speed countercurrent chromatography

Tianyou Zhang

8.1 INTRODUCTION

Medicinal herbs are an important source of natural products for medicine. These include various chemical components ranging from fat-soluble to water-soluble compounds. The isolation of the biologically active components is the starting point for further research in chemistry and pharmacology as well as for the utilization of these compounds.

Traditional Chinese medicine is extremely rich in experience acquired over a long period of time. In order to make greater use of traditional Chinese medicine, modern scientific methods are being used to find the bioactive compounds in traditional drugs and to use them as lead-compounds for new drug design. New drugs developed in this way include anisodamine and the antimalarial agent Qinghaosu (artemisin).

For separating and purifying bioactive compounds from medicinal herbs, modern chromatographic techniques—such as gas chromatography, high performance liquid chromatography, thin-layer chromatography and electrophoresis—have significantly raised the technical level and shortened the time required for research projects.

High-speed countercurrent chromatography (HSCCC) has been recognized as an effective means for the separation and purification of a wide variety of bioactive components [1–5]. It is a liquid–liquid partition chromatography system based on a coil planet centrifuge system without the use of any solid support. This technique has developed rapidly during the past decade. It has been shown to have preparative capabilities and unique properties for fractionating a variety of natural products and medicinal herbs.

In the Beijing Research Center for Separation & Purification Technologies of Natural Products, a hi-tech R & D organization led by the Beijing Municipal Science Commission, we have developed a series of HSCCC instruments and separation methods such as semi-preparative HSCCC, analytical HSCCC, multidimensional HSCCC, pH-zone refining CCC, and cross-axis CCC. In the field of natural products and Chinese herbs, we have worked on the application of separations of the bioactive components of *Ginkgo biloba* L., *Taxus chinensis*, *Polygonum multiflorum*, *Coptis chinensis* Franch, *Pueraria lobata*, *Sophora flavescens* Ait., *Salvia miltiorrhiza* Bunge, and others. Various bioactive chemicals including alkaloids, flavone aglycones, isoflavones, stilbene glucosides, glycoproteins, and proteins, were separated, purified and prepared. Some of these methods and typical applications are described in this chapter.

The results of these studies were obtained by the author and his colleagues, Dr. Fuquan Yang, Dr. Xueli Cao, Dr. Yun Wei, Yu Tian, Guilian Tian, Xiao Han and others. Dr. Yoichiro Ito contributed greatly to these works.

8.2 APPARATUS FOR HIGH-SPEED CCC

8.2.1 Semi-preparative HSCCC

The semi-preparative HSCCC instrument is a multilayer coil planet centrifuge referred to as Model GS10A2. Its separation column was prepared by winding about 130 m of a 1.6 mm i.d. PTFE tube coaxially on two spools in the column holder hub. Each spool was made by 14 layers of 17 or 18 turns (a total of 245 turns). The β ratio varied from 0.5 for the inner layer (r = 4 cm) to 0.75 for the visible external layer (r = 6 cm) ($\beta = r/R$, where r is the distance from the holder axis to the coil and R, 8 cm, is the distance from the holder axis to the centrifuge axis). The two spools made a total column capacity measured as 260 ml with 490 tube turns. The speed of revolution of the centrifuge is adjustable with a speed controller, in the range of 0–1000 rpm. The separation system was equipped with an NS-1007 constant-flow pump, 8823A UV monitor, a Yokogawa model 3057 recorder and a sample injection valve with a 10 ml loop.

8.2.2 Analytical HSCCC

The analytical HSCCC instrument is referred to as Model GS20. It holds a pair of spool holders symmetrically on the rotary frame at a distance of 5 cm from the central axis of the centrifuge. The multilayer coiled separation column was prepared by winding a 50 m × 0.85 mm i.d. PTFE (polytetrafluoroethylene) tube directly onto each of the two holder hubs to form the two spools. Each spool was made by 145 tube turns coiled in 10 layers of 9 or 10 turns. The β value varied from 0.4 at the internal terminal ($r = 2$ cm) to 0.7 at the external terminal ($r = 3.5$ cm). The total GS20 machine capacity was 30 ml. The revolution speed of the apparatus could be regulated with a speed controller in a range between 0 and 2000 rpm.

8.2.3 Two-dimensional HSCCC system

The two-dimensional HSCCC system was developed in our laboratory for rapid preparative separations. It is set up with two sets of semi-preparative HSCCC equipment, Model GS10A2, and some additional devices as shown in Fig. 8.1.

Two valve positions, A and B, are shown in Fig. 8.1. The switching valve is initially set in position A. As an example of applications, HSCCC systems 1 and 2 are simultaneously filled with the upper aqueous stationary phase using pumps 1 and 2, respectively. Then, both apparatuses are rotated at 800 rpm, while the lower organic phase is eluted through HSCCC systems 1 and 2 using their respective pumps at a flow rate of 2.0 ml/min. After the mobile phase front emerges and the two phases have established hydrodynamic equilibrium in each column, the sample solution is injected into HSCCC System 1 through the injection valve while Pump 2 is stopped. The effluent from the outlet of HSCCC 1 is continuously monitored with UV detector 1 at 254 nm, and collected with fraction collector 1, according to the chromatogram traced by Recorder 1. When the target peak appears, the effluent from HSCCC 1 is cut and introduced into HSCCC 2 by turning the switching valve to position B. After the target peak has eluted from HSCCC 1, the switching valve is returned to position A and the target peak fraction introduced in HSCCC 2 is eluted by restarting Pump 2 together with Detector 2, Recorder 2, and Fraction Collector 2.

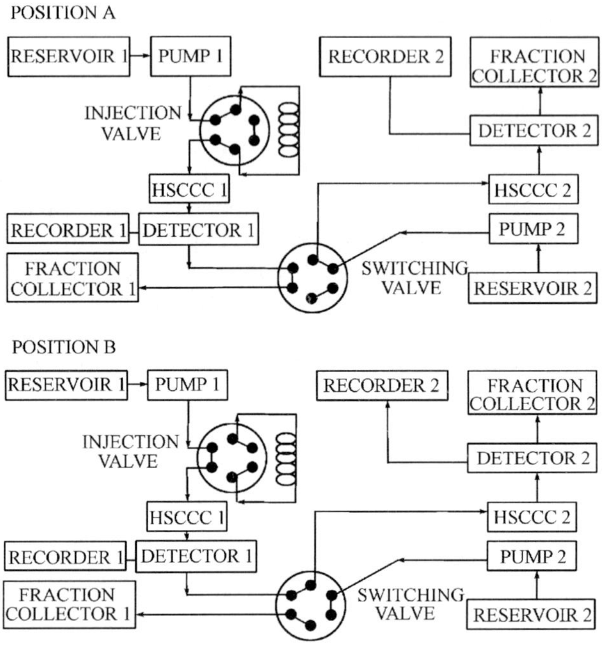

Fig. 8.1. Schematic diagram of our multidimensional countercurrent chromatography (MDCCC) system with two sets of high-speed countercurrent chromatography (HSCCC) system Model GS20 (30 ml), a six-port injection valve and a six-port switching valve.

8.2.4 pH-Zone-refining CCC system

pH-Zone-refining countercurrent chromatography is a recently developed preparative method which provides many important advantages over the conventional CCC method, including a more than 10-fold increase in sample loading capacity, high concentration of fractions, concentration of minor impurities, etc. The method uses a retainer base (or acid) in the stationary phase to retain the analytes in the column, and an eluting acid (or base) to elute the analytes according to their pK_a values and hydrophobicities. It produces a succession of highly concentrated rectangular peaks with minimum overlap similar to those observed in displacement chromatography. The method has been applied successfully to the separation of a variety of compounds including acidic and basic amino acid derivatives, many hydroxyxanthene dyes, alkaloids from natural sources, geometrical and structural isomers from synthetic products, chiral compounds, etc.

The pH-zone-refining CCC system was performed with the semi-preparative HSCCC manufactured in our laboratory as the one introduced before (260 ml machine, Model GS10A2). The solvent was pumped into the column with a Model NS-1007 constant-flow pump (Beijing Institute of New Technology Application). The continuous monitoring of the effluent was achieved with a Model 8823A UV monitor (Beijing Institute of New Technology Application) operating at 254 nm and a Model 333 pH meter (ATI Orion Research, Boston, MA, USA). A manual sample-injection valve with a 20 or 30 ml loop (Tianjin High-New Science and Technology, Tianjin, China) was used to introduce the sample into the column. A portable recorder (Yokogawa Model 3057, Sichuan Instrument Factory, Chongqin, China) was used to draw the chromatogram. A Model RE-90 rotary evaporator and a Model FC-95 auto fraction collector (both from Beijing Institute of New Technology Application) were also used.

8.2.5 Cross-axis CCC

The cross-axis Coil Planet Centrifuge, called XLL-CCC, has emerged as a universal countercurrent chromatograph. It is able to work with high flow rates of the mobile phase maintaining a large volume of stationary phase in the machine, and this even with highly polar two-phase solvent systems. Furthermore, the column efficiency is at least comparable to, or even higher than, that of the type J CCC machines (Models GS10A2 and GS20). The drawback is some mechanical weakness, owing to the two axes of rotation being at right angles. This design produces a high mechanical load on the ball bearings and gears, inducing some wear. However, this system has been improved to perform countercurrent chromatography with viscous polar solvent systems. In the XLL-CCC, the column holder is positioned laterally along the rotary shaft to generate a strong lateral force field that effectively stabilizes retention of the stationary phase for the polymer aqueous/aqueous phase system.

The polymer phase systems, useful for partitioning proteins, may be divided into two types of aqueous-aqueous biphasic systems: the polyoxyethylene glycol (PEG)–inorganic salt system and the PEG–dextran system. The PEG–salt system has relatively high interfacial tension and low viscosity, hence producing an efficient separation for many proteins.

It is essential to select the proper composition of the polymer-phase system used for separation.

The cross-axis countercurrent chromatograph, Model XLL-CCC 96 X was made in the Beijing Institute of New Technology Application, China. It holds a pair of horizontal rotary shafts symmetrically, one on each side of the rotary frame, at a distance of 12.5 cm from the central axis of the centrifuge. A spool-shaped column holder is mounted on each rotary shaft at a lateral distance of 14.5 cm away from the midpoint. Each multilayer spool was prepared by winding 14 m of 2.6 mm i.d. PTFE tubing onto a 7.6 cm-diameter holder forming six layers of seven or eight turns between a pair of flanges spaced 5 cm apart. The total column (two spools) had a volume of 150 ml with a total tube length of 28 m and a total number of 90 turns.

The solvent was pumped into the column using a Model NS-1007 constant-flow pump (Beijing Institute of New Technology Application). The continuous monitoring of the effluent was achieved with a Model 8823A UV monitor (Beijing Institute of New Technology Application) operating at 280 nm. A manual sample injection valve with a 2 or 20 ml loop (Tianjin High-New Science and Technology, Tianjin, China) was used to introduce the sample into the column. A portable recorder (Yokogawa Model 3057, Sichuan Instrument Factory, Chongqing, China) was used to plot the chromatogram. A Model FC-95 auto fraction collector (Beijing Institute of New Technology Application) was also used.

8.3 APPLICATIONS OF SEMI-PREPARATIVE HSCCC

8.3.1 Separation of 10-deacetylbaccatin III from the needles of *Taxus chinensis* [6]

Purification steps

The anticancer drug taxol is a natural product obtained from the bark of the Pacific yew, *Taxus brevifolia*. The quantity of taxol in the bark is so low, however, that a large amount of bark is required to meet the increasing demand for taxol. Since the removal of the bark kills the whole tree, intensive efforts have been devoted to searches for alternative sources, such as leaves and branches. At present the most promising approach is the semi-synthesis of paclitaxel or its analogs from 10-deacetylbaccatin III, a compound available in relatively high quantities from the foliage of several yew species. 10-Deacetylbaccatin

III is also used for synthesizing taxotere (an analog of taxol) that possesses improved pharmaceutical properties. Consequently, considerable interest exists in the isolation of the compound and its analogs.

In the present study, the isolation and purification are described of 10-deacetylbaccatin III (10-DAB) from the needles of Chinese yew (*T. chinensis*), directly from the crude water extract, by semi-preparative HSCCC.

A 500 mg sample of ground needles of *T. chinensis* was extracted twice with 2.5 l of deionized water at 50°C. The filtered aqueous solution was then extracted three times with 3 l of ethyl acetate. The organic phases containing the target compound were combined and washed with 2 l of 0.1 *M* sodium carbonate solution and then twice with 0.9 l of deionized water. The extract was finally concentrated to dryness.

The crude extract of taxus needles, the partially purified fraction from the first-step CCC separation (Fig. 8.1) and the final fraction after the second-step CCC separation were analyzed by HPLC. The analysis was performed with a Rainin (Woburn, MA, USA) HPX HPLC system equipped with a Microsorb-MC C_8 column (150×4.6 mm i.d., 5 μm diameter). The mobile phase composed of acetonitrile–water (25:75 v/v) was eluted isocratically at a flow rate of 1.0 ml/min, and the effluent monitored at 230 nm by a UV detector.

In the first separation, a quaternary solvent system composed of *n*-hexane–ethyl acetate–ethanol–water (2:5:2:5 v/v/v/v) was used. The fraction containing 10-deacetylbaccatin III was collected and concentrated, and then subjected to the second separation by HSCCC using another solvent system composed of *n*-hexane-chloroform–methanol–water (5:25:34:20 v/v/v/v).

The purified fraction of 10-deacetylbaccatin III obtained from the HSCCC separation was analyzed by FAB mass spectrometry (MS) and ^{1}H nuclear magnetic resonance (NMR). FAB-MS were taken on a Finnigan MAT711, Tabspec instrument, and ^{1}H NMR spectra on a Bruker AM-500 spectrometer (in D_2O).

Results

It has been reported that 10-deacetylbaccatin III was separated from a crude extract of European yew (*T. baccata*) by centrifugal partition chromatography [21]. In the present study, the crude needle extract of *T. chinensis* contained a much greater number of components, some of which elute close to the target component of 10-deacetylbaccatin III, as shown by HPLC analysis (Fig. 8.2).

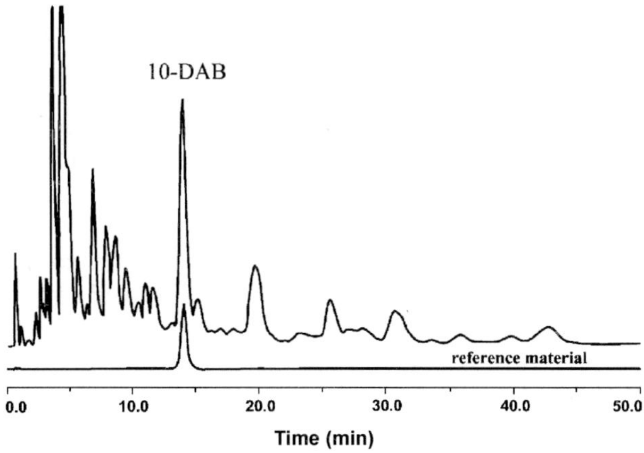

Fig. 8.2. HPLC analysis of the crude needle extract of *Taxus chinensis*. Column: Spherisorb-MV C_8 (150×4.6 mm i.d., 5 μm, Rainin); mobile phase: acetonitrile–water (25:75, v/v); flow rate: 1 ml/min: detection: UV, 230 nm. 10-DAB = deacetylbaccatin III.

In the first step of HSCCC separation, 500 mg of the crude extract was injected in 5 ml of sample solution. There was obtained 120 mg of a partially purified fraction (shaded portion of chromatogram in Fig. 8.3). HPLC analysis of this fraction showed that many interfering compounds which were present in the original sample solution, had been removed (Fig. 8.3, HPLC analysis).

This semi-pure product (120 mg in a 5 ml sample solution) was subjected to a second HSCCC purification. This second step yielded 20 mg of the final product. The HPLC analysis indicated that the purity of 10-deacetylbaccatin III in this fraction was over 98% (Fig. 8.4).

The final fraction was identified by both FAB-MS and ^{1}H NMR. FAB-MS showed three fragments of $[MH]^+$ 545, 527 and 509, which correspond to $[M+1]^+$, $[M+1–H_2O]^+$ and $[M+1–2H_2O]^+$, respectively. This indicates that the compound has a molecular mass of 544, which is identical to that of 10-deacetylbaccatin III. ^{1}H NMR produced a spectrum identical to that of a 10-deacetylbaccatin III standard. These findings confirmed that our final product is pure 10-deacetylbaccatin III.

The overall results of the present studies led us to conclude that HSCCC is a suitable method for the semi-preparative separation and purification of 10-deacetylbaccatin III from a crude extract of Taxus needles, and may also be applied to other parts of Taxus plants. The present method can produce a pure fraction of 10-deacetylbaccatin III

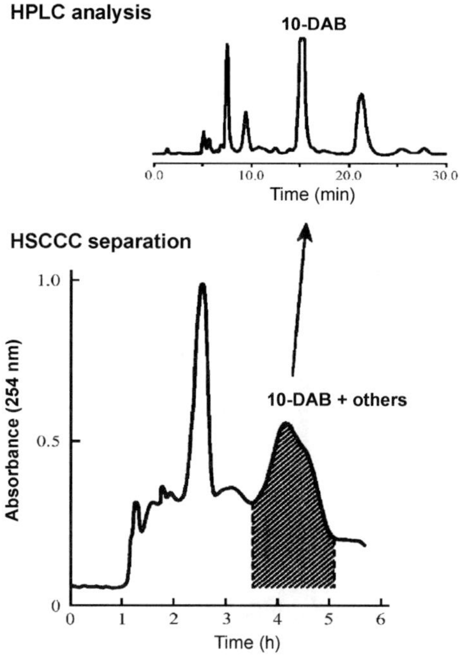

Fig. 8.3. The first-step HSCCC separation of the crude needle extract of *Taxus chinensis* and HPLC analysis of the 10-DAB fraction. Experimental conditions: apparatus: a Model GS10A2 mutilayer coil planet centrifuge with 8 cm revolution radius; column: mutilayer coil of 1.6 mm i.d. and 260 ml capacity. 800 rpm. Solvent system: *n*-hexane–ethyl acetate–ethanol–water (2:5:2:5 v/v/v/v); mobile phase: lower phase; sample: 500 mg of crude needle extract of Taxus chinensis in 5 ml solvent; flow rate: 2 ml/min in the head to tail direction; detection: UV at 254 nm. HPLC conditions, see Fig. 8.2. 10-DAB = 10-deacetybaccatin III.

with a high recovery rate, through a relatively simple separation procedure, in less than 10 hours.

8.3.2 Separation of a stilbene glycoside from *Polygonum multiflorum* Thunb [7]

Sample preparation

The dried roots of *Polygonum multiflorum* Thunb have been used for treating a variety of inflammatory disorders such as suppurative dermatitis, gonorrhoea, favus athlete's foot, and hyperlipidemia, in Chinese and Japanese traditional medicine. The roots contain 2,3,5,4′-tetra-

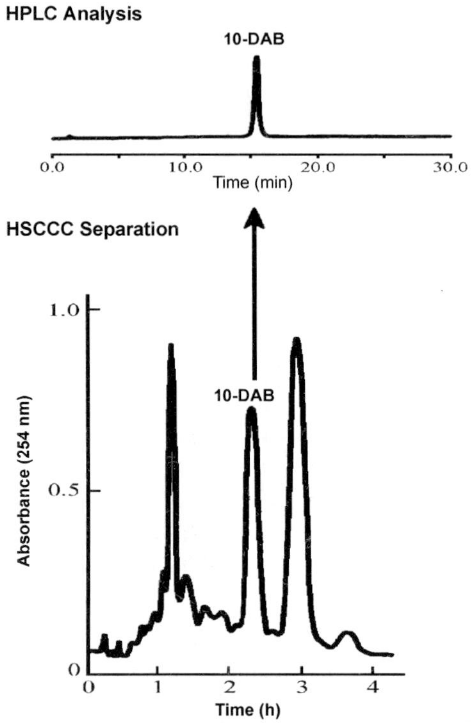

Fig. 8.4. The second-step HSCCC separation of the partially purified fraction and HPLC analysis of the 10-DAB fraction. Experimental conditions: apparatus: a Model GS10A2 multilayer coil planet centrifuge (see Fig. 8.3); solvent system: *n*-hexane–chloroform–methanol–water (5:25:34:20 v/v/v/v); mobile phase: lower phase; sample: 120 mg of first-step HSCCC fraction (shaded portion of chromatogram in Fig. 8.3) in 5 ml solvent; flow rate: 2 ml/min in the head to tail direction; detection: UV at 254 nm. For HPLC conditions see Fig.8.2. 10-DAB = 10-deacetylbaccatin III.

hydroxystilbene-2-O-D-glucoside, two new stilbene glycoside gallates, and proanthocyanidins, in addition to various anthraquinones including chrysophanol, emodin, rhein, and physcion. These stilbene components (Fig. 8.5) are known to have useful pharmaceutical activities: 2,3,5,4′-tetrahydroxystilbene can reduce liver injury in peroxidized-oil-fed rats.

The dried milled roots (2.7 g) of *P. multiflorum* collected from Yunnan, China were extracted three times with 30 ml of chloroform, under ultrasonication, to remove anthraquinones. Then, the residue was similarly extracted three times with 30 ml ethanol. The ethanol

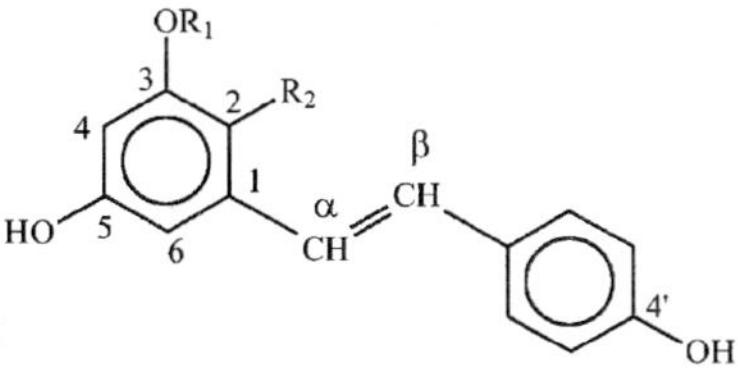

Resveratrol: R_1=R_2=H

Piceid: R =D-Glucose, R_2=H

2,3,5,4' - Tetrahydroxystilbene-2-O-D-glucoside: R_1=H, R_2=O-D-Glucose

Fig. 8.5. The structure of stilbene components.

extracts were combined and concentrated to dryness. The dried extract was dissolved in the mobile phase for HSCCC separation.

In the present study, a solvent system composed of ethyl acetate–ethanol–water (10:1:10 v/v/v) was used in the first separation to harvest an enriched fraction of stilbene glycoside from the crude extract. Then, this partially purified fraction was concentrated and subjected to HSCCC separation using the same solvent composition, at a different volume ratio of 50:1:50.

The HSCCC peak fractions were analyzed by HPLC. The HPLC instrument used for analysis was a Rainin SD-2000 HPLC system (Rainin Instrument Company, Mack Road, Woburn, MA, USA). The analysis was performed with a Spherisorb-MV C_{18} column (250×4.6 mm i.d., 5 μm diameter, Rainin Instrument Company). The mobile phase composed of methanol and water (40:60 v/v) was eluted at a flow rate of 1.0 ml/min, and the effluent monitored by a UV detector at 254 nm.

The stilbene glycoside purified by HSCCC was identified by FAB-MS and ^{1}H-NMR. The FAB-MS were taken on a Finnigan MAT711, Tabspec instrument, and ^{1}H-NMR spectra on a Bruker AM-500 spectrometer (in D_2O).

Results

In the present study, 240 mg of the crude ethanol extract (10 ml) from the dried roots of *P. multiflorum* was separated by HSCCC using a solvent system composed of ethyl acetate–ethanol–water at a volume ratio of 10:1:10 (Fig. 8.6). In this separation, the fraction corresponding to the shadowed peak in the HSCCC chromatogram contained stilbene glycoside at about 80% purity (by HPLC analysis). This partially purified fraction was further subjected to HSCCC separation using the same set of solvents at a modified volume ratio of 50:1:50 (Fig. 8.7). In

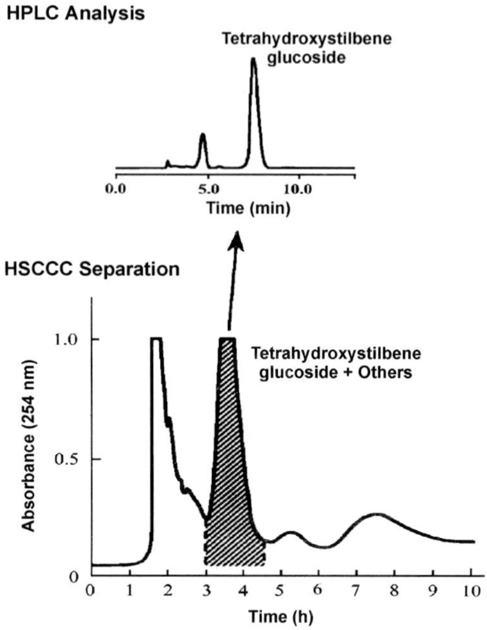

Fig. 8.6. HSCCC separation of crude root extract of *P. multiflorum*. Experimental conditions: apparatus: model GS10A2 (260 ml, see Fig. 8.3); sample: 240 mg of crude root extract of *P. multiflorum* in 10 ml solvent; solvent system: ethyl acetate–ethanol–water (10:1:10 v/v/v); mobile phase: lower phase; flow rate: 2 ml/min in the head to tail direction; revolution: 800 rpm; detection: UV at 254 nm. HPLC analysis: Rainin SD-200 HPLC system; column: Spherisorb-MV C_{18} column (250×4.6 mm i.d., 5 μm); mobile phase: methanol–water (40:60 v/v); flow rate: 1.0 ml/min; detection: UV at 254 nm.

this second separation the stilbene glycoside fraction showed a high purity of over 96% by HPLC analysis. The recovery of the final product was 30 mg, or 1.1% (w/w) from the dried roots (2.7 g) of Yunnan *P. multiflorum*.

The stilbene glycoside isolated by HSCCC, emits bright fluorescence under UV light, and was identified by both FAB-MS and ^{1}H-NMR. FAB-MS showed two peak fragments with [MH^+] of 407 and 244, which correspond to tetrahydroxystilbene glycoside and tetrahydroxystilbene parent, respectively. The ^{1}H-NMR spectrum exhibited in the aromatic- and olefinic range a pair of *meta*-coupled doublets at δ 6.09 (1H, d, J = 2 Hz, 4-H), δ 6.29 (1H, d, J = 2 Hz, 6-H), *trans*-olefinic proton signals at δ 6.36 (1H, d, J + 16 Hz, β-H), δ 7.01 (1H, d, J = 2 Hz, β-H), and an

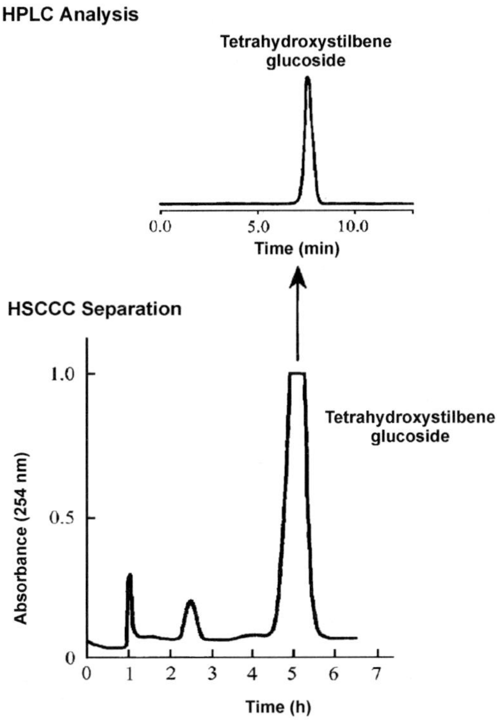

Fig. 8.7. HSCCC separation of partially purified extract of *P. multiflorum*. Experimental conditions: apparatus: model GS10A2 (260 ml, see Fig. 8.3); sample: partially purified fraction of root extract of *P. multiflorum* (all shaded portion of chromatogram in Fig. 8.6); solvent system: ethyl acetate–ethanol–water (50:1:50 v/v/v); mobile phase: lower phase in the head to tail direction; flow rate: 2 ml/min; revolution: 800 rpm; detection: UV at 254 nm. HPLC analysis: Rainin SD-200 HPLC system; column: Spherisorb-MV C_{18} column (250×4.6 mm i.d., 5 μm); mobile phase: methanol–water (40:60, v/v); flow rate: 1.0 ml/min; detection: UV at 254 nm.

A_2B_2-type signal δ 6.25 (2H, d, J = 8 Hz, 2^1,6^1-H), δ 6.91 (2H, d, J= 8 Hz, 3′,5′-H). At high field, the spectrum exhibited C-H signals of glucose, δ 3.99 (1H, d, J = 8 Hz, C_1-H), δ 2.88–3.52 (6H, *m*, other C-H).

Considering these spectra, together with the biological origin of the molecule, it is concluded that the compound is 2,3,5,4′-tetrahydroxy-stilbene-2-O-D-glycoside.

The sample size processed by HSCCC can be greatly increased using a large multilayer column: hence, the method is ideal for preparing the standard of stilbene glycoside for quality control.

8.3.3 Separation of Resveratrol and Piceid from extract of *Polygonum cuspidatum* Sieb. et Zucc

Resveratrol and Piceid exist in more than 72 species of plants, such as grape, peanut and some Chinese medicinal herbs. They have obvious bioactivities for cardiovascular diseases, nervous system diseases, inflammation, and abscesses, among others. Some recent investigations showed that they appear in the inhibition of the initial stage, proliferation stage and expansion stage of cancer.

HSCCC was used to separate these compounds from a medicinal herb called *Polygonum cuspidatum* (*Huzhang* in Chinese) which is rich in them. An ethyl acetate extract (1.5 g) was prepared from 20 g of raw material of *P. cuspidatum*. Then 252 mg of this extract was separated with HSCCC using a solvent system composed of ethyl acetate–ethanol–water (10:1:10, v/v/v). The result is shown in Fig. 8.8. Collection of the effluent corresponding to Peak 5 and evaporation of the solvents produced 72.5 mg of purified Resveratrol.

A 1.88 g water extract was prepared similarly from 20 g of raw material. A 310 mg extract was then separated by a two-step HSCCC experiment. Two solvent systems composed of ethyl acetate–ethanol–water (50:1:50 and 10:1:10 v/v/v) were used. The chromatograms are shown in Fig. 8.9. By this procedure, 35.45 mg of ~99% pure Piceid was

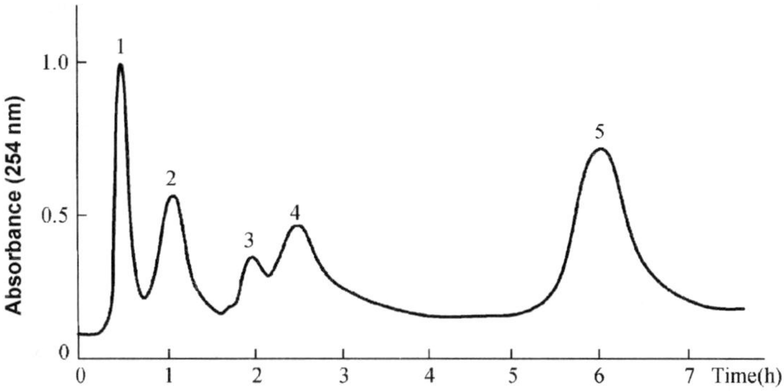

Fig. 8.8. HSCCC Separation of Resveratrol from *Polygonum cuspidatum* extract. Experimental conditions: apparatus: GS10A2 (260 ml, see Fig. 8.3); sample: 252 mg of ethyl acetate extract of *Polygonum cuspidatum* in 4 ml lower phase. Solvent system: ethyl acetate–ethanol–water (10:1:10 v/v/v); mobile phase: lower phase in the head-to-tail direction; flow rate: 2 ml/min; revolution: 800 rpm; detection: UV at 254 nm. Peak 5 = Resveratrol (72.5 mg).

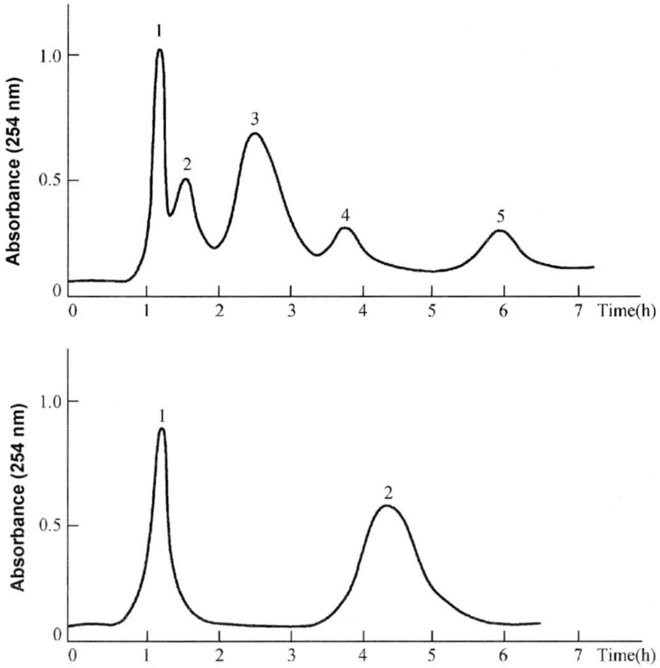

Fig. 8.9. HSCCC separation of Piceid from *Polygonum cuspidatum* extract. Experimental conditions: apparatus: model GS10A2 (260 ml, see Fig. 8.3); sample: 310 mg of water extract of *Polygonum cuspidatum* in 10 ml lower phase; solvent system: ethyl acetate–ethanol–water (50:1:50 v/v/v, top chromatogram, and 10:1:10 v/v/v, bottom chromatogram); mobile phase: lower phase in the head to tail direction; flow rate: 2 ml/min; revolution: 800 rpm; detection: UV at 254 nm. Peak 2 of the top chromatogram was reinjected giving the bottom chromatogram whose Peak 2 was Piceid (35.5 mg).

obtained. The calculated data showed that the contents of Resveratrol and Piceid in raw material of Huzhang were about 2.18% and 1.07% w/w, respectively.

8.3.4 Separation of gallic acid from *Cornus officinalis* Sieb. et Zucc [8]

The fruit of *Cornus officinalis* Sieb. et Zucc has been used as a traditional Chinese medicine in China and Japan. It represents one of the eight-component drugs in Hachimi-gan that is one of the most popular traditional Chinese medicines in Japan. A study of the constituents of the Zucc fruit by Okuda et al. revealed the presence of various organic acids and hydrolysable tannins [22]. However, separation of these components by conventional chromatographic methods is difficult, because these tannins adsorb strongly onto any silica-containing solid supports.

High-speed countercurrent chromatography (HSCCC), being a support-free partition chromatography, eliminates the adsorption problem and can be an excellent alternative method for separating these tannins. In our study for developing a new anti-aging preparation, the n-butanol extract of fruits of *Cornus officinalis* Sieb. et Zucc was separated by HSCCC in two steps, to purify gallic acid that is known to be an antioxidant with a high anti-aging activity.

All organic solvents used for HSCCC were of analytical grade and purchased from Beijing Chemical Factory, Beijing, China. The gallic acid standard sample was purchased from the National Institute for the Control of Pharmaceutical & Biological Products, Ministry of Health, Beijing, China.

About 1.0 kg of fruits of *Cornus officinalis* Sieb. et Zucc was extracted three times in a glass bottle (2500 ml capacity) with 1.3 l of 70% acetone (total volume 4 l) at room temperature. The extracts were combined, evaporated to dryness under reduced pressure, and redissolved in 500 ml of water. The aqueous solution was again extracted successively with chloroform (3×500 ml) and n-butanol (10 times 500 ml). The n-butanol extract, which showed a high anti-aging activity, was concentrated to yield 172.8 g. Of this extract, 1 g was dissolved in 10 ml of each phase of the solvent system used for HSCCC separation. The sample solution was sonicated for several minutes before loading into the column of the HSCCC Model GS10A2 machine (260 ml column capacity).

The following two solvent systems were prepared: ethyl acetate–n-butanol–water (5:1.8:6, v/v/v) and ethyl acetate–ethanol–water (5:0.5:6, v/v/v). Each solvent mixture was thoroughly equilibrated in a separatory funnel at room temperature and the two phases were separated shortly before use. For each separation, the column was first entirely filled with the organic stationary phase, the upper phase. Then the sample solution was injected through the sample port and the aqueous mobile phase was pumped through the column at a flow rate of 2.0 ml min^{-1} while the column was rotated at 800 rpm. The effluent from the outlet of the column was monitored with a UV detector at 254 nm. Peak fractions were manually collected according to the chromatogram. After the desired peaks were eluted, the rotation and elution were stopped and the column contents were collected in a graduated cylinder by N_2 at a pressure of approx. 0.5 MPa. The retention of the stationary phase relative to the total column capacity was computed from the volume of the stationary phase collected from the column.

HPLC analysis was performed with a Shimadzu LC-10AVP system including an LC-10AT liquid chromatograph, an SPD-M10AVP photodiode array detector, an SCL-OAVP system controller, a CTO-10SVP column oven, and a DGU-12A degasser (Shimadzu, Kyoto, Japan) using an Intersil ODS-3 column (150×4.6 mm i.d.) at a column temperature of 40°C. The mobile phase was composed of acetonitrile–0.1 *M* H_3PO_4 (linear gradient from 0:100 to 40:60 v/v in 30 min). The flow rate was set at 1.0 ml min^{-1} and a photodiode array detector continuously monitored the effluent. Identification of HSCCC peak fractions was carried out by comparing the retention times and UV spectra with standard samples.

In the present study, 350 g of crude acetone extract from the dried fruit of *Cornus officinalis* Sieb. et Zucc was successively extracted with chloroform and *n*-butanol. Pharmacodynamic tests done on 100 mg each of acetone, chloroform, *n*-butanol, and water extracts revealed a high anti-aging activity in the *n*-butanol extract only. Consequently, the *n*-butanol extract was separated by HSCCC using a solvent system composed of ethyl acetate–*n*-butanol–water (5:1.8:6, v/v/v), yielding six peak fractions (I–V, and the residue in the column) (Fig. 8.10) among which fraction V showed a high anti-aging activity. The HPLC analysis of this fraction indicated that it contained gallic acid at about 70% purity. This partially purified fraction was further subjected to HSCCC using the second solvent system composed of ethyl acetate–ethanol–water (5:0.5:6, v/v/v) (Fig. 8.11). HPLC analysis of the fraction obtained after the second separation showed that the gallic acid purity was over 97%.

The two-step HSCCC separation described above yielded 60 mg of gallic acid from 1 g of *n*-butanol extract (6% w/w). The gallic acid thus obtained was a white needle-shaped crystal. The fast atom bombardment (FAB) MS date are: *m/z* 171 $[M^+]$: UV, λ_{EtOH} 217, 273: ^1H NMR (D_2O, δ ppm, TMS) 7.5 (2 H, s, H-2, H-6): ^{13}C NMR, δ ppm; 112.6 (C_2, C_6), 124.3 (C_3, C_5), 140.4 (C_4), 147.1 (C_1), 173.4 (COOH).

The results of our studies show that HSCCC is a useful tool for isolating the bioactive fraction from a crude extract from *Cornus officinalis* Sieb. et Zucc. The quantity of the sample can be increased several times using a larger multilayer coil separation column, suggesting that the method is also ideal for preparation of the standards of gallic acid and other components.

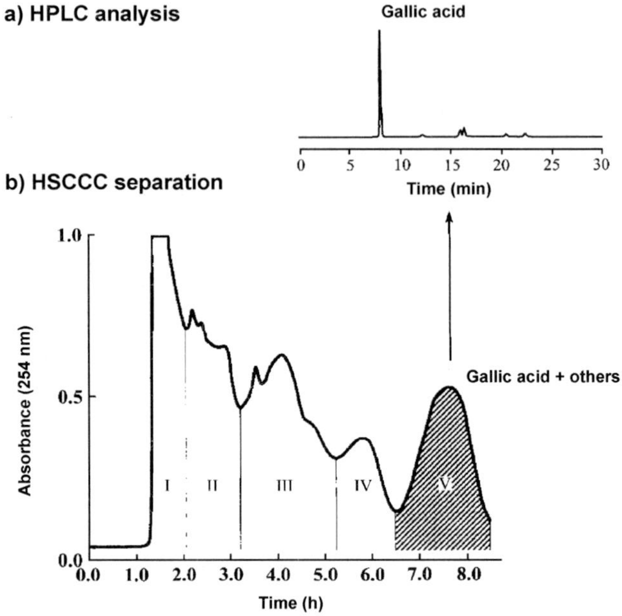

Fig. 8.10. Preparative HSCCC separation of *n*-butanol extract of *Cornus officinalis* Sieb. et Zucc. (a) HPLC analysis: column: Intersil ODS-3 (150×4.6 mm i.d.); column temperature: 40°C; mobile phase: acetonitrile-0.1 *M* H_3PO_4 (linear gradient, % v/v from 0:100 to 40:60 in 30 min); flow rate: 1.0 ml min^{-1}; detection: 254 nm. (b) CCC experimental conditions: Model GS10A2, 260 ml machine, rotation speed: 800 rpm; solvent system: ethyl acetate–*n*-butanol–water (5:1.8:6 v/v/v); mobile phase: lower aqueous phase; flow rate: 2 ml min^{-1} in the head to tail direction; sample size: 1.0 g; retention of the stationary phase: 51%.

8.4 APPLICATIONS OF ANALYTICAL HSCCC

8.4.1 Separation of Alkaloids from *Coptis chinensis* Franch [9]

Coptis chinensis Franch is an important traditional Chinese medicinal herb, which has been effectively used for reducing fevers, relieving headache and fidgetiness, decreasing flu-related sweating, and detoxicating. The major active constituents of this herb are alkaloids, including berberine, epiberberine coptisine, palmatine, jateorhizine and columbamine, and their structures are illustrated in Fig. 8.12. The berberine molecule, present at about 10% in *C. chinensis* Franch, has a strong anti-bacterial action on *shigella dysenteriae*, *staphylococci* and *strepto-*

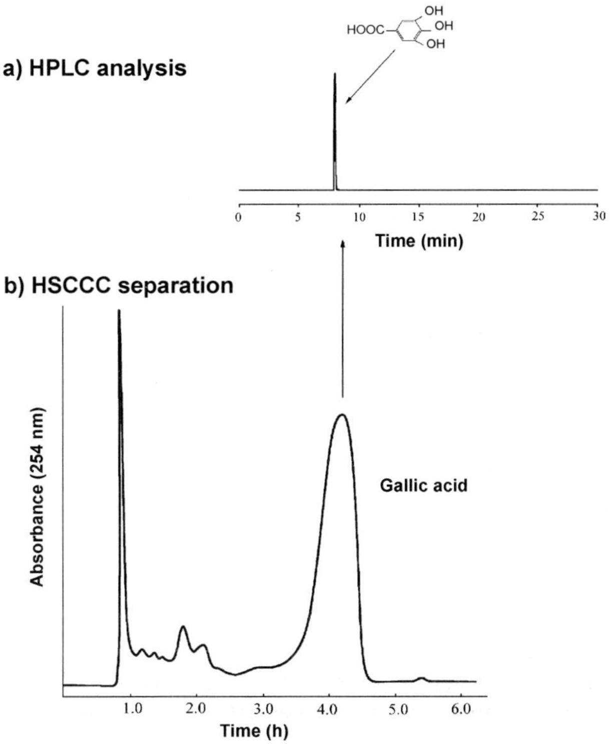

Fig. 8.11 Preparative HSCCC separation of partially purified extract of *Cornus officinalis* Sieb. et Zucc. (a) HPLC analysis: see Fig. 8.10. (b) CCC experimental conditions: see Fig. 8.10. Solvent system: ethyl acetate–ethanol–water (5:0.6:5 v/v/v); mobile phase lower aqueous phase; flow rate: 2 ml min^{-1} in the head to tail direction; retention of stationary phase: 56%.

cocci, and has been used for the treatment of dysentery. All other alkaloids from *C. chinensis* Franch possess a distinct anti-inflammatory action.

In this experiment, the analytical HSCCC machine, Model GS20, 30 ml, is used for the systematic selection and optimization of the two-phase solvent system for the separation of alkaloids from *C. chinensis* Franch. Using the optimized solvent system, the preparative HSCCC separation of the crude alkaloids is performed with the 260 ml Model GS10A2 machine from larger quantities of the crude extract of *C. chinensis* Franch.

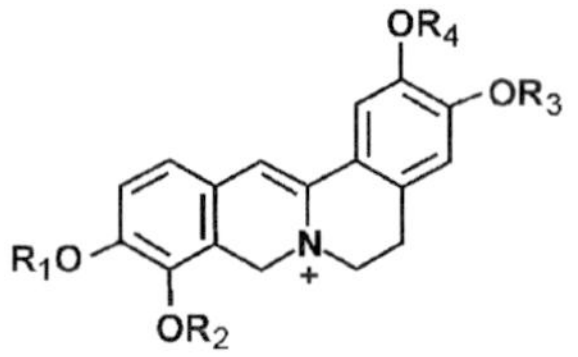

	R_1	R_2	R_3	R_4
palmatine	CH_3	CH_3	CH_3	CH_3
berberine	CH_3	CH_3	CH_2	
epiberberine	CH_2		CH_3	CH_3
coptisine	CH_2		CH_2	
jateorhizine	CH_3	CH_3	H	CH_3
columbamine	CH_3	CH_3	CH_3	H

Fig. 8.12. Molecular structures of alkaloids from *C. chinensis* Franch.

Sample preparation

Raw roots of *C. chinensis* Franch (ca. 1.0 kg) were extracted with methanol three times at room temperature. The extracts were combined and evaporated to dryness under reduced pressure. The above crude extracts containing alkaloids as major components were directly subjected to HSCCC.

The sample solutions were prepared by dissolving the crude alkaloid extract of *C. chinensis* Franch in the lower organic phase at suitable concentrations for analytical and preparative purposes.

After HSCCC separation, all peak fractions were analyzed using silica gel-G thin-layer chromatography (TLC) plates and developing with a solvent mixture composed of C_6H_6–EtOAc–CH_3OH–$(CH_3)_2$CHOH–aqueous NH_3 (12:6:3:3:1, v/v/v/v) and staining with a Dragendorff reagent to detect alkaloids. The alkaloids purified by HSCCC were identified by mass spectrometry (MS), nuclear magnetic resonance (NMR), IR and UV detection.

Results

During the selection of the two-phase solvent system, a solvent system composed of $CHCl_3$–MeOH–water (4:3:2, v/v/v) was first applied for the analytical separation of the crude alkaloids from *C. chinensis* Franch. In this separation, no alkaloids were eluted in 2.5 h. In the following separation with a solvent system composed of $CHCl_3$–MeOH–0.05 *M* HCl (4:3:2, v/v/v), all major alkaloid components were eluted within 2 h but with poor resolution. Further studies were carried out using the above solvent system composed of chloroform, methanol and dilute HCl, by systematically modifying the relative volumes of methanol (4:2:2–4:1.5:2, v/v/v) and the concentration of HCl (0.3–0.1 *M*).

Figure 8.13 shows the results of separations. Nine chromatograms are arranged in such a way that the effects of methanol and HCl concentrations on the separation of alkaloids are each clearly visualized. As seen from the top to the bottom rows, decreasing the relative volumes

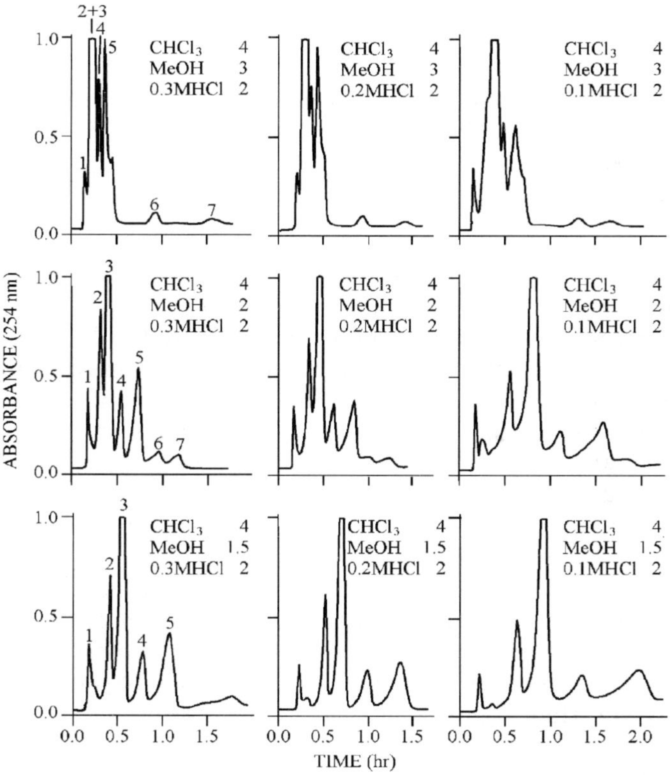

Fig. 8.13. Chromatograms of the crude alkaloid from C. chinensis Franch by the 30 ml GS20 analytical machine. Experimental conditions: sample: 2.5 mg of crude alkaloid methanol extract; solvent system: shown in each chromatogram; mobile phase: lower organic phase; flow rate: 1 ml/min (head to tail); revolution: 1500 rpm. Retention of the stationary phase was 77% for all biphasic liquid systems.

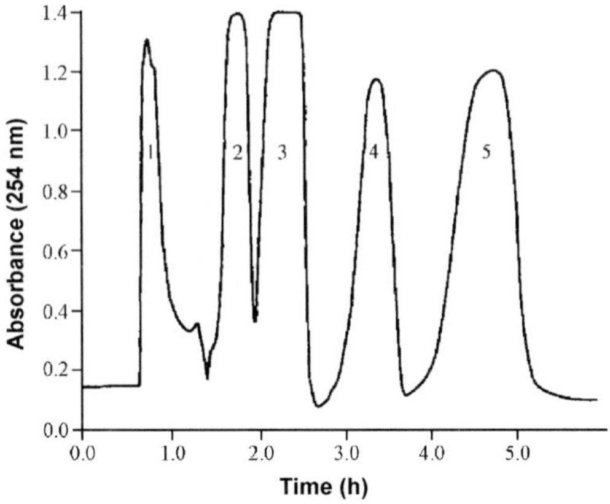

Fig. 8.14. Chromatogram of the crude alkaloids from *C. chinensis* Franch by preparative HSCCC. Model GS10A2, 260 ml machine with $CHCl_3$–MeOH–0.2 *M* HCl (4:1.5:2 v/v). Mobile phase: lower phase; flow rate: 2.0 ml/min (head to tail); sample size: 200 mg dissolved in 10 ml mobile phase. 1: unknown compound; 2: palmatine; 3: berberine; 4: epiberberine; 5: coptisine. Retention of the stationary phase was 73.3%.

of methanol in the solvent system from 4:3:2 to 4:1.5:2 increases the retention time of alkaloids. A similar effect is also produced by decreasing the concentration of HCl from 0.3 *M* to 0.1 *M* in the solvent system (from left to right in each row). Among those, the solvent system composed of $CHCl_3$–MeOH–0.2 *M* HCl (4:1.5:2, v/v/v) (bottom row, center) produced the best separation of all four major components of alkaloids (Peaks 2–5) within 1.5 h, while minor components (Peaks 6 and 7) were still retained in the column. On the other hand, the solvent systems containing a large volume of methanol (top row) produced a good separation for Peaks 6 and 7 within 2 h.

To do a large-scale separation, the 260 ml GS10A2 machine was used with the solvent system composed of $CHCl_3$–MeOH–0.2 *M* HCl (4:1.5:2, v/v/v). A 200 mg sample of the crude alkaloids was dissolved in 10 ml mobile phase and injected into the column. Figure 8.14 shows the results of this separation. Four main alkaloids, berberine, epiberberine, coptisine and palmatine, were well resolved within 5 h.

Figure 8.15 shows the result of the TLC analyses of various samples including the crude alkaloid extract and HSCCC fractions corresponding to Peaks 2, 3, 4 and 5 (Fig. 8.14). Silica gel-G TLC plates were developed with a solvent composed of C_6H_6–EtOAc–CH_3OH–$(CH_3)_2CHOH$–aqueous NH_3 (12:6:3:3:1, v/v/v/v/v) and stained with a Dragendorff

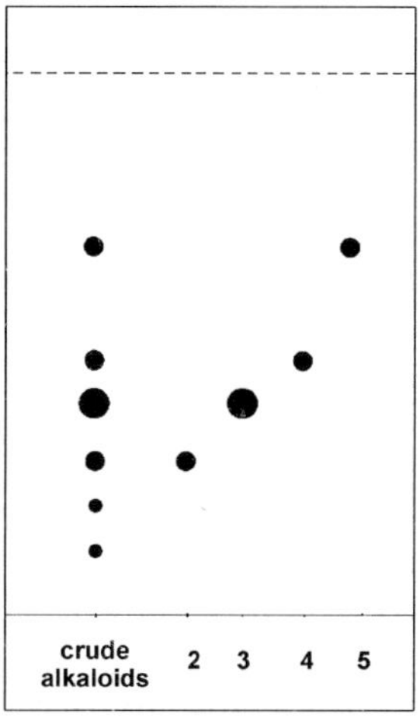

Fig. 8.15. The result of TLC analyses of the crude alkaloid extract of *C. chinensis* Franch and HSCCC fractions corresponding to Peaks 2, 3, 4 and 5 made by a silica gel G TLC plate developed with the solvent system: C_6H_6–EtOAc–CH_3OH–$(CH_3)_2$–CHOH– aqueous NH_3 (12:6:3:3:1 v/v/v/v).

reagent to detect the alkaloids. Alternatively, spots on the TLC plates were observed under a UV lamp. The result indicated that HSCCC fractions corresponding to Peaks 2, 3, 4 and 5 each contained a single alkaloid species, identified as palmatine, berberine, epiberberine and coptisine, respectively, by MS, NMR, IR and UV.

8.4.2 Separation of isoflavones from *Pueraria lobata* (willd.) Ohwi [10]

Pueraria lobata (willd.) Ohwi is a traditional Chinese medicinal herb. Its crude extract has been used for the treatment of hypertension and angina pectoris in China. Studies in pharmacology and clinical practice have shown that the active constituents in the extract are isoflavones, mainly puerarin, that significantly dilate coronary arteries, decrease myocardial oxygen consumption, and improve micro-circulation in both animals and human patients suffering from cardiovascular disease. Further studies on pharmacological and clinical effects of puerarin and other active constituents necessitate the development of an efficient preparative separation method for these drugs. Such a method will also facilitate quality control and improvement of the quality of existing *pueraria radix* products. Except for puerarin, pure standards of these isoflavones are not yet available.

Sample preparation
The separation and purification of puerarin using the conventional methods such as column chromatography and HPLC requires several steps, resulting in low recoveries of products.

In our laboratory, fresh roots of *P. lobata* collected in Tibet were chopped and extracted with ethanol. The extract was concentrated to dryness and redissolved in water. The water-soluble extract was concentrated and then chromatographed on polystyrene resin (NKA-0, 0.3–1.25 mm; Nankai Chemical Factory, Tianjing, China) with 20% ethanol to yield the crude sample for CCC separation. The sample solution was prepared by dissolving the crude sample in the mobile phase of the solvent system selected for HSCCC separation. The two-phase solvent system composed of ethyl acetate–*n*–butanol–water in various volume ratios such as: 10:1:10, 1:4:5, 1:2:3, 1:1:2, and 2:1:3, v/v/v were used.

In each analytical separation, the column was first entirely filled with the organic stationary phase, and then the apparatus was rotated at 1,800 rpm, while the aqueous mobile phase was pumped into the column at a flow rate of 1.0 ml/min. About 1 ml of the sample solution containing 2 mg of the crude extract was injected through the injection valve. The effluent of the column was continuously monitored with a UV detector at 254 nm. Peak fractions were collected according to the elution profile.

The preparative separation based on the information from the analytical separation was performed similarly using a 10 ml sample volume (80 mg crude extract) at a flow rate of 2.0 ml/min and a revolution speed at 800 rpm.

Results
Figure 8.16 shows an HPLC separation of the crude extract from *P. lobata*. Peak 2 corresponds to puerarin, the major component comprising approximately 30% of the total extract. Peaks 1, 4, 5 were identified as 3′-hydroxypuerarin, and two puerarin xylosides, respectively, by FAB-MS, ^{1}H NMR and ^{13}C NMR. Peaks 3 and 6 were identified as 3′-methoxypuerarin and daidzin, respectively. The structures of these isoflavones are shown in Fig. 8.17.

The performance of the two-phase solvent systems composed of ethyl

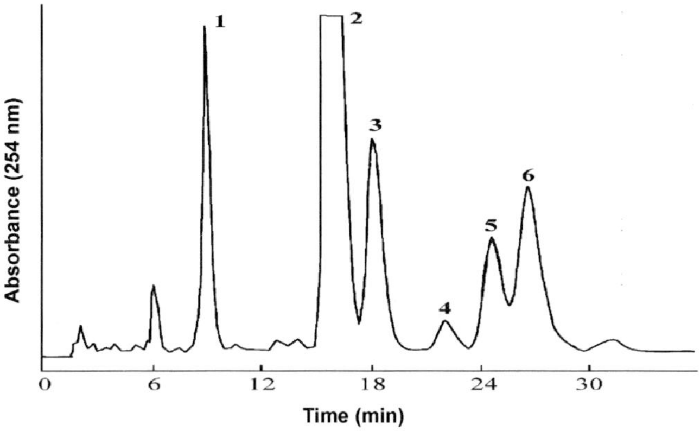

Fig. 8.16. HPLC separation of the crude extract from *Pueraria lobata*. Experimental conditions: apparatus: Rainin SD-200 HPLC system with a Supelcosil LC-18 column (250×4.6 mm i.d., 5 μm); mobile phase: methanol–water (25:75 v/v); flow rate: 1.0 ml/min; detection: UV at 254 nm. Peak identification, see Fig. 8.17.

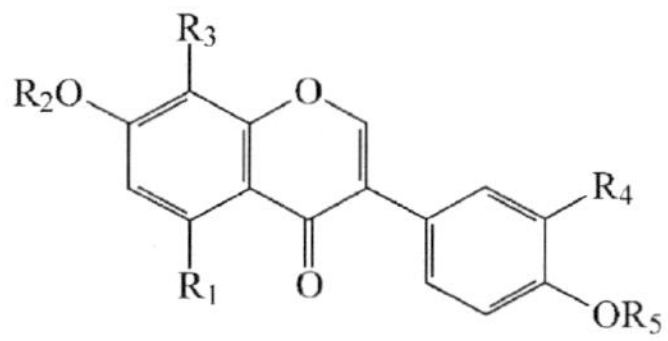

	R_1	R_2	R_3	R_4	R_5
1. 3'-hydroxyl-puerarin	H	H	glc	OH	H
2. puerarin	H	H	glc	H	H
3. 3'-methoxy-puerarin	H	H	glc	OCH_3	H
4. puerarin-xyloside	H	H	glc–6xyl	H	H
5. puerarin-xyloside	H	H	glc–2xyl	H	H
6. daidzin	H	glc	H	H	H

Fig. 8.17. The structures of six isoflavones.

acetate–*n*–butanol–water at various volume ratios (10:1:10, 1:4:5, 1:2:3, 1:1:2, and 2:1:3) were evaluated by analytical HSCCC in terms of peak resolution of various isoflavones. The results indicated that all of these solvent systems except for 10:1:10, v/v/v, could be used to separate puerarin, 3′-methoxypuerarin, and daidzin from the crude sample. However, for the resolution of the other three components, and for the separation time, the solvent-system at a volume ratio of 2:1:3 was found to be most suitable.

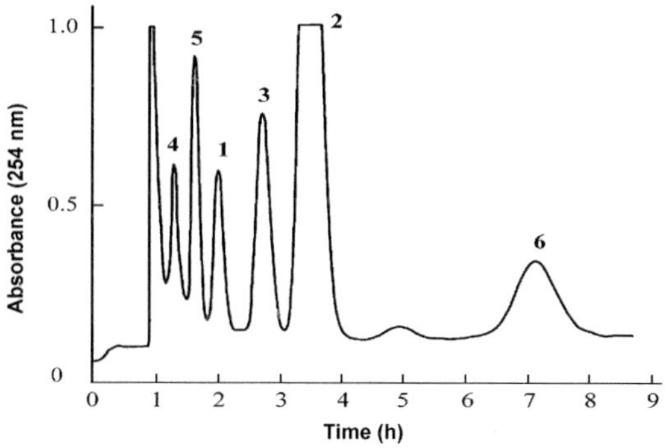

Fig. 8.18. Preparative HSCCC separation of the crude extract of *Pueraria lobata*. Experimental conditions: apparatus: Model GS10A2; 260 ml column; rotation speed: 800 rpm; solvent system: ethyl acetate–*n*-butanol–water (2:1:3 v/v/v); mobile phase: lower aqueous phase; flow rate: 2 ml/min head to tail; detection: UV at 254 nm; sample size: 80 mg; retention of the stationary phase: 56%. Peak identification, see Fig. 8.17.

Figure 8.18 shows the preparative HSCCC separation of 80 mg of the crude sample using the selected solvent system. HPLC analysis of each peak fraction of this separation revealed that puerarin corresponding to Peak 2 was over 98% pure, while other isoflavones also showed high purities over 95% except for those corresponding to Peaks 1 and 4 which were about 90% pure.

8.4.3 Separation of catechins from tea polyphenols [11]

Among numerous types of polyphenols contained in green tea leaves, flavan-3-ols and their gallates, referred to as "catechins" (Fig. 8.19), are the most important major components. It has been recognized that these catechins possess many biological activities, such as antioxidant, anti-mutagenic and anti-carcinogenic properties, anti-HIV activities, prevention of dental caries, and reduced risk of cardiovascular problems. Because of these properties, catechins are seriously considered as potent compounds for medical use. However, catechins with high purity are currently only commercially available as their standards, offered at high price. This may be mainly due to a lack of an efficient purification method for these compounds. The isolation of catechins using conventional methods such as column chromatography requires complicated multiple steps, resulting in low recoveries of products owing to irreversible adsorption and decomposition on the solid support.

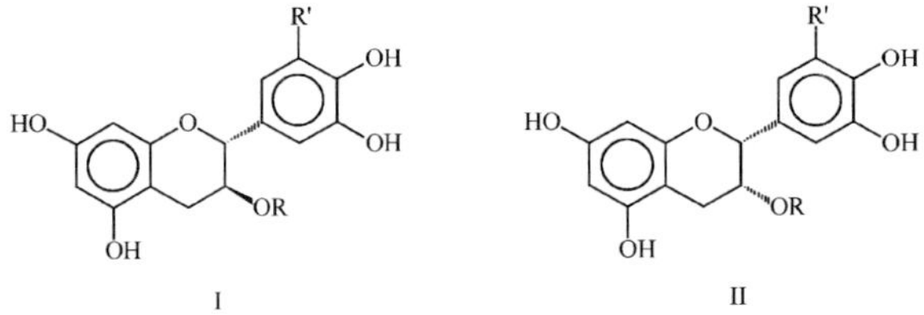

I	II	R	R'
C	EC	H	H
GC	EGC	H	OH
CG	ECG	G	H
GCG	EGCG	G	OH

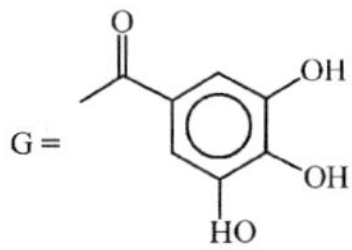

Fig. 8.19. The structure of catechins. C = catechin, E = epi, G = gallo or gallate.

The present study describes the application of analytical and semi-preparative HSCCC to the separation and purification of three individual catechins: epigallocatechin-3-*o*-gallate (EGCG), gallocatechin-3-*o*-gallate (GCG), and epicatechin-3-*o*-gallate (ECG) from a crude sample of tea polyphenol mixture.

All solvents used for HSCCC were of analytical grade and purchased from Beijing Chemical Factory, Beijing, China. The Qun Li Company, Hainan, China, provided the tea polyphenol sample mixture.

Fraction preparation

A series of preliminary experiments was performed to determine suitable two-phase solvent systems using the analytical HSCCC Model GS20 (30 ml) machine. With its speed and minimal solvent consumption this machine offers a very promising way to carry out method development for preparative HSCCC separations. Two sets of solvent systems composed of ethyl acetate–ethanol–water and hexane–methanol–water were examined by varying the mutual volume ratios. Each solvent mixture was thoroughly equilibrated in a separatory funnel and the two phases were separated just before use. The sample solution was prepared by dissolving the polyphenol mixture in the mobile phase of the solvent system to be used for separation.

In the analytical separation, the coiled column was first entirely filled with the organic stationary phase, and the HSCCC machine was rotated at 1800 rpm, while the aqueous mobile phase was pumped into the column at a flow rate of 1.0 ml/min in the head to tail direction. After the mobile phase front emerged and hydrodynamic equilibrium was established, 1 ml of the sample solution containing about 16 mg of sample was injected through the injection valve. The effluent from the column was monitored with a UV detector at 254 nm. Peak fractions were collected according to the elution profile.

The preparative separation was similarly performed with the GS10A2 260 ml machine on 1 g of sample (20 ml) at a flow rate of 2.0 ml/min and 800 rpm. During the separation, a nitrogen circulation protected the collected fractions from any oxidation, and, after separation, the fractions were freeze-dried under dark conditions.

The fractions corresponding to the HSCCC peaks were analyzed using a Shimadzu Class-VP HPLC system equipped with a Phenomenex LUNA C_{18} column (150×4.6 mm i.d.), a C_{18} precolumn (4×2.0 mm i.d.), and a photodiode array (PDA) detector. The mobile phase composed of A and B at a 1:1 volume ratio was pumped at a flow rate 1 ml/min where A was acetic acid–methanol–water (1:1:98, v/v/v) and B was acetic acid–methanol–water–*N,N*-dimethylformamide (DMF) (1:1:48:50, v/v/v/v).

Each catechin component purified by HSCCC was identified by fast-atom-bombardment (FAB) MS and ^{1}H and ^{13}C NMR. FAB-MS were taken on a Finnigan MAT711, Tabspec instrument, while ^{1}H and ^{13}C NMR spectra were obtained using a Bruker AM-500 spectrometer (in acetone-d_6).

Results

As mentioned above, the starting material for CCC separation was a tea polyphenol mixture. According to the result of HPLC analysis (Fig. 8.20), this mixture mainly contained three catechins that were later identified as EGCG, GCG and ECG.

During previous work aimed at obtaining a pure EGCG standard from different sources of material by HSCCC, it was found that the quaternary solvent system composed of hexane–ethyl acetate–methanol–water was not suitable for the separation, because EGCG was eluted too quickly and contaminated with closely eluting impurities.

Our purpose in the present studies is to separate EGCG, GCG, and ECG, all at high purity, in a single HSCCC run. In order to achieve this goal the performance of two different solvent systems composed of ethyl acetate–methanol–water and hexane–ethyl acetate–water at various volume ratios was examined by analytical HSCCC.

In the ethyl acetate–methanol–water system, the retention time of three catechins can be adjusted conveniently by changing the relative volume of methanol in the solvent system. When the relative volume of

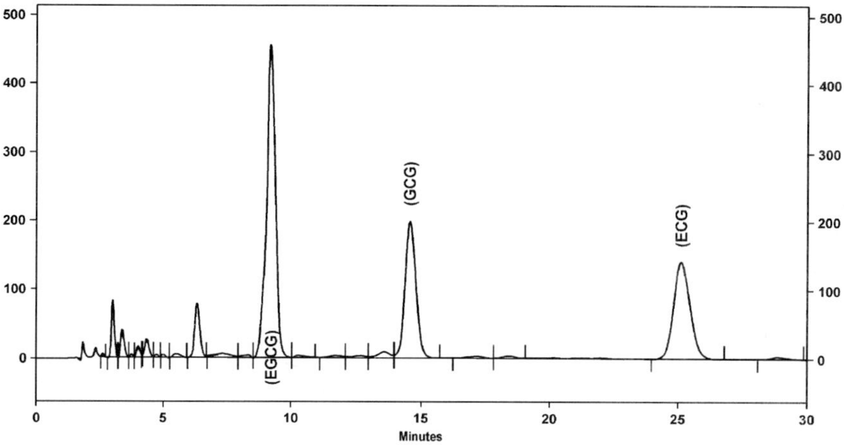

Fig. 8.20. HPLC chromatogram of tea polyphenol mixture. Conditions: Shimadzu Class-VP, Phenomenex LUNA C18 column (150×4.6 mm i.d.); mobile phase: water–DMF–methanol–acetic (73:25:1:1 v/v/v/v); flow rate: 1 ml/min; detection: UV at 280 nm.

methanol was small, ranging from 20:1:20 to 25:1:25, 50:1:50, or even 1:0:1 v/v/v, all catechins were well resolved, whereas the retention time of GCG and ECG became too long. As the relative volume of methanol was increased, the retention time of all catechins was reduced, but the early eluting EGCG peak was contaminated with polar impurities. However, these problems were solved by a stepwise elution with three different compositions of the mobile phase, by increasing the relative volume of methanol as shown in Fig. 8.21. In this separation the column was first eluted with the first mobile phase (25:1:25 v/v/v) until all polar impurities were eluted. Then the mobile phase was switched to the second mobile phase (10:1:10) at P1 (Fig. 8.21) to elute the EGCG peak. Finally the column was eluted with the third mobile phase (5:1:5) at P2 to elute the remaining two catechins.

In the hexane–ethyl acetate–water system, the polarity of the solvent system is conveniently adjusted by changing the volume ratio between hexane and ethyl acetate. For example, all catechins including EGCG can be well resolved with polar phase compositions of 1:10:10 (Fig. 8.22), 1:20:20, or 1:30:30 v/v/v.

As shown in Figs. 8.21 and 8.22, three catechins can be successfully separated with two different solvent systems by analytical HSCCC. However, if these solvent systems were applied directly to preparative

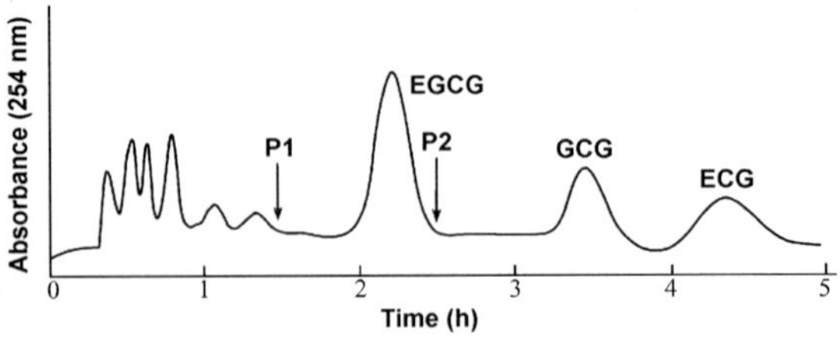

Fig. 8.21. Analytical HSCCC separation of EGCG, GCG and ECG from tea polyphenol mixture. Apparatus: Model GS20 analytical CCC machine; column: 30 ml; rotation speed: 1800 rpm; solvent system: ethyl acetate–ethanol–water (25:1:25 v/v/v), next at P1: 10:1:10 v/v/v, then at P2 5:1:5 v/v/v; mobile phase: lower aqueous phase; flow rate: 1 ml/min head to tail; detection: UV at 254 nm; sample size: 16 mg; retention of stationary phase: 51%.

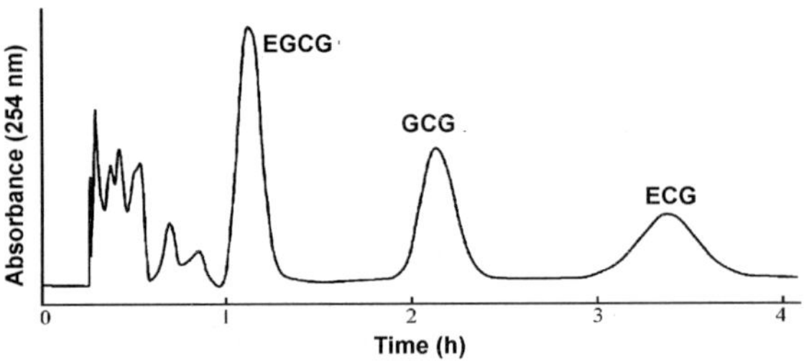

Fig. 8.22. Analytical HSCCC separation of EGCG, GCG, and ECG from tea polyphenol mixture. Conditions: solvent system: hexane–ethyl acetate–water (1:10:10 v/v/v); sample size: 16 mg; retention of stationary phase: 62%. Other experimental conditions are listed in the caption to Fig. 8.21.

HSCCC, the elution time for GCG and ECG would become excessively long. Therefore, the following two-step separation was performed. In the first step, EGCG was purified from the mixture using the first solvent system composed of ethyl acetate–methanol–water by stepwise elution from 25:1:25 to 10:1:10 v/v/v as shown in Fig. 8.23. Then, the residues in the column were pushed out, concentrated and subjected to the second step of separation using the second solvent system composed of hexane–ethyl acetate–water at 1:4:5 v/v/v, which resulted in elution of pure fractions of GCG and ECG (Fig. 8.24). This preparative HSCCC was applied to 1 g of a polyphenol mixture yielding 275 mg of EGCG, 140 mg of GCG and 130 mg of ECG.

Each fraction was analyzed by HPLC with a PDA detector at 280 nm as shown in Fig. 8.25, where the first peak in each chromatogram was a solvent peak. From the relative peak area, the purity of each fraction

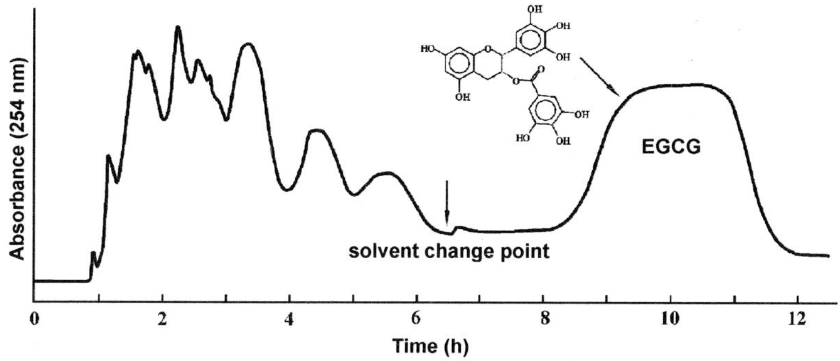

Fig. 8.23. Preparative HSCCC separation of EGCG from the tea polyphenol mixture. Conditions: apparatus: Model GS10A2 preparative MLCPC; column volume: 260 ml; rotation speed: 800 rpm; solvent system: ethyl acetate–ethanol–water (25:1:25 v/v/v) to 10:1:10 (arrow); mobile phase: lower aqueous phase; flow rate: 2 ml/min head to tail; detection: UV@254 nm; sample size: 1 g; retention of stationary phase: 70%.

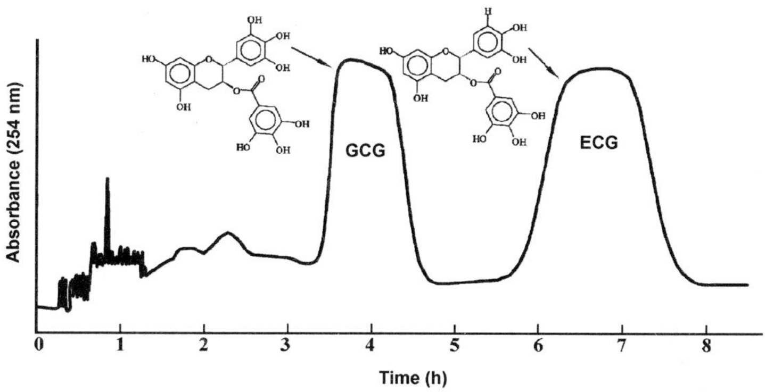

Fig. 8.24. Preparative HSCCC separation of GCG and ECG from the residual in the column of the first separation (Fig. 8.23.). Solvent system: hexane–ethyl acetate–water (1:4:5 v/v/v); retention of stationary phase: 83%. Other conditions, see caption to Fig. 8.23.

was computed as about 98%. The chromatograms showed no measurable impurity peaks at wavelengths between 200 to 500 nm.

The structure of each catechin, obtained by preparative HSCCC, was identified by FAB-MS, ^{1}H NMR and ^{13}C NMR. The fragments [MH^+] in the MS spectrum indicate that two of the compounds have the identical molecular mass of 458, which corresponds to the EGCG and GCG diastereoisomers, whereas the other gallate has a molecular mass of 442

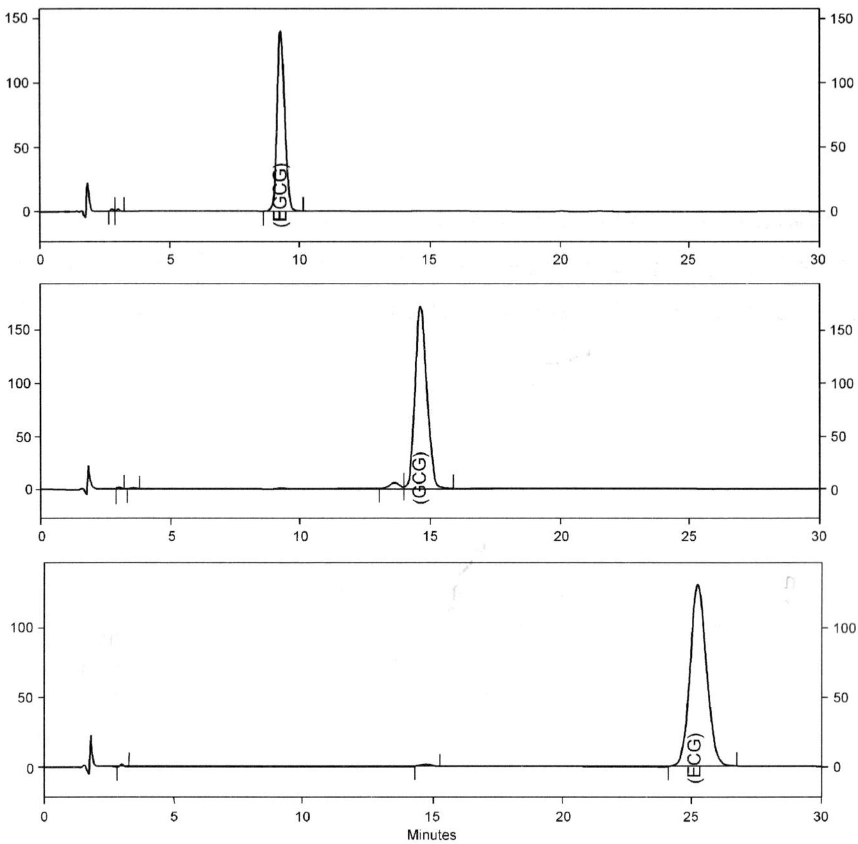

Fig. 8.25. HPLC chromatogram of EGCG, GCG and ECG purified by preparative HSCCC. Conditions: Shimadzu Class-VP, Phenomenex LUNA C18 column (150×4.6 mm i.d.); mobile phase: water-DMF-methanol-acetic (73:25:1:1 v/v/v/v); flow rate: 1 ml/min; detection: UV at 280 nm.

which corresponds to ECG. The chemical shifts and coupling constants obtained in ^{1}H and ^{13}C NMR spectra confirmed the assignments of each peak fraction against the reference data published elsewhere.

The overall results of the present studies demonstrate that HSCCC is a powerful tool for preparative isolation of individual catechins with high purity from a polyphenol mixture. The sample-loading capacity in this preparative separation is greater than that of the standard CCC separation.

8.4.4 Separation of notopterol and isoimperatorin from *Notopterygium forbessi* Boiss [12]

Notopterygium forbessi Boiss (*Qianghuo* in Chinese) is a traditional Chinese medicinal herb that has been used as a diaphoretic, an antipyretic and an anodyne. The major active constituents of this herb are coumarins, including notopterol and isoimperatorin, whose chemical structures are given in Fig. 8.26. Notopterol and isoimperatorin are used as standards in the quality control of Qianghuo products. The present work describes HSCCC separation of notopterol, isoimperatorin and two unknown compounds from a crude ethyl acetate extract of *N. forbessi* Boiss. Analytical HSCCC is first used for selecting a pair of two-phase solvent systems, and their combined use in stepwise elution led to the successful preparative separation of notopterol and isoimperatorin.

The high-performance liquid chromatography (HPLC) equipment used was a Shimadzu LC-10AVP system including two LC-10ATVP solvent delivery units, a SPD-M10AVP UV–VIS photodiode array detector, a Model 7726 injection valve with a 1 μl loop, an SCL-10AVP system controller, a CTO-10ASVP column oven, a DGU-12A degasser, and a Class-VP-LC workstation (Shimadzu, Kyoto, Japan).

All organic solvents used for HSCCC were of analytical grade and purchased from Beijing Chemical Factory, Beijing, China. Methanol and acetonitrile used for HPLC analysis were of chromatographic grade and purchased from Tianjin Huaxi Special Reagent Factory, Tianjin, China. The crude ethyl acetate extract of *N. forbessi* was provided by Professor Youfu Sun of the Institute of Chinese Materia Medica, China Academy of Traditional Chinese Medicine.

Three slightly different two-phase solvent systems were prepared:

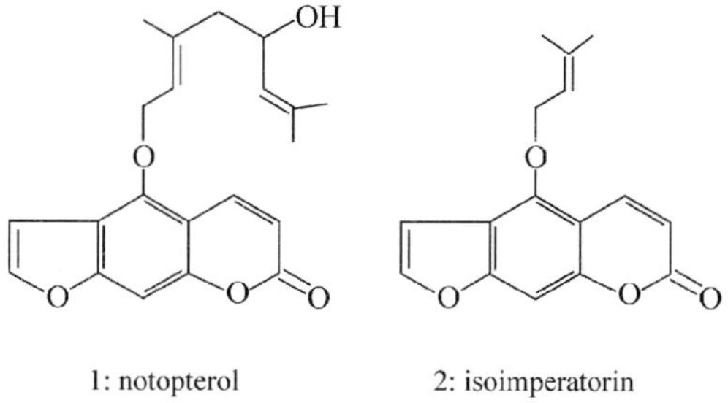

Fig. 8.26. Chemical structures of coumarins from *N. forbessi* Boiss.

– Solvent System 1: light petroleum ether (b.p. 30–60°C)–EtOAc–MeOH–water (5:5:5:5, v/v/v/v)
– Solvent System 2: light petroleum ether (b.p. 30–60°C)–EtOAc–MeOH–water (5:5:4.8:5, v/v/v/v)
– Solvent System 3: light petroleum ether (b.p. 30–60°C)–EtOAc–MeOH–water (5:5:5:4, v/v/v/v).

Each solvent system was thoroughly equilibrated in a separatory funnel at room temperature and the two phases were separated shortly before use.

The sample solutions were prepared by dissolving the crude extract in the organic upper phase at suitable concentrations according to the analytical or preparative purpose.

Analytical HSCCC was performed using solvent systems 1–3 and the Model GS20 30 ml machine as follows: the multilayer-coiled column was first entirely filled with the upper organic stationary phase. Then, the lower aqueous mobile phase was pumped into the head end of the column at a flow rate of 1.0 ml/min, while the apparatus was rotated at 1500 rpm. After hydrodynamic equilibrium was reached, as indicated by a clear mobile phase eluting at the tail outlet, the sample solution (10 mg dissolved in 1 ml of the upper organic phase) was injected through the sample port. The effluent from the tail end of the column was continuously monitored with a UV detector at 254 nm. The retention of the stationary phase relative to the total column capacity was computed from the volume of the stationary phase collected from the column after the separation was completed.

Preparative HSCCC was performed using solvent systems 2 and 3 and the model GS10A2 260 ml machine as follows: the multilayer-coiled column was first entirely filled with the upper organic phase of Solvent System 2 as a stationary phase. Then, the lower aqueous phase was pumped into the head end of the column at a flow rate of 2.0 ml/min, while the apparatus was rotated at 800 rpm. After hydrodynamic equilibrium was reached, as indicated by a clear mobile phase eluting at the tail outlet, the sample solution was injected through the injection valve. The effluent from the outlet of the column was continuously monitored with a UV detector at 254 nm. Peak fractions were manually collected into test tubes according to the chromatogram. After five hours, the mobile phase was changed to the lower aqueous phase of Solvent System 3, and elution, detection and fraction collection were resumed.

The crude ethyl acetate extract of *N. forbessi* Boiss and each peak fraction obtained by HSCCC were analyzed by HPLC. The analyses were

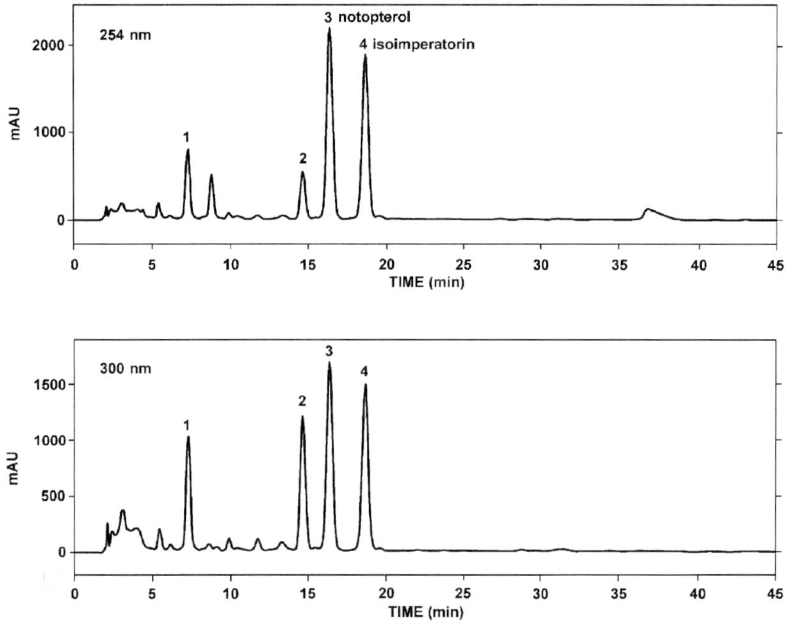

Fig. 8.27. The result of HPLC analyses of the ethyl acetate extract from *N. forbessi* Boiss. Column Shim-pack CLC-ODS column (150×4.6 mm i.d.); temperature: 40°C; mobile phase: methanol–acetonitrile–water (30:30:40 v/v/v); flow rate: 1.0 ml/min. 1: Unknown 1; 2: Unknown 2; 3: notopterol; 4: isoimperatorin.

performed with a Shim-pack CLC-ODS column (150×4.6 mm i.d.) at a column temperature of 40°C. The mobile phase composed of MeOH–MeCN–water (30:30:40, v/v/v) was eluted at a flow rate of 1.0 ml/min and the effluent monitored by the PDA detector.

Identification of HSCCC peak fractions was carried out by MS (Finnigan MAT711), ^{1}H NMR and ^{13}C NMR spectra (Varian).

Results

The crude ethyl acetate extract of *N. forbessi* Boiss was first analyzed by HPLC. As indicated in Fig. 8.27, it contained four major compounds, including notopterol and isoimperatorin. In order to achieve an efficient resolution of the target compounds, three different solvent systems were examined using the 30 ml analytical HSCCC machine. The results indicated that the Solvent System 1 could only partially separate notopterol. Solvent System 2 could resolve notopterol and Unknown 1 well,

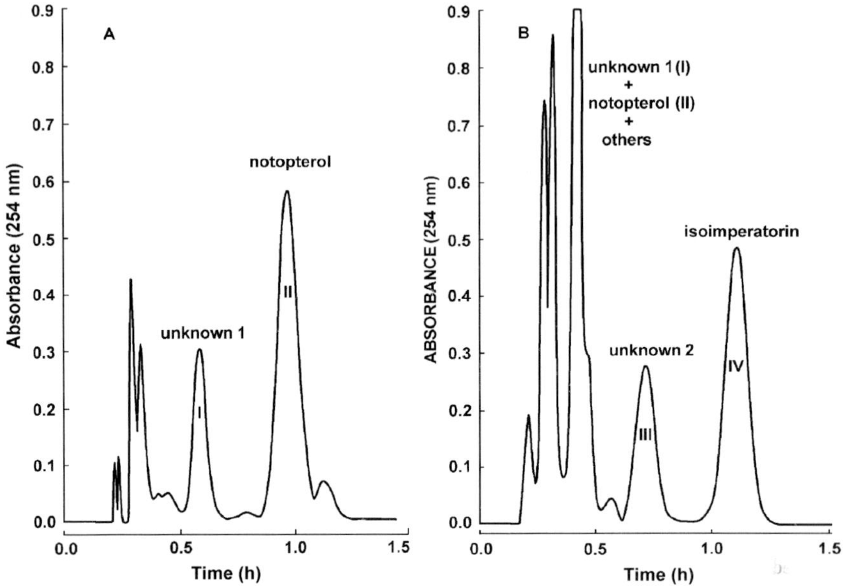

Fig. 8.28. Chromatograms of crude ethyl acetate extract from *N. forbessi* Boiss by the analytical GS20 30 ml CCC machine. Solvent systems: (A) light petroleum ether–ethyl acetate–methanol–water, 5:5:4.8:5 v/v/v/v; (B) light petroleum ether–ethyl acetate–methanol–water, 5:5:5:4 v/v/v/v; stationary phase: upper organic phase; mobile phase: lower aqueous phase; flow rate: 1.0 ml/min head to tail; revolution speed: 1500 rpm; sample size: 6–10 mg dissolved in 1 ml organic phase; retention of stationary phase: ~55%.

while iso-imperatorin and Unknown 2 were retained in the column (Fig. 8.28A). Solvent System 3, on the other hand, could separate iso-imperatorin and Unknown 2 well, but failed to resolve notopterol and Unknown 1 (Fig. 8.28B). Consequently, the combined use of solvent systems 2 and 3 in stepwise elution resulted in a successful separation of notopterol, isoimperatorin, and unknown compounds in a single run. Figure 8.29 shows the result obtained from 200 mg of the crude ethyl acetate extract of *N. forbessi* Boiss by preparative HSCCC. After five hours of elution using Solvent System 2. The mobile phase was switched to the lower phase of Solvent System 3 to elute the retained two peaks. This separation yielded four pure peak fractions, each at over 98% purity as determined by HPLC. The results of HPLC analysis and the UV spectra characteristic of each fraction were given in Fig. 8.30 A–D.

Here, it is worthwhile discussing the advantage of the stepwise elution applied in the present studies. After some desired peaks eluted,

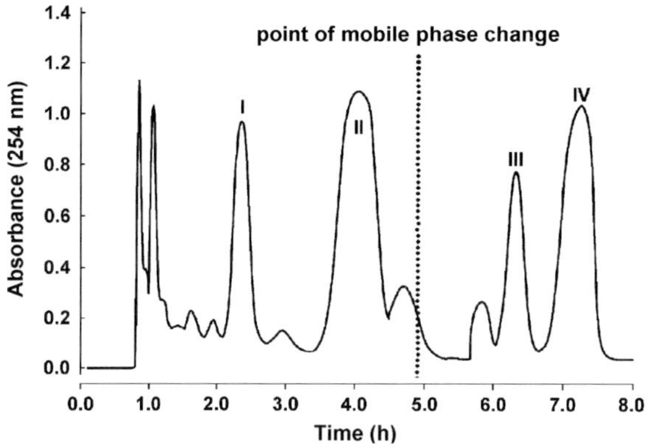

Fig. 8.29. Chromatograms of crude ethyl acetate extract of *N. forbessi* Boiss with the GS10A2 260 ml preparative CCC machine using stepwise elution with solvent systems 2 and 3, i.e., light petroleum ether–ethyl acetate–methanol–water (5:5:4.8:5 v/v/v/v) and (5:5:5:4 v/v/v/v), respectively; stationary phase: upper organic phase of System 2; mobile phase: 590 ml of lower phase of solvent System 2 and 360 ml of lower phase of Solvent System 3; flow rate: 2.0 ml/min head to tail; revolution speed: 800 rpm; sample size: 200 mg dissolved in 5 ml organic phase of System 2; retention of the stationary phase: 57%. I: Unknown 1; II: Notopterol; III: Unknown 2; IV: Isoimperatorin.

other peaks with much greater partition coefficients will be retained in the column for a long period of time. In this case, one can change the chromatographic conditions to facilitate elution of the retained peaks without loss of peak resolution. This can be done in the following two different ways: the first method is a dual elution method in which the column is eluted in the opposite direction, with the phase other than that used originally as the stationary phase. Although this method is simple, the target peaks still present near the original inlet of the column owing to its large partition coefficient (K_D) may be immediately eluted without further separation from the surrounding impurities. The second method is the stepwise elution applied to the present method. This method requires a careful choice of the second mobile phase in such a way that the volume of the stationary phase is not altered significantly after the two phases re-establish the hydrodynamic equilibrium in the column. However, the method allows a choice of the second mobile phase with suitable K_D values for the target compounds so that the peak resolution is further increased.

The structural identification of HSCCC peak fractions was carried out by MS, ^{1}H-NMR and ^{13}C-NMR spectra. The results demonstrate that

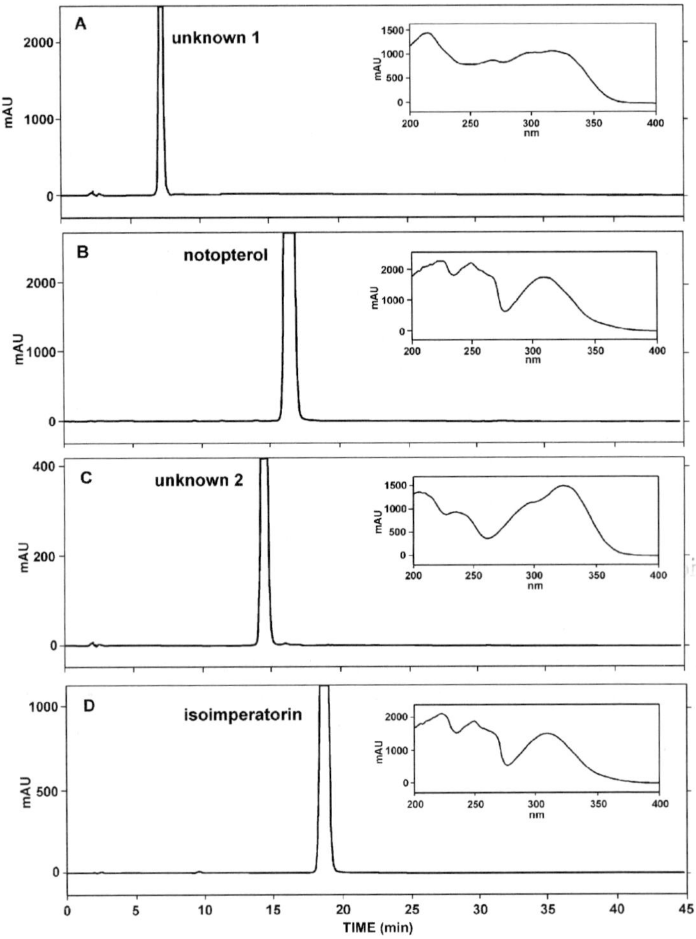

Fig. 8.30. The result of HPLC analyses and UV spectrum of the purified CCC peak fractions. (A) Peak I, (B) Peak II, (C) Peak III, and (D) Peak IV of the preparative HSCCC separation shown in Fig. 8.29. Column: Shim-pack CLC-ODS column (150×4.6 mm i.d.); temperature: 40°C; mobile phase: methanol–acetonitrile–water (30:30:40 v/v/v); flow rate: 1.0 ml/min.

a combined use of two solvent systems in stepwise elution by HSCCC provides highly efficient preparative separation of notopterol and isoimperatorin from a crude extract of *N. forbessi* Boiss.

8.4.5 Separation of tanshinones from *Salvia miltiorrhiza* Bunge

Salvia miltiorrhiza Bunge (*Dan-Shen* in Chinese) is a well-known traditional Chinese medicinal herb used for the treatment of various kinds of diseases, particularly for coronary disorders. The major active constituents of this herb are tanshinones, including tanshinone I, tanshinone IIA, and cryptotanshinone. Pharmacological tests revealed that all these components have an anti-cancer effect, and among these tanshinones, tanshinone IIA is most effective and has been used as a quality controller for some medicines. The chemical structures of these tanshinones are given in Fig. 8.31.

The present study describes the separation and purification of tanshinones from a crude extract of *Salvia miltiorrhiza* Bunge by analytical and semi-preparative HSCCC [13]. All organic solvents used for HSCCC and HPLC were of analytical grade and purchased from Beijing Chemical Factory, Beijing, China. All standard samples were purchased from the National Institute for the Control of Pharmaceutical and Biological Products, Ministry of Health, Beijing, China.

Sample preparation

About 1.0 kg of roots of *Salvia miltiorrhiza* Bunge was extracted three times in a glass bottle (2.5 l capacity) with 1.5 l of methanol (a total volume of 4.5 l) at room temperature. The extracts were combined,

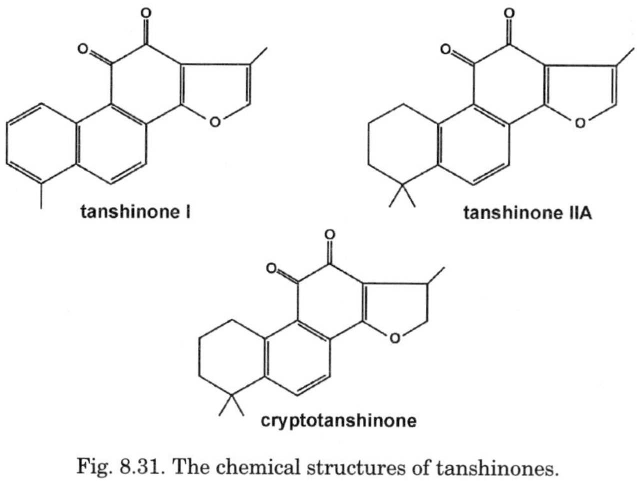

Fig. 8.31. The chemical structures of tanshinones.

evaporated to dryness under reduced pressure, and then redissolved in 500 ml of water. The aqueous solution was again extracted with light petroleum ether (500 ml six times), and these extracts were combined and evaporated to dryness to yield 201 g of the crude sample. Then, 5 mg of the extract was dissolved in 1 ml of each phase of the solvent system used for analytical HSCCC separation. The sample solutions were sonicated for several minutes before loading into the column.

The *n*-hexane–EtOH–water two-phase solvent systems were prepared with increasing amounts of ethanol in the lower aqueous phase:

- Solvent system A: 4:1.8:2, v/v/v
- Solvent system B: 4:2.3:2, v/v/v
- Solvent system C: 4:3:2, v/v/v

The upper phase is almost pure hexane. Each solvent mixture was thoroughly equilibrated in a separatory funnel at room temperature and the two phases were separated shortly before use.

Because of its speed of separation and minimal solvent consumption, analytical HSCCC is an ideal tool for optimizing the key parameters for preparative HSCCC separation. In each analytical separation with the 30 ml Model GS20 apparatus, the coiled column was first entirely filled with the hexane stationary phase, and the aqueous mobile phase was pumped into the column at a flow rate of 1.0 ml/min, while the apparatus was rotated at 1700 rpm. After the mobile phase emerged from the outlet of the column and hydrodynamic equilibrium was established, about 1 ml of the sample solution containing 5 mg of the light petroleum extract was injected through the injection valve. The effluent from the column was continuously monitored with a UV detector at 254 nm. Peak fractions were collected manually according to the chromatogram.

Preparative separation was performed with the 260 ml Model GS10A2 machine using stepwise elution with solvent systems A, B and C in sequence, as follows: the coiled column was first entirely filled with the upper hexane phase, and the lower aqueous phase of solvent system A was pumped into the column at a flow rate of 2.0 ml/min, while the apparatus was rotated at 800 rpm. After the mobile phase emerged from the tail outlet and hydrodynamic equilibrium was established in the column, 10 ml of the sample solution containing 50 mg of the light petroleum extract was injected through the injection valve. The effluent of the column was monitored continuously as above. Peak fractions were collected manually according to the chromatogram. After 2 h of elution with the lower phase of solvent system A, the mobile phase was switched to the lower phase of system B to elute the column for 50 min, followed by elution with the lower phase of system C to complete the separation.

HPLC analyses were performed with a Shimadzu LC-10AVP system including a LC-10AT liquid chromatograph, an SPD-M10AVP PDA detector, an SCL-OAVP system controller and a CTP-10SVP column oven using an Inertsil ODS-3 column (150 mm, 4.6 mm i.d.) at 30°C. The mobile phase composed of methanol–water (80:20, v/v) was eluted at a flow rate of 1.0 ml/min and the effluent monitored by the PDA detector. Identification of HSCCC peak fractions were carried out by comparing retention times and UV spectra with the standard samples.

Results

The HPLC analyses revealed that tanshinones were present mainly in the light petroleum extract from the dried roots of *Salvia miltiorrhiza* Bunge. Consequently, the light petroleum extract was separated by analytical HSCCC to select suitable solvent systems and the other experimental conditions. The solvent systems A–C were evaluated in terms of peak resolution of various tanshinones. The results indicated that cryptotanshinone and tanshinone I were well resolved by solvent system A (Fig. 8.32a), but the profile of unknown components was irregular, and tanshinone IIA was retained in the column. When the relative volume of ethanol in the solvent system was increased (solvent systems B and C), the retention time of the unknown component decreased and tanshinone IIA was eluted from the column, however the resolution between cryptotanshinone and tanshinone I became worse (Fig. 8.32 c).

In this situation, we selected a stepwise elution as follows, to improve the resolution of target compounds. The crude sample was first separated by solvent system A for 40 min, then the mobile phase was changed to the lower phase of system B, and 20 min later the mobile phase was again changed to the lower phase of system C to complete the separation. This stepwise elution produced resolution of all four target peaks (I, II, III and IV) as shown in Fig. 8.32 d.

Figure 8.33 shows the preparative HSCCC separation of 50 mg of the light petroleum extract using the optimized stepwise elution determined by analytical HSCCC. The crude sample was first separated by solvent system A for 2 h, and the mobile phase was changed to the lower phase of system B to elute for 50 min, followed by elution with the lower phase of system C to complete the separation. Although the sample size was increased ten times compared to the analytical HSCCC run, an equivalent separation was achieved. Each peak fraction of this preparative

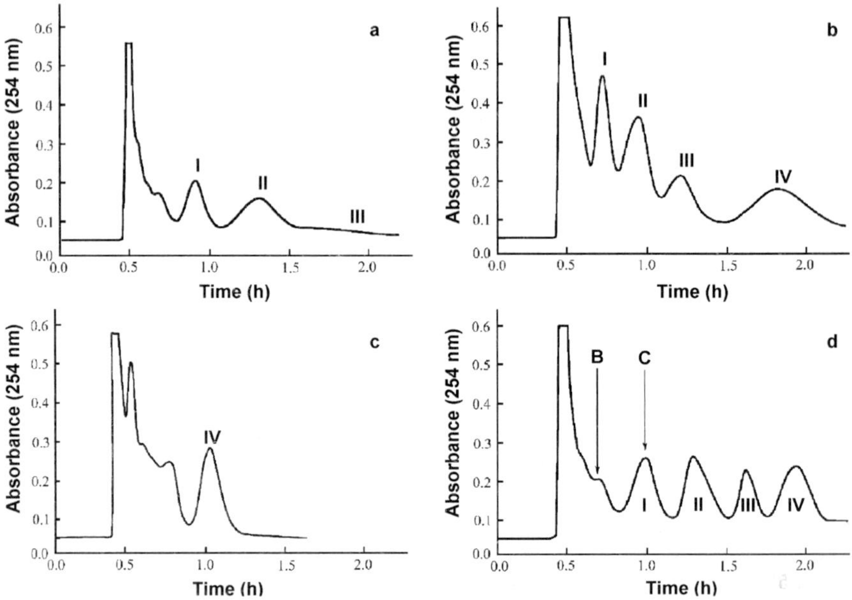

Fig. 8.32. Analytical HSCCC separation of the light petroleum extract of *Salvia miltiorrhiza* Bunge. (a) Solvent system A (*n*-hexane–ethanol–water, 4/1.8/2 v/v/v); (b) solvent system B (*n*-hexane–ethanol–water, 4/2.3/2 v/v/v); (c) solvent system C (*n*-hexane–ethanol–water, 4/3/2 v/v/v); (d) stepwise elution with the solvent systems A, B and C in sequence, i.e. the stationary phase was the upper hexane phase of solvent system A, and the mobile phase was the sequence: 60 ml of the lower phase of solvent system A, 20 ml of the lower phase of solvent system B, and 90 ml of the lower phase of solvent system C. Other experimental conditions: GS20 30 ml machine, revolution speed: 1700 rpm; flow rate: 1 ml/min (head to tail); sample size: 5 mg; retention of the stationary phase: 51%; detection: UV at 254 nm. Peaks: I = cryptotanshinone; II = tanshinone I; III = unknown component; IV = tanshinone II A.

HSCCC was analyzed by HPLC (Fig. 8.34), which indicates that the purity of tanshinone IIA corresponding to Peak IIA was over 97% and other tanshinones also gave high purities of over 95%. The yields of tanshinone IIA, tanshinone I, and cryptotanshinone were 7 mg (14%), 4 mg (8%) and 3 mg (6%, all w/w), respectively.

Once again, these results indicate that CCC is a powerful tool for both analytical and preparative separations of tanshinones from a crude extract of *Salvia miltiorrhiza* Bunge. Using a stepwise elution with three solvent systems, preparative HSCCC can yield several milligrams of four compounds at high purity in a single run.

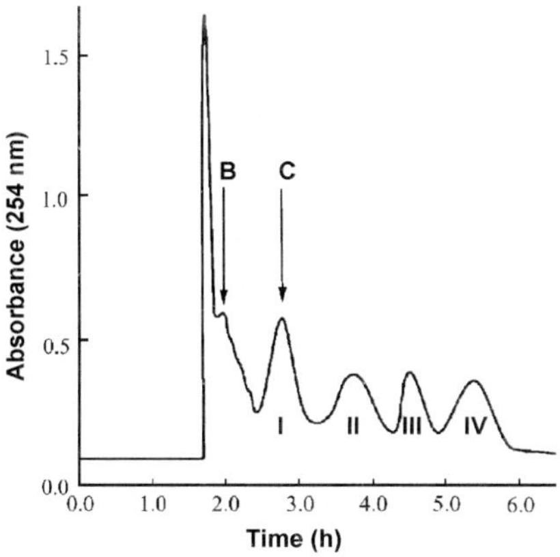

Fig. 8.33. Preparative HSCCC separation of the light petroleum extract of *Salvia miltiorrhiza* Bunge using stepwise elution with solvent systems A, B and C. Experimental conditions: GS10A2 260 ml machine; revolution speed: 800 rpm; stationary phase: the upper hexane phase of solvent systems A; mobile phase: 250 ml of the lower phase of system A, 100 ml of the lower phase of system B and 450 ml of the lower phase of system C eluted in this sequence; flow rate: 2 ml/min; sample size: 50 mg; retention of the stationary phase: 55%; detection: 254 nm. Peaks: I = cryptotanshinone; II = tanshinone I; III = unknown component; IV = tanshinone II A.

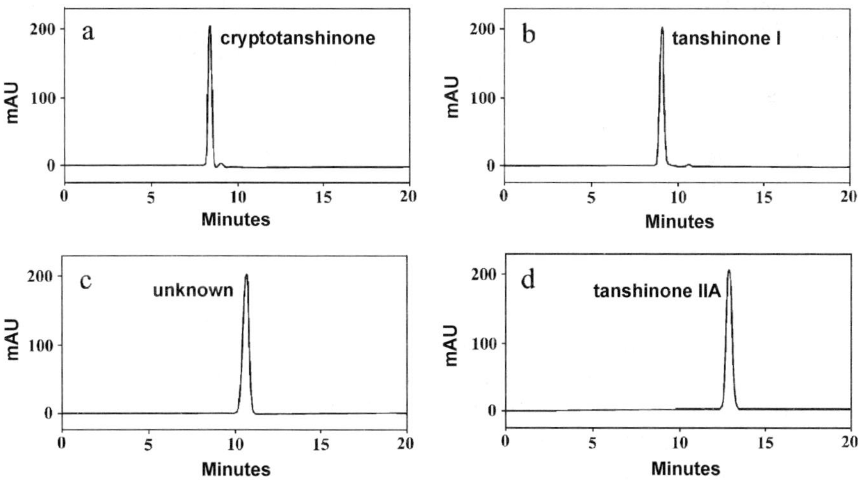

Fig. 8.34. HPLC analyses of the HSCCC fractions of tanshinones from *Salvia miltiorrhiza* Bunge. (a) Peak I; (b) Peak II; (c) Peak III; (d) Peak IV of the preparative HSCCC separation shown in Fig. 8.33. Column: Intersil ODS-3 (4.6×150 mm i.d.); column temperature: 30°C; mobile phase: methanol–water (80:20 v/v); flow rate: 1.0 ml/min; detection: UV at 254 nm.

8.5 APPLICATION OF TWO-DIMENSIONAL HSCCC SYSTEMS

8.5.1 Separation of flavone aglycones from *Hippophae rhamnoids* L. [14]

Multidimensional chromatography is a frontier of contemporary chromatographic science. Its applications are abundant and varied, involving all branches of chromatography. Multidimensional chromatography styles developed so far include gas chromatography (GC)–GC, liquid chromatography (LC)–GC, LC–LC, supercritical fluid chromatography (SFC)–SFC, SFC–GC, LC–SFC, LC-capillary zone electrophoresis (CZE), field flow fractionation (FFF)–FFF and all the couplings with thin layer chromatography (TLC). According to the theory of Poisson statistics, overlapping of the components on a complex chromatogram is more serious than was thought. Consequently, the best strategy to improve the chromatographic resolution is to increase the peak capacity. This is readily achieved by a multidimensional separation.

CCC is a special liquid–liquid partition chromatography with no solid support matrix, which reduces tailing of solute peaks and sample loss owing to irreversible adsorption. Consequently the method is ideal for the preparative separation of natural products such as flavonoids. In the past HSCCC was applied to the analytical and preparative separations of isorhamnetin, kaempferol and quercetin, which are important standard samples used for controlling the quality of the Ginkgo extracts and its preparations [4].

This study describes the application of the multidimensional countercurrent chromatography (MDCCC) system shown in Fig. 8.1 to the preparative separation of isorhamnetin, kaempferol and quercetin from a crude sample from *Hippophae rhamnoides* L.

All organic solvents were of analytical grade and purchased from Beijing Factory (Beijing, China). A two-phase solvent system composed of chloroform–methanol–water at a volume ratio of 4:3:2 was used. The solvent mixture was thoroughly equilibrated in a separatory funnel at room temperature and the two phases separated shortly before use.

A crude mixture of the flavone aglycones from *Hippophae rhamnoides L.* was provided by the National Engineering Research Center for Traditional Chinese Medicine (Shanghai, China). The sample solution was prepared by dissolving the crude sample in an equal amount of the two solvent phases used for the separation.

Figure 8.1 shows our MDCCC system made of two GS20 30 ml

machines, two pumps and two detectors interconnected by a six-port valve. The switching valve is initially set in position A, and the two HSCCC systems (1 and 2) are simultaneously filled with the upper aqueous stationary phase using Pumps 1 and 2, respectively. Then, both apparatuses were set in rotation at 800 rpm, while the lower organic phase is eluted through HSCCC Systems 1 and 2 using their respective pumps at a flow rate of 2.0 ml/min. After the mobile phase front emerges and the two phases have established hydrodynamic equilibrium in each column, the sample solution is injected into HSCCC 1 through the injection valve while Pump 2 is stopped. The effluent from the outlet of HSCCC 1 is continuously monitored with UV Detector 1 at 254 nm, and collected with Fraction Collector 1 according to the chromatogram traced by Recorder 1. When the target peak appears, the effluent from HSCCC 1 is cut and introduced into HSCCC 2 by turning the switching valve to position B. After the target peak has eluted from HSCCC 1, the switching valve is returned to position A and the target peak fraction introduced in HSCCC 2 is eluted by restarting Pump 2 together with Detector 2, Recorder 2, and Fraction Collector 2.

The analysis of each major HSCCC peak fraction containing the target analytes was performed with a Microsorb-MV ODS column (150×4.6 mm i.d., 5 μm, Rainin). The mobile phase composed of methanol–buffer solution (0.04% H_3PO_4) (50:50, v/v) was eluted at a flow rate of 1.0 ml/min and the effluent monitored at 254 nm.

Figure 8.35 shows the separation of 200 mg of the crude flavone aglycones from *Hippophae rhamnoides* L. by the MDCCC system. Figure 8.35A shows the chromatogram obtained from HSCCC 1 and Recorder 1. The first peak in Fig. 8.35A contained some unknown lipophilic impurities. The second large peak between the broken lines, contained a large amount of isorhamnetin, a small amount of kaempferol, and unknown impurities. It was cut and introduced into HSCCC 2. The chromatogram in Fig. 8.35B was obtained by HSCCC 2 from the cut fraction eluted from HSCCC 1. It shows that isorhamnetin was well separated from kaempferol and lipophilic impurities.

During the separation of the crude flavone aglycones from *Hippophae rhamnoides* L. by HSCCC 1, the initial loss from the stationary phase was about 65 ml. After injection, carry-over of the stationary phase began and continued until the lipophilic impurities had eluted from the column. This additional loss of the stationary phase was 35–45 ml, which may be the main cause of the inefficient peak resolution obtained from HSCCC 1. If HSCCC Systems 1 and 2 were simply

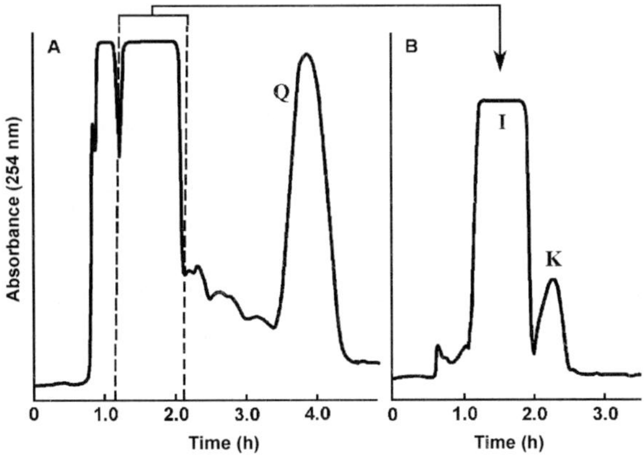

Fig. 8.35. Multidimensional countercurrent chromatograms of the crude flavone aglycones from *Hippophae rhamnoides* L. I, isorhamnetin; K, kaempferol; Q, quercetin. Solvent system, chloroform–methanol–water (4:3:2 v/v/v); mobile phase, lower organic phase; flow rate, 2.0 ml/min; sample size: 200 mg. (A) Chromatogram obtained by HSCCC 1 (B) chromatogram of the cut fraction (between the broken lines) obtained by HSCCC2. See Fig. 8.1.

connected in series, the loss of the stationary phase and its adverse influence on the separation would spread to HSCCC 2. Furthermore, a large amount of lipophilic impurities would adversely affect the separation of the target compounds in HSCCC 2. This may also be the case in the separation of the crude Ginkgo flavone aglycones by MDCCC.

By using the MDCCC system, the serious loss of the stationary phase caused by lipophilic impurities was limited to HSCCC 1. The peak resolution of the MDCCC system may be improved by the use of the two-phase solvent system with two different volume ratios, one for HSCCC 1 and the other for HSCCC 2.

8.5.2 Separation of flavone aglycones from the leaves of *Ginkgo biloba* L

Using the two-dimensional HSCCC system, isorhamnetin, kaempferol and quercetin can be separated from the extract of *Ginkgo biloba* L. The separation method was the same as in Section 8.5.1.

Figure 8.36 shows the separation of 200 mg of the crude Ginkgo flavone aglycones obtained by MDCCC. Figure 8.36A shows a chromatogram obtained from HSCCC 1. The first large peak contained several unknown impurities. The major component of the second peak is

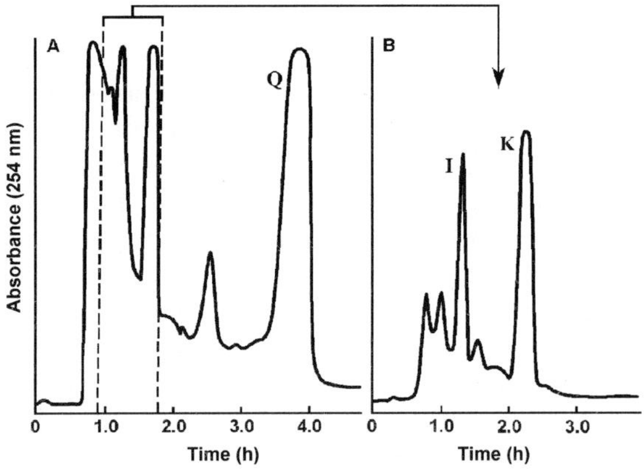

Fig. 8.36. Multidimensional countercurrent chromatograms of the crude Ginkgo flavone aglycones. I, isorhamnetin; K, kaempferol; Q, quercetin. Solvent system: chloroform–methanol–water (4:3:2 v/v/v); mobile phase, lower organic phase; flow rate, 2.0 ml/min head to tail; sample size: 200 mg. (A) Chromatogram obtained by HSCCC 1; (B) chromatogram of the cut fraction (between the broken lines) obtained by HSCCC 2. See Fig. 8.1.

isorhamnetin, and that of the third peak is kaempferol. The second and the third peaks between the broken lines were cut and introduced together into HSCCC 2 for further separation. Figure 8.36B is a chromatogram of the cut fraction by HSCCC 2, showing a good separation between isorhamnetin and kaempferol.

8.6 APPLICATIONS OF THE PH-ZONE-REFINING CCC SYSTEM

8.6.1 Preparative separation of alkaloids from the root of *Sophora flavescens* Ait [15,16]

Sophora flavescens Ait, a typical traditional Chinese medicinal herb, is used for the treatment of various diseases in China as an antipyretic, diuretic, anthelmintic, and antidote to poisons. The roots of *S. flavescens* Ait contain more than ten kinds of alkaloids, among which sophocarpine and matrine are the main effective constituents used for cancer treatment. pH-Zone-refining CCC is applied to the separation of these alkaloids from *S. flavescens* Ait.

Experimental part

Raw roots of *S. flavescens* Ait (ca. 1.5 kg) were extracted three times each with 2 l of 95% ethanol. Then, the extracts were combined and evaporated to dryness under reduced pressure. The residue obtained from the combined extracts was dissolved with 500 ml of 2% HCl. After filtration, the acidic aqueous solution was extracted three times, each with 500 ml portions of chloroform. The pH of the aqueous solution was adjusted to 9.6 with NaOH solution and the total alkaloids were extracted three times again, each with 500 ml of chloroform. The chloroform extracts were combined and evaporated to dryness. Portions of this crude alkaloid extract of *S. flavescens* Ait were subjected to pH-zone-refining CCC.

The solvent pairs used for CCC separations were prepared as follows. About equal volumes of methyl *tert*-butyl ether and distilled water were thoroughly equilibrated in a separatory funnel at room temperature and the two phases were separated. Then triethylamine (retainer) was added to the upper organic phase at a concentration of 10 m*M*, while the lower aqueous phase was acidified with hydrochloric acid (eluter) at 5 or 10 m*M*. The sample solutions were prepared by dissolving the crude alkaloid extract of *S. flavescens* Ait in 20–30 ml of a phase mixture consisting of equal volumes of each phase. The sample solutions were sonicated for several minutes before injection into the column.

For each separation, the 260-ml GS10A2 machine was first entirely filled with the basic upper organic stationary phase. Then the sample solution was injected through the sample port and the acidic aqueous mobile phase was pumped in the head-to-tail direction through the column at a flow rate of 1.5 ml/min while the column was rotated at 800 rpm. The effluent from the outlet of the column was continuously monitored with a UV detector at 254 nm and a pH monitor, and then collected in test tubes at 2 min intervals (3.0 ml/tube) with a fraction collector. After the desired peaks were eluted, the rotation and elution were stopped and the column contents were collected into a graduated cylinder by connecting the inlet of the column to a nitrogen line pressurized at approximately 80 p.s.i. (6 bar or 0.6 MPa). The retention of the stationary phase relative to the total column capacity was computed from the volume of the stationary phase collected from the column.

All collected fractions were analyzed by using silica gel G thin-layer chromatography (TLC) plates (100×50 cm, 0.20–0.25 mm thick) (Qingdao Haiyang Chemical Factory, Qingdao, China) by developing with a solvent mixture composed of C_6H_6–EtOAc–CH_3COCH_3–25% aqueous NH_3 (2:3:4:0.2, v/v/v/v) and staining with a Dragendorff reagent to detect the alkaloids.

Results

Figure 8.37A shows a typical pH-zone-refining countercurrent chromatogram obtained for the separation of 1.0 g of crude alkaloid extract of *S. flavescens* Ait. Two pure alkaloids were obtained in this

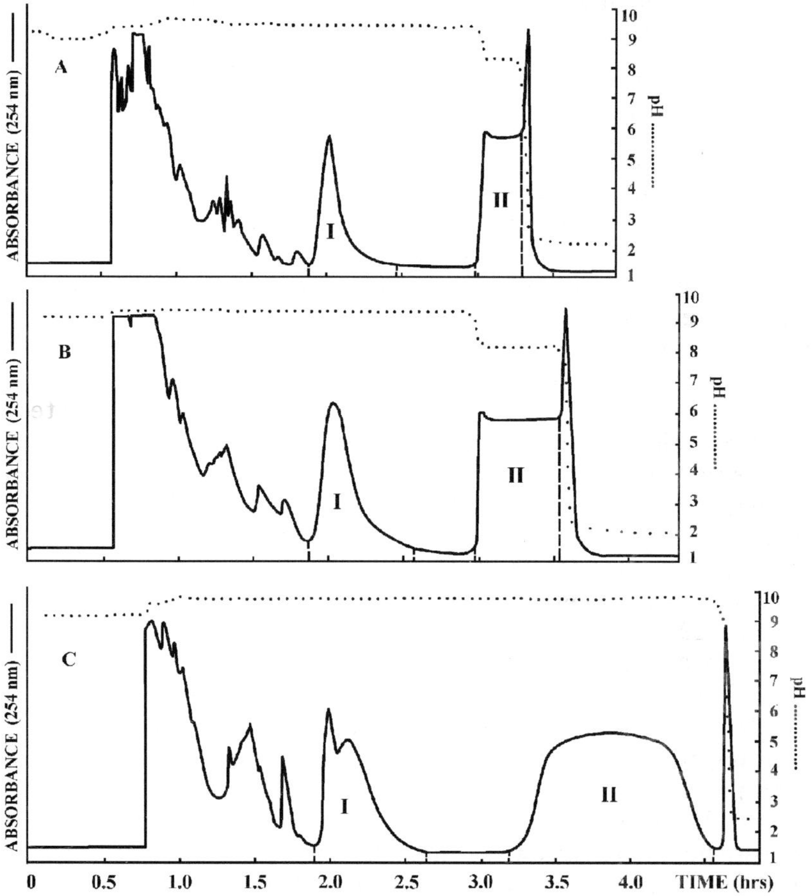

Fig. 8.37. Chromatograms of the crude alkaloid extract from the root of *Sophora flavescens* Ait obtained by pH-zone-refining CCC. Experimental conditions: GS10A2 260 ml machine, solvent system: methyl *tert*-butyl ether–water (1:1); stationary phase: upper basic phase (10 m*M* triethylamine); mobile phase: lower acid phase (10 m*M* HCl for A and B, and 5 m*M* for C); flow rate: 1.5 ml/min (head to tail); sample size: A = 1.0 g in 20 ml; B = 2.0 g in 30 ml and C = 1.0 g in 20 ml of an equal amount of each phase; revolution speed: 800 rpm; retention of stationary phase: 66%. Peak I = matrine; Peak II = sophocarpine.

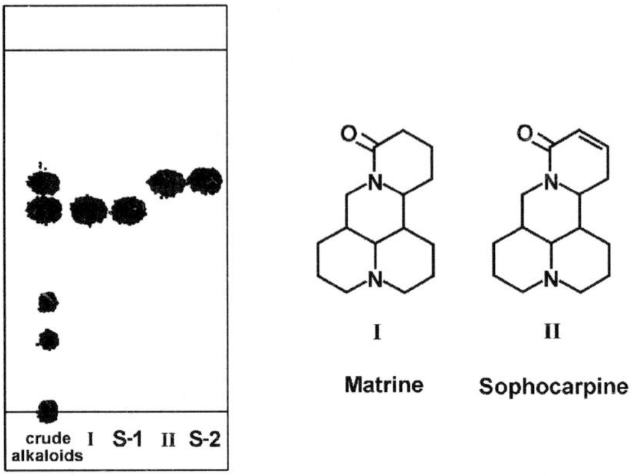

Fig. 8.38. The results of TLC analyses. Left: the crude alkaloid extract of *S. flavescens* Ait; I: Peak I; S-1: matrine standard; II: Peak II; S-2: sophocarpine standard. The analyses were made by a silica gel G TLC plate developed with a solvent system composed of C_6H_6–EtOAc–CH_3COCH_3–25% aqueous NH_3 (2:3:4:0.2 v/v/v/v) and stained with a Dragendoff reagent to detect the alkaloids.

separation: matrine (I in Fig. 8.38) from Peak I and sophocarpine (II in Fig. 8.38) from Peak II. Peak I showed a normal Gaussian shape and no pH change was observed when it eluted. In contrast, a sharp drop in pH coincided with Peak II eluting with a rectangular shape. A flat pH zone followed the elution of Peak II. As shown in the chromatogram, Peak II is associated with a sharp impurity peak (1–2 ml) at its rear boundary where the pH drops abruptly from 8.3 to 2.2.

Figure 8.37C shows a chromatogram obtained for the separation of 1.0 g of the same extract using a mobile phase containing a lower concentration of HCl (5 m*M*). The reduction of the eluter-acid concentration in the mobile phase produced a remarkable change in the elution profile of Peak II that became broader, losing its rectangular shape, and being totally separated from the sharp impurity peak. However, pure alkaloids were still obtained from each of these two peaks: matrine (600 mg) and sophocarpine (170 mg), both more than 98% pure.

Figure 8.38 shows the result of the TLC analysis of various samples including the crude alkaloid extract of *S. flavescens* Ait, combined fractions of Peak I, matrine standard (S-1), combined fractions of Peak II, and sophocarpine standard (S-2), using silica gel-G TLC developed with a solvent mixture composed of C_6H_6–EtOAc–CH_3COCH_3–25%

aqueous NH_3 (2:3:4:0.2, v/v/v/v) and stained with a Dragendorff reagent to detect the alkaloids. The result indicates that Peaks I and II each contained one alkaloid compound (R_F values: 0.63 for sophocarpine and 0.55 for matrine). Although Peak I contained minor brown impurities (as it was only partially resolved from the main peak in Fig. 8.21C), they were easily removed by recrystallization using a mixture of chloroform–methanol (10:1, v/v).

These results demonstrated that pH-zone-refining CCC produced efficient separations of two main alkaloids from gram quantities of crude alkaloid extracts from *S. flavescens* Ait. The present method may be applied to purification of various other alkaloids from natural products.

8.6.2 Preparative separation of lappaconitine by pH-zone-refining CCC

Lappaconitine is the main diterpene alkaloid in the root of *Aconitum sinomontanum* Nakai. Lappaconitine hydrobromide is a new drug used to alleviate pains. The molecule was found not to be addictive. The drug is used for the treatment of obstinate pains, especially cancer pains, and the effective success rate reaches about 90%. It is also effective for chronic pains such as gastric and intestinal ulcer, gastritis, hepatitis, cholecystitis, pulmonary tuberculosis, rheumatic arthritis, sciatica, and others. Lappaconitine hydrobromide can be used to treat addictions to drugs or narcotics (including opium, morphine, heroine, cocaine, marijuana, amphetamine or barbiturates).

Experimental part

The crude alkaloid sample was provided by Lanzhou Pharmaceutical Factory, Lanzhou, China. In separations, two solvent systems were selected:

- Solvent System I: methyl *tert*-butyl ether–water. Stationary phase: the organic phase (with 10 m*M* triethylamine as retainer): mobile phase, the acidic aqueous phase (10 m*M* HCl as eluter).
- Solvent System II: chloroform–water. Stationary phase: the acidic aqueous phase (10 m*M* HCl as retainer); mobile phase, the basic organic phase (10 m*M* triethylamine as eluter).

Results

Figure 8.39 shows the chromatogram of the separation of Lappaconitine (III) from 2.0 g of the crude extract of *A. sinomontanum* Nakai. TLC

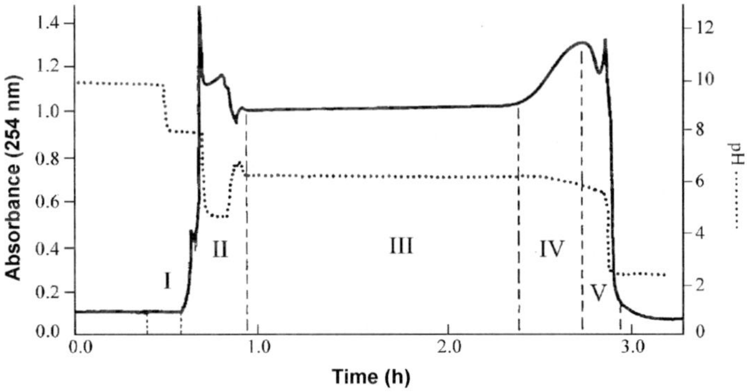

Fig. 8.39. Chromatogram of the preparative separation of Lappaconitine from the crude extract of *A. sinomontanum* Nakai by pH-zone-refining CCC. Solvent system: methyl *tert*-butyl ether–water (1:1 v/v); stationary phase: upper basic organic phase (10 m*M*, triethylamine); mobile phase: lower acid phase (10 m*M* HCl); flow rate: 2.0 ml/min; sample size: 2.0 g.

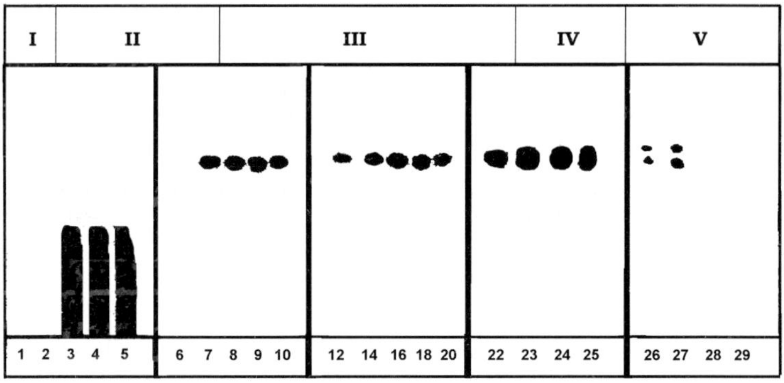

Fig. 8.40. The result of TLC analyses of all collected fractions. Silica gel G TLC plates (100×50 mm, 0.20–0.25 mm thickness); developing reagent: a solvent mixture composed of C_6H_6–$EtOA_C$–CH_3COCH_3–25% aqueous NH_3 (10:20:15:1 v/v/v/v); staining with a Dragendorff reagent to detect the alkaloids.

identifications of these fractions are shown in Fig. 8.40. Solvent System I was used in this separation.

Using Solvent System II, 2.0 g (A) and 5.0 g (B) of crude extract were completely separated; the chromatograms are shown in Fig. 8.41. The chemical structural formula of Lappaconitine was identified as shown in Fig. 8.42.

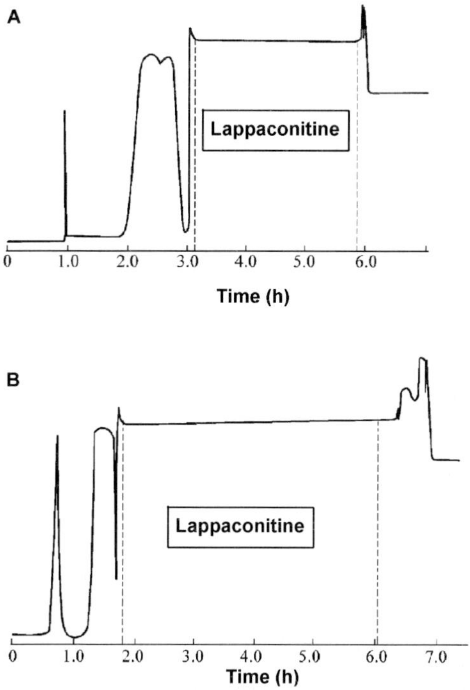

Fig. 8.41. Chromatograms of the preparative separation of Lappaconitine from crude extract of *A. sinomontanum* Nakai by pH-zone-refining CCC. Solvent system: chloroform–water (1:1 v/v); stationary phase: upper aqueous phase (10 m*M*, triethylamine); flow rate: 2.0 ml/min; sample size: 2.0 g (A) and 5.0 g (B).

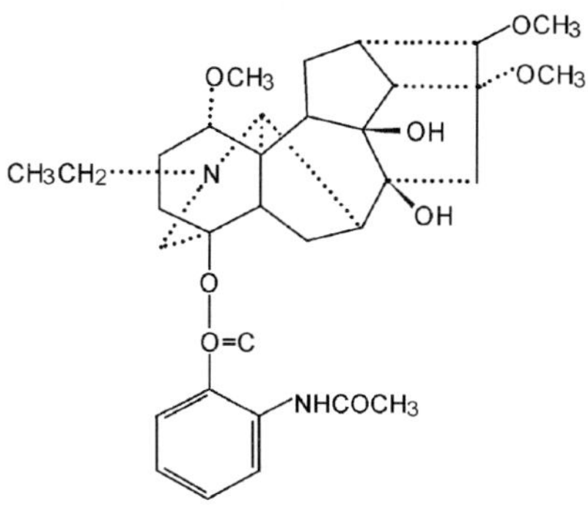

Fig. 8.42. Chemical structural formula of lappaconitine, $C_{32}H_{44}N_2O_8$, m.w. 584.7.

8.7 APPLICATIONS FOR A CROSS-AXIS CCC MACHINE [17–20]

8.7.1 Partition of protein in aqueous two-phase systems

The reagents used in our applications were as follows. Polyethylene glycol 8000, dibasic potassium phosphate, potassium dihydrogen phosphate, were of analytical grade and purchased from Beijing Chemical Factory (Beijing, China). Acetonitrile was of HPLC grade: lysozyme, pepsin, albumin, haemoglobin were purchased from Sigma.

The aqueous polymer-phase systems were prepared by dissolving 125 g of PEG 8000 and 250 g dibasic potassium phosphate in 1625 ml distilled water adjusted to the desired pH, or dissolving 125 g of PEG 8000 and 300 g potassium dihydrogen phosphate. The mixture was thoroughly equilibrated in a separatory funnel at room temperature, and allowed to separate completely into the two layers before use. Sample solutions were prepared by dissolving each standard protein in 1 ml of the lower phosphate phase of the two-phase solvent system used for separation.

The partition coefficient of each protein was determined spectrophotometrically using a simple test-tube procedure. Two ml of each phase of the equilibrated two-phase solvent system were delivered into a test tube to which 1 mg of the sample was added. The contents were thoroughly mixed and allowed to settle at room temperature. After the two clear layers were formed, a 1-ml aliquot of each phase was diluted with 2 ml of distilled water and the absorbance was measured at 280 nm using a spectrophotometer (Model UV-2501PC, Shimadzu Corporation).

The partition coefficient was obtained by dividing the absorbance value of the upper phase by that of the lower phase. Table 8.1 shows partition coefficients of four proteins in PEG–phosphate salt solvent system.

TABLE 8.1

Partition coefficients of four proteins

Protein	Partition coefficient
Pepsin	1.849
Albumin (egg)	0.298
Lysozyme	0.530
Haemoglobin	0.980

Partition coefficients were calculated from the absorbance of the PEG upper phase divided by that of the phosphate lower phase.

8.7.2 Standard proteins separation

For CCC separation, the coil of the 150-ml XLLCCC machine was first completely filled with the stationary phase, which is the upper PEG phase. This was followed by injection of the sample solution through the sample port. Then the phosphate mobile phase was eluted through the coil at 1.0 ml/min while the apparatus was rotated at 500 rpm. The effluent from the outlet of the column was continuously monitored with an 8823A UV detector at 280 nm and recorded by a Yokogawa 3057 recorder, collected into test tubes (5.0 ml/tube) with a FC-95 collector.

Figure 8.43 shows a chromatogram of a mixture of pepsin, albumin (egg), lysozyme and haemoglobin. Pepsin, lysozyme and haemoglobin were separated on the baseline. Figure 8.44 shows a chromatogram of three standard proteins by different pH polymer-phase countercurrent chromatography. Figure 8.44C shows that the three proteins were baseline-separated at pH 4.9. The resolution factors were 3.1 (pepsin–albumin) and 3.5 (albumin–lysozyme). Mobile phase pH values of 4.9, 6.0, 6.5, 7.0, and 9.0 of the phosphate buffer solution were tried with the PEG8000 stationary phase. The effect of pH on retention is shown in Fig. 8.45. There is a sharp change in the retention of lysozyme and a slight decrease in the retention of albumin with increases in pH values. However, the retention of pepsin is less affected. These phenomena are due to the different pH values of the proteins. The polarity on the surface of the proteins changes as the pH is adjusted so that they have a variety of chromatographic characteristics.

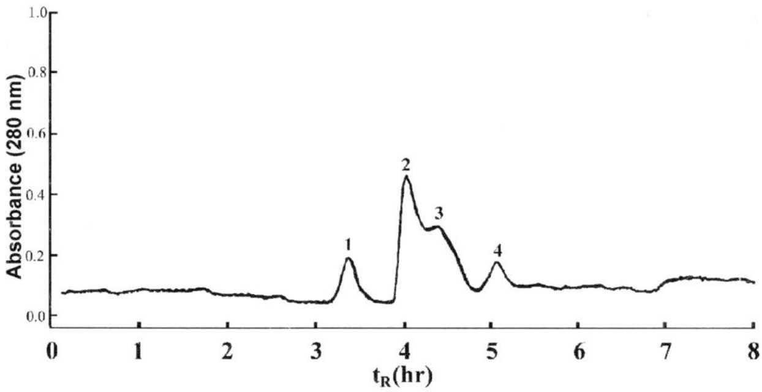

Fig. 8.43. Chromatogram of proteins obtained by the XLL CCC 150 ml machine. Experimental conditions: 10–15 mg each in 1 ml solvent; solvent system: 12.5% (w/w) PEG 8000–25% (w/w) K_2HPO_4 (pH = 9.0) in distilled water; mobile phase: lower phase; flow rate: 1 ml/min head to tail; revolution: 500 rpm. Sample 1: Pepsin; 2: albumin (egg); 3: lysozyme; 4: haemoglobin.

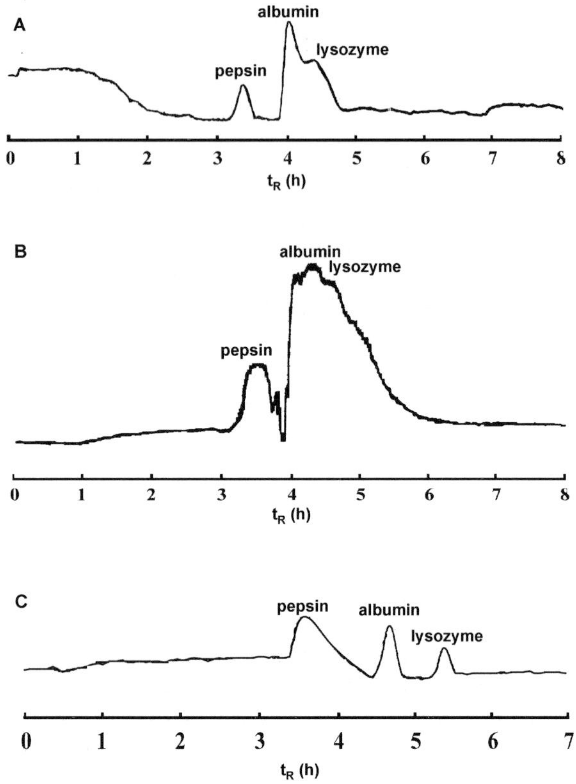

Fig. 8.44. Chromatogram of standard proteins by different pH polymer phase. Experimental conditions: XLLCCC 150 ml machine; 5–15 mg each protein in 0.5 ml solvent, flow rate: 1.0 ml/min, revolution: 500 rpm, mobile phase: phosphate lower phase. Solvent System A: 12.5% (w/w) PEG8000–25% (w/w) K_2HPO_4 (pH = 9.0) in distilled water. Solvent System B: 12.5% (w/w) PEG8000–25% (w/w) K_2HPO_4+KH_2PO_4 (pH = 6.5) in distilled water. Solvent System C: 12.5% (w/w) PEG8000–30% (w/w) KH_2PO_4 (pH = 4.9) in distilled water.

8.7.3 Preparation of glycoprotein from *Morchella esculenta* (L.)

Morchella esculenta (L.), morel, is an edible mushroom. A glycoprotein in *Morcella esculenta* (L.) has a strong anticancer activity. It can increase the leukocyte number as well as improving cancer immunity. So, it is very important to obtain the glycoprotein in a purified form.

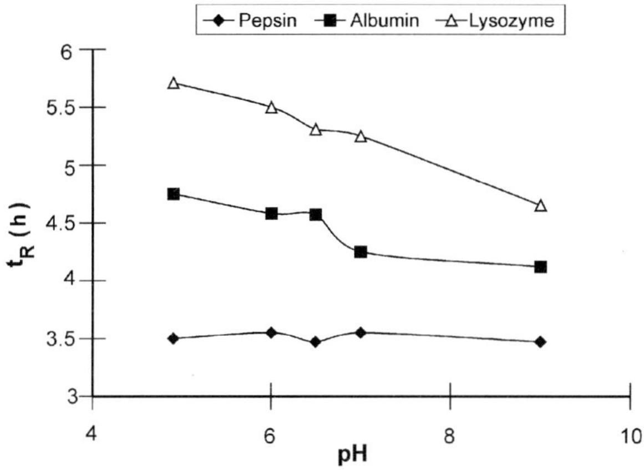

Fig. 8.45. Effect of pH on the retention of proteins by cross axis counter current chromatography. Experimental conditions: 5–15 mg of each protein in 0.5 ml solvent, flow rate: 1.0 ml/min, revolution: 500 rpm, mobile phase: lower phosphate phase. Solvent system: pH = 9: 12.5% (w/w) PEG8000–25%(w/w) K_2HPO_4 in distilled water, pH = 6, 6.5 and 7: 12.5% (w/w) PEG8000–25% (w/w) K_2HPO_4+KH_2PO_4 in distilled water; pH = 4.9: 12.5% (w/w) PEG800—30% (w/w) KH_2PO_4 in distilled water.

The nutrient liquid of *Morchella esculenta* (L.), obtained from liquid deep-layer fermentation, was treated by ultrafiltration to remove low-molecular-weight components, defatted, decolorized, and concentrated under reduced pressure and low temperature. It was further purified by ultracentrifugation. The crude glycoprotein was obtained by repeated precipitations and purifications with ethanol.

The purification of the glycoprotein fractions was done by utilizing the cross-axis countercurrent chromatograph. The separation was performed with a polymer phase composed of 12.5% (w/w) polyethylene glycol 8000 and 25% (w/w) dibasic potassium phosphate in distilled water, by eluting the lower phosphate phase (pH = 9) at a flow rate of 1.0 ml/min in the head to tail direction.

Figure 8.46 shows a chromatogram of *Morchella esculenta* (L.) glycoprotein. Three glycoprotein fractions (called Glycoprotein 1, Glycoprotein 2, and Glycoprotein 3) were obtained within 6 h and their resolutions between neighbour glycoproteins were 2.0 and 0.6 respectively. The yields were respectively 49.1 mg, 10.2 mg and 22.3 mg, obtained from 100 mg of the crude glycoprotein fraction.

All collected fractions were analyzed by using a home-made methyl linoleate silica-based polymer-bonded column (150×4.6 mm i.d.). A

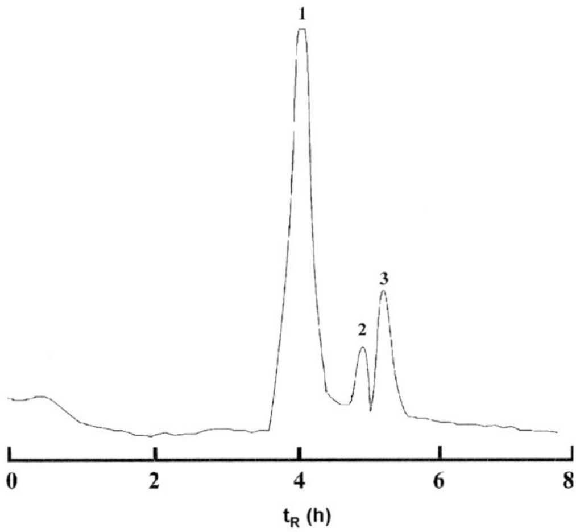

Fig. 8.46. Chromatogram of glycoprotein obtained by the XLLCCC 150 ml machine. Experimental conditions: 100 mg *Morchella esculenta* glycoprotein in 1 ml solvent. Solvent system: 12.5% (w/w) PEG 8000–25% (w/w) K_2HPO_4 (pH = 9.0) in distilled water; mobile phase: lower phase head to tail; flow rate: 1 ml/min; revolution: 500 rpm. Peak 1: Glycoprotein 1; Peak 2: Glycoprotein 2; Peak 3: Glycoprotein 3.

gradient elution mode was used with mobile phases A, 0.1% TFA in 100% deionized water and B, 0.1% TFA in 100% acetonitrile (20-min gradient from 95% A to 100% B). Figure 8.46 shows the HPLC chromatogram of the crude glycoprotein extract containing three different glycoproteins.

8.8 CONCLUSION

High-speed countercurrent chromatography has been recognized as an effective technique with the advantages of good purification, high recovery, large preparative scale, excellent reproducibility, and so on. It is very suitable for use in the fields of natural products, natural medicine, agricultural products and other bioorganic products. The applications described in this chapter are some typical examples.

The separation efficiency of HSCCC is not determined by a selection of supporting material or filled column as in the HPLC methods. Users can perform a successful separation by designing for themselves suitable two-phase solvent system according to the properties of a particular sample. For this reason, HSCCC can be used widely and easily for different applications. The selections of solvent systems introduced in this chapter should serve as possible reference systems for newcomers to CCC.

The different types of HSCCC instruments and ways of using them, e.g., analytical HSCCC, semi-preparative HSCCC, two-dimensions HSCCC, pH-zone-refining CCC, cross-axis CCC, gave us a series of effective tools for separations and purification. By choosing a simple type or a combination of these types of apparatus, the high-speed and high-efficiency purification and preparation of various kinds of mixtures can be achieved on different quantity scales. It is expected that a great contribution to separation science and technology will be realized by the spread of HSCCC, associated with other modern chromatographic techniques and detection modes.

REFERENCES

1 Y. Ito, *CRC Crit. Rev. Anal. Chem.*, 17 (1986) 65.
2 T.Y. Zhang, X. Hua, R. Xiao and S. Kong, *J. Liq. Chromatogr.*, 11(1) (1988) 233.
3 T.Y. Zhang, R. Xiao, Z.Y. Xiao, L.K. Pennell and Y. Ito, *J. Chromatogr.*, 445 (1988) 199.
4 F.Q. Yang, T.Y. Zhang, B.X. Mo, L.J. Yang, Y.Q. Gao and Y. Ito, *J. Liq. Chromatogr. Relat. Technol.*, 21 (1&2) (1998) 209.
5 T.Y. Zhang, In: Y. Ito and W.D. Conway, *High-Speed Countercurrent Chromatography*. Wiley-Interscience, New York, 1996, Ch. 8, pp. 225–263.
6 X.L. Cao, Y. Tian, T.Y. Zhang and Y. Ito, *J. Chromatogr. A*, 813 (1998) 397.
7 X.L. Cao, Y. Tian, T.Y. Zhang, P.N. Shen, L.H. Zhu and Y. Ito, *J. Liq. Chromatogr. Relat. Technol.*, 21 (18) (1998) 2897.
8 G.L. Tian, T.Y. Zhang, F.Q. Yang and Y. Ito, *J. Chromatogr. A*, 886 (2000) 309.
9 F.Q. Yang, T.Y. Zhang, R. Zhang and Y. Ito, *J. Chromatogr. A*, 289 (1998) 137.
10 X.L. Cao, Y. Tian, T.Y. Zhang and Y. Ito, *J. Chromatogr. A*, 855 (1999) 709.
11 A. Berthod, K. Talabardon, F. De La Poype and A.J. Erdelmeir, *J. Liq. Chromatogr. Relat. Technol.*, 21 (1998) 3003.
12 F.Q. Yang, T.Y. Zhang, Q.H. Lui, G.Q. Xu, Y.B. Zhang, S. Zhangand Y. Ito, *J. Chromatogr. A*, 883 (2000) 67.
13 G.L. Tian, Y.B. Zhang, T.Y. Zhang, F.Q. Yang and Y. Ito, *J. Chromatogr. A*, 904 (2000) 107.
14 F.Q. Yang, J. Quan, T.Y. Zhang and Y. Ito, *J. Chromatogr. A*, 803 (1998) 298.

15 Y. Ito, In: Y. Ito and W.D. Conway (Eds.), *High-Speed Countercurrent Chromatography*. Wiley-Interscience, New York, 1996, Ch. 6, pp. 121–178.

16 F.Q. Yang, J. Quan, T.Y. Zhang and Y. Ito, *J. Chromatogr. A*, 822 (1998) 316.

17 K. Shinomiya, T. Komatsu, T. Murata, Y. Kabssawa and Y. Ito, *J. Liq. Chrom. Relat. Technol.*, 23 (9) (2000) 1403.

18 Y. Shibusawa, Y. Ito, *J. Liq. Chromatogr.,* 15 (1992) 2787.

19 Y. Shibusawa, Y. Ito, *J. Chromatogr.*, 550 (1991) 695.

20 Y. Wei, T.Y. Zhang and Y. Ito, *J. Chromatogr. A*, 917 (2001) 347.

21 R. Margraff, In: A.P. Foucault (Ed.), *Centrifugal Partition Chromatography*, Chromatographic Science Series, Vol. 68. Marcel Dekker, New York, 1995, Ch. 12, pp. 335–350.

22 T. Okuda, T. Hatano and N. Ogawa, *Chem. Pharm. Bull.*, 32 (1984) 4664.

Chapter 9

Separation of inorganic compounds: theory, methodology and applications

Yu-Ren Jin, Li-Xing Zhang, Lian-Zhong Zhang and Shi-Jun Han

9.1 INTRODUCTION

Countercurrent chromatography (CCC) is a liquid–liquid partition chromatography in which centrifugal or gravitational force is employed to maintain one liquid as the stationary phase in a coil or train of chambers, while the second immiscible liquid (the mobile phase) is passed through the stationary phase [1]. Hitherto, CCC has been used successfully for the separation of organic and bio-organic substances, while its applications to inorganic separation are more limited. The major applications are, for example, analytical and radiochemical separations, preconcentration, and purification of inorganic substances in solutions on a laboratory scale [2].

It has been noted that the merits of CCC, mainly derived from its absence of a solid support, are valuable for the separation of both organic and inorganic compounds. However, the chromatographic behaviour of inorganic compounds is rather different from that of organic substances, and several reports [3–7] expose these differences.

Two types of CCC techniques, namely centrifugal partition chromatography and the coil planet centrifuge, are mainly used in inorganic separation. Their constructions and phase-retaining mechanisms are markedly different. However, they both use the same principles of separation. We will not distinguish between them in this chapter while the fundamentals of inorganic separation are discussed; the abbrevi-

Comprehensive Analytical Chemistry XXXVIII
A. Berthod (Ed.)

ation CPC refers to centrifugal partition chromatography. This chapter mainly deals with the theoretical investigations, methodology and applications of CCC in inorganic separation.

9.2 THEORY

9.2.1 Definitions and expressions

The theory of CCC as a chromatographic technique has been established on the basis of the classical expression used in extraction chromatography [1,8]. The volume of the column (V_c) is the sum of the volume of the stationary phase (V_s) and the volume occupied by the mobile phase (V_m).

$$V_c = V_s + V_m \tag{9.1}$$

The retention ratio of the stationary phase, *Sf*, is equal to V_s divided by V_c.

$$Sf = \frac{V_s}{V_c} \tag{9.2}$$

The retention volume of a solute is based on the distribution coefficient (K_D) in the two phases. K_D is also called the partition coefficient. It is generally defined as the ratio of the concentration of a solute in the stationary phase (c_s) to the concentration in the mobile phase (c_m):

$$K_D = \frac{c_s}{c_m} \tag{9.3}$$

The retention factor, k, which is very important in any chromatographic process, is defined as the ratio of the quantity of the solute in the stationary phase, Q_s, to the quantity in the mobile phase, Q_m.

$$k = \frac{Q_s}{Q_m} \tag{9.4}$$

The relationship between k and K_D is:

$$k = K_D \frac{V_s}{V_m} \tag{9.5}$$

The retention time of a solute is expressed by Eq. (9.6):

$$t_R = t_m(1 + k) \tag{9.6}$$

where t_m is the retention time in the mobile phase. The retention value of a solute is frequently expressed by its retention volume, V_R, which can be obtained by:

$$V_R = V_m(1 + k) = V_m + K_D V_s \tag{9.7}$$

The separation or selectivity factor, α, of two solutes is defined as the ratio of their distribution coefficients:

$$\alpha = \frac{K_{D1}}{K_{D2}} \tag{9.8}$$

The efficiency of the CCC process is measured by the number of theoretical plates, N, which is calculated according to the peak retention volume (or retention time) and peak-width at the base ($W = 4\sigma$, in which σ is the standard deviation of the peak position):

$$N = 16\left(\frac{V_R}{W}\right)^2 \tag{9.9}$$

The resolution, *Rs,* is also quoted from other chromatography techniques to characterize the separation result of two adjacent solutes:

$$Rs = 2\left(\frac{V_{R,2} - V_{R,1}}{W_2 + W_1}\right) \tag{9.10}$$

9.2.2 Retention of inorganic compounds

Since the absence of a solid support precludes adsorption, and solute retention depends only on its distribution coefficient, CCC can be a valuable alternative to batch equilibrium or RP-HPLC for measuring distribution coefficients [9]. Equation (9.7) shows that the V_R volume of a solute can be predicted from its K_D value provided V_m and V_s are known. CCC is more convenient and reliable than batch methods for measurement of K_D values [10,11]. The separation factor can be estimated by Eq. (9.8). CCC has been developed into a highly efficient chromatographic method for organic separation. The number of theoretical plates can reach the low thousands. This is quite encouraging for scientists engaged in inorganic separation. They now have a new technique, which avoids the problems caused by the solid support, and gives them an alternative way to solve many daunting problems, for example the mutual separation of trivalent rare earth ions. If Eq. (9.7) is also valid for inorganic separations, the retention volume of the inorganic compounds will be predictable, and we will be able to judge which extraction system is most effective by using the large number of available solvent extraction data.

Unfortunately, many separations of inorganic compounds have shown that V_R volumes deviated from the values predicted by Eq. (9.7) [5,12–15]. At the same time, the distribution coefficients obtained from the CCC profile, K_D^{dyn}, were also different from those obtained from batch equilibrium, K_D^{bat}. These results indicate that Eq. (9.7) is not always right for inorganic separation. We cannot impute the discrepancies between the K_D^{dyn} and K_D^{bat} coefficients to experimental errors. They should be attributed to the existence of complex reactions in the two phases [6,7,13,14].

Unlike non-ionizable organic substances that normally exist in the same chemical forms in the two liquid phases, inorganic compounds exist in different chemical forms during the CCC process. The distribution species in the organic phase are normally in the form of neutral extraction complexes, while they exist in the form of hydrated ions in the aqueous phase. The injected inorganic compound is carried by the mobile phase, e.g., an aqueous phase, into the column or a train of cartridges where the extraction complex is extracted into the stationary phase, i.e., the organic phase. When the mobile phase is depleted in inorganic compounds, further flow of the mobile phase will cause the back-extraction of the inorganic compounds. It needs several steps of chemical reactions to form the extraction complex from ions, and *vice versa*. The retention behaviour is determined not only by the distribution equilibrium but also by the mechanism of the related chemical reactions. For example, Fe(III) was eluted in two peaks when the two-phase liquid system, *n*-decane containing the complexing agent

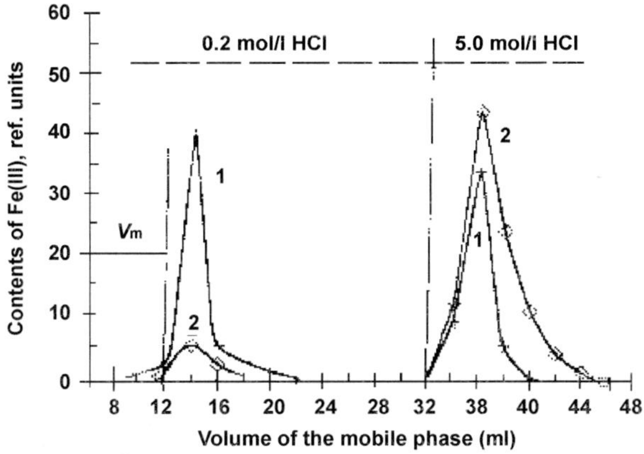

Fig. 9.1. Chromatographic behaviour of Fe(III) in DEHPA/n-decane/HCl systems. $Sf = 0.5$. Stationary phases: (1) 0.05 M DEHPA; (2) 0.5 M DEHPA. Step elution, F = 1 ml/min (adapted from Ref. [5]).

diethylhexyl phosphoric acid (DEHPA), and water containing HCl, was used in a CCC process, as shown in Fig. 9.1. A similar chromatographic behaviour (two peaks) was observed with the Ta(V) and Nb(V) ions and the 0.5 M DEHPA in n-decane/2.0 M HNO_3 system [5].

In order to elucidate the influence of chemical reactions on the retention behaviour, we choose a neutral complexing agent (E) as the extracting agent to illustrate the mass-transfer process in CCC of a metal ion (M^{n+}) [6,7]. The mobile phase is the aqueous phase and the stationary phase is the organic phase. The overall reaction in the CCC separation process can be expressed as:

$$M^{n+}_{(a)} + nL^{-}_{(a)} + bE_{(o)} \underset{k_r}{\overset{k_f}{\rightleftharpoons}} ML_nE_{b(o)} \tag{9.11}$$

where k_f and k_r are the forward and back-extraction reaction rate constants, and subscripts (a) and (o) denote the aqueous and organic phase, respectively. The M^{n+} complexes with E that diffuses from the organic phase and reacts with the inorganic anion (L^-) from the aqueous phase and form the extraction complex (ML_nE_b) in the interface. ML_nE_b is distributed in the two phases according to the partition law. The process described above is reversible. The metal ions leave the column, eventually, via a series of extraction and back-extraction processes between the two liquid phases.

Assuming that the cross-section of the CCC column is q, the linear

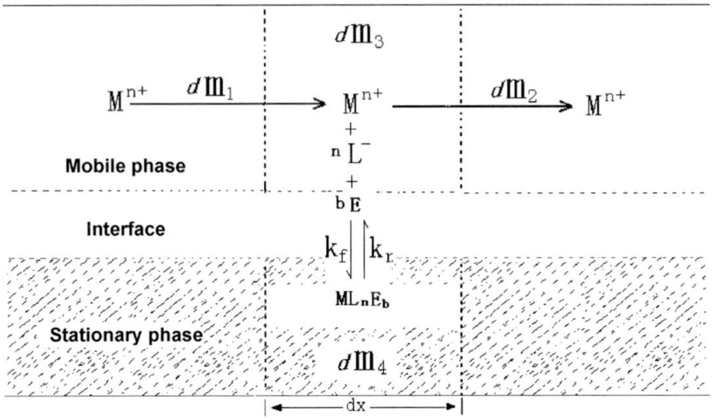

Fig. 9.2. The model for the mass transfer of inorganic ions during the CCC process.

flow rate of the mobile phase is μ: considering the dx unit of the column in Fig. 9.2, the concentration of metal ion in the mobile phase is c, and the concentration of extraction complex in the stationary phase is a. With the flow of mobile phase, the quantity of M^{n+} entering into a dx unit is dm_1, and the quantity leaving the dx unit is dm_2, in the time interval dt. The amount of M^{n+} ions in the dx unit of mobile phase is dm_3, and the variation in stationary phase is dm_4.

Then

$$dm_1 = c\mu(1-Sf)q\,dt \tag{9.12}$$

$$dm_2 = \mu(1-Sf)q\left[c+\left(\frac{\partial c}{\partial x}\right)_t dx\right]dt \tag{9.13}$$

$$dm_3 = (1-Sf)q\left(\frac{\partial c}{\partial t}\right)dt\,dx \tag{9.14}$$

$$dm_4 = Sf\,q\left(\frac{\partial a}{\partial t}\right)dt\,dx \tag{9.15}$$

According to the equilibrium of materials,

$$dm_1 - dm_2 = dm_3 + dm_4 \tag{9.16}$$

Substituting Eqs. (9.12), (9.13), (9.14), and (9.15) into Eq. (9.16), we obtain:

$$(1-Sf)\,\mu\left(\frac{\partial c}{\partial x}\right)+(1-Sf)\left(\frac{\partial c}{\partial t}\right)+Sf\left(\frac{\partial a}{\partial t}\right)=0 \tag{9.17}$$

Assuming that the reaction of the metal ion is of pseudo-first order, the variation of the concentration in the organic phase could be expressed as:

$$\frac{\partial a}{\partial t}=k_{\mathrm{f}}c[\mathrm{L}^-]^i_{(\mathrm{a})}\,[\mathrm{E}]^j_{(\mathrm{o})}-k_{\mathrm{r}}a \tag{9.18}$$

Defining:

$$k'_{\mathrm{f}}=k_{\mathrm{f}}[\mathrm{L}^-]^i_{(\mathrm{a})}\,[\mathrm{E}]^j_{(\mathrm{o})} \tag{9.19}$$

then

$$\frac{\partial a}{\partial t}=k'_{\mathrm{f}}\,c-k_{\mathrm{r}}a \tag{9.20}$$

where i and j are exponential orders, which are related to the chemical mechanism of Reaction (9.11). Assuming that the sample is injected instantaneously and that the amount of solute is small, the concentrations of extractant and L^- remain constant during the CCC process. The expression for the retention time, t_{R}, is derived by a mathematical treatment similar to that in the rate theory of chromatography [16].

$$t_{\mathrm{R}}=\frac{L}{u}\left[1+\frac{k'_{\mathrm{f}}}{k_{\mathrm{r}}}\varphi_{\mathrm{v}}\right] \tag{9.21}$$

where L is the CCC column length, u is the mobile phase linear velocity and ϕ_{v} is the phase ratio.

$$\varphi_{\mathrm{v}}=\frac{V_{\mathrm{s}}}{V_{\mathrm{m}}}=\frac{Sf}{1-Sf} \tag{9.22}$$

According to the definition of Sf and the relationship between t_{R} and V_{R}, Eq. (9.21) can be rewritten as:

$$V_R = V_M + \frac{k'_f}{k_r} V_s \tag{9.23}$$

K_D^{dyn} can be expressed in Eq. (9.24), according to its definition.

$$K_D^{dyn} = \frac{k'_f}{k_r} \tag{9.24}$$

Substituting Eq. (9.19) into Eq. (9.24), then

$$K_D^{dyn} = \frac{k_f}{k_r}[L^-]^i_{(a)}[E]^j_{(o)} \tag{9.25}$$

K_D^{bat} can be expressed in Eq. (9.26) according to the extraction reaction (9.11),

$$K_D^{bat} = \frac{k_f}{k_r}[L^-]^n_{(a)}[E]^b_{(o)} \tag{9.26}$$

Thus the relationship between K_D^{dyn} and K_D^{bat} is:

$$K_D^{dyn} = K_D^{bat}[L^-]^{i-n}_{(a)}[E]^{j-b}_{(o)} \tag{9.27}$$

Generally, the reaction orders i and j are not equal to the stoichiometric coefficients n and b. Therefore, K_D^{dyn} is not equal to K_D^{bat} in most cases and, consequently, the retention volume is not equal to that predicted by Eq. (9.7) using K_D^{bat}. However, K_D^{bat} is still useful for optimizing the separation conditions. It can always be used to predict the elution order of the metal ions. A lower K_D^{bat} will lead to a smaller V_R, a larger K_D^{bat} will induce a greater V_R and need a longer time for separation.

Equation (9.27) shows that K_D^{dyn} is linearly related to K_D^{bat} when the two-phase system and other experimental parameters are known. Figure 9.3. is the chromatogram of the fourteen stable lanthanide ions using 25% dihexyl (*N,N*-diethyl)methylene carbamoyl phosphonate (DHDECMP) in cyclohexane as stationary phase and 2.91 *M* HNO_3 as mobile phase [6]. The K_D^{dyn} of the fourteen lanthanides calculated from Fig. 9.3, as well as the K_D^{bat} in the corresponding separation system, are listed in Table 9.1. Figure 9.4. shows that K_D^{dyn} is proportional to K_D^{bat} as indicated by Eq. (9.27).

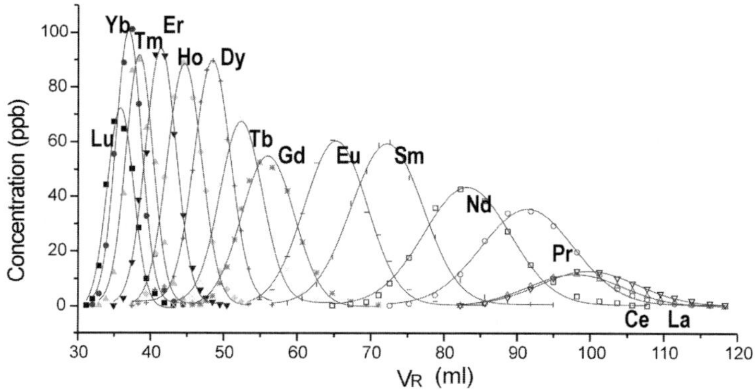

Fig. 9.3. Chromatograms of the 14 stable rare earth ions. Mobile phase, 2.91 *M* HNO_3: stationary phase, 25% DHDECMP in cyclohexane: flow rate, 1 ml/min: rotor speed, 1200 rpm: temperature 22°C: *Sf* retention ratio, 20.4%; β = 0.4–0.72, V_c = 30.5 ml.

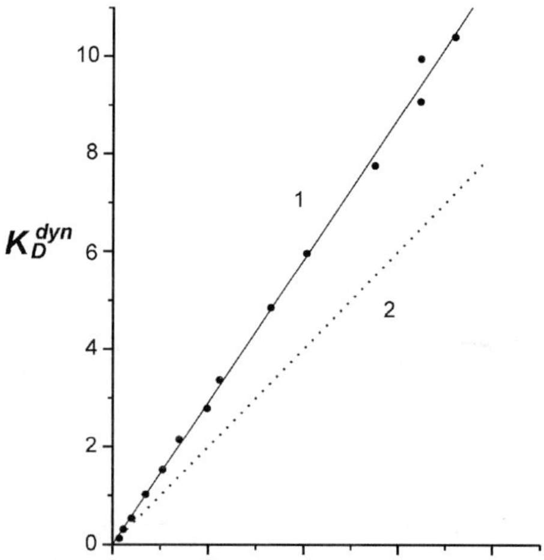

Fig. 9.4. Dependence of K_D^{dyn} on K_D^{bat}. (1) K_D^{dyn} to K_D^{bat} values corresponding to Fig. 9.3 and Table 9.1. r = 0.9986, SD = 0.22. (2) Line of slope 1 corresponding to $K_D^{dyn} = K_D^{bat}$.

The case in which K_D^{dyn} equals K_D^{bat} occurs when all the steps related in Reaction (9.11) and Fig. 9.2. are very fast [17]. Only then, is the expression for K_D^{dyn} derived from the mechanism of reaction (9.11), using the stationary state method [18], the same as K_D^{bat}.

TABLE 9.1

Results of chromatographic separation

Metal ion	K_D^{bat} (SD)	K_D^{dyn}	α^{bat}	α^{dyn}	*Rs*	*N*
La	7.19 (0.27)	10.39	1.11	1.04	0.102	130
Ce	6.47 (0.24)	9.947	1.00	1.10	0.202	131
Pr	6.46 (0.24)	9.074	1.17	1.17	0.306	155
Nd	5.50 (0.21)	7.756	1.35	1.30	0.469	163
Sm	4.06 (0.15)	5.965	1.23	1.23	0.346	185
Eu	3.31 (0.12)	4.848	1.48	1.44	0.533	192
Gd	2.24 (0.08)	3.372	1.13	1.21	0.250	206
Tb	1.98 (0.15)	2.788	1.42	1.30	0.302	214
Dy	1.39 (0.10)	2.152	1.33	1.41	0.324	231
Ho	1.05 (0.08)	1.529	1.51	1.49	0.283	241
Er	0.69 (0.05)	1.024	1.72	1.90	0.300	257
Tm	0.40 (0.05)	0.538	1.79	1.74	0.162	285
Yb	0.23 (0.04)	0.309	1.56	2.50	0.138	300
Lu	0.14 (0.04)	0.124				309

9.2.3 Separation efficiency

High separation efficiency of inorganic compounds using CCC has been reported. Alkali metals were separated with high efficiency using dibenzo-18-crown-6 in $CHCl_3$/1-BuOH as stationary phase, in the presence of picrate anion in the mobile phase [19]. The separation efficiency of the ion-pair $(tetraheptylammonium)_2IrCl_6$ was very similar to that obtained with the organic compound 3-picoline, at different flow rates and phase-volume ratios for the solvent pairs examined [4]. More investigations, however, reported that the peaks were significantly broadened [13,14,20–22] and the efficiencies of inorganic separations were four to five times lower than those of organic separations [14]. These results puzzled the scientists and stimulated them to explore the fundamental factors that affect the efficiencies of inorganic separations.

Freiser et al. [3,4,11,14] conducted separations of trivalent lanthanide ions, platinum group elements, and nickel by CPC with the trioctylphosphine oxide (TOPO) chelating agent. They found that the peak width for the $PdCl_2$ $(TOPO)_2$ complex was greater than that of 3-picoline, a molecular solute, with identical retention volume and flow

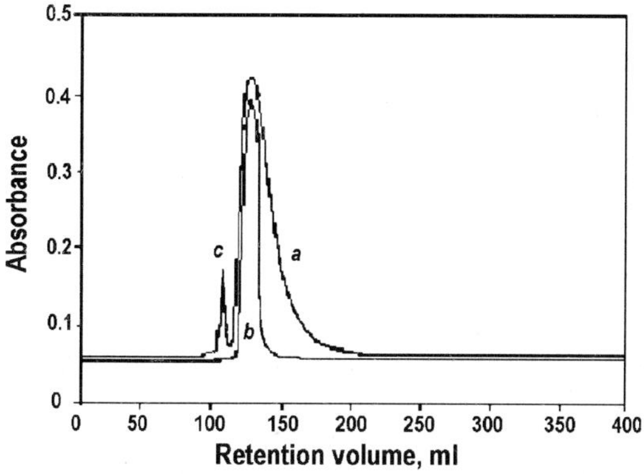

Fig. 9.5. Chromatograms of 3-picoline and $PdCl_2$ $(TOPO)_2$ at identical distribution ratios in the heptane–H_2O phase pair: $\varphi_v = 0.2$, flow rate = 2 ml/min. (a) 10^{-3} *M* Pd(II), 0.3 *M* TOPO, 0.1 *M* Cl^-, pH = 3; (b) 3-picoline, pH = 6.1; (c) dead-volume peak (adapted from Ref. [14]).

rate (Fig. 9.5). Namely, the efficiency for the Pd(II) complex was significantly lower than that for 3-picoline. They defined the CETP parameter, channel equivalent to a theoretical plate (number of channels needed to make one plate), which is the parameter equivalent to the reduced plate height characterizing the separation efficiency in HPLC. The CETP variations with flow rate were studied for the Pd(II) complex and 3-picoline with the heptane/water system. The CETP for the Pd(II) complex was always higher than that for 3-picoline and it exhibited a greater sensitivity to flow rate. Simple diffusion alone cannot explain the significantly lower efficiency for Pd(II), even if the viscosity of the heptane solution, increased by TOPO, was taken into account. The metal complex formation reactions were rapid when the ligand concentration was in large excess compared to the metallic ion concentration during the CPC separations. The factor responsible for low efficiency was most likely the slow kinetics of back-extraction of the metal ions into the aqueous phase. The additional inefficiency was characterized by $CETP_{ck}$, which could be isolated from the observed $CETP_{obs}$ for Pd(II), subtracting the $CETP_{dif}$ value obtained with 3-picoline. The $CETP_{ck}$ was correlated to the half-life, $t_{1/2}$, for the kinetics of back-extraction of the $PdCl_2$ $(TOPO)_2$ complex and it was directly proportional to $t_{1/2}$. This strong correlation between $CETP_{ck}$ and this kinetic parameter proved unequivocally that the band-broadening of the Pd(II) peak is due to the slow kinetics of the complex dissociation [14].

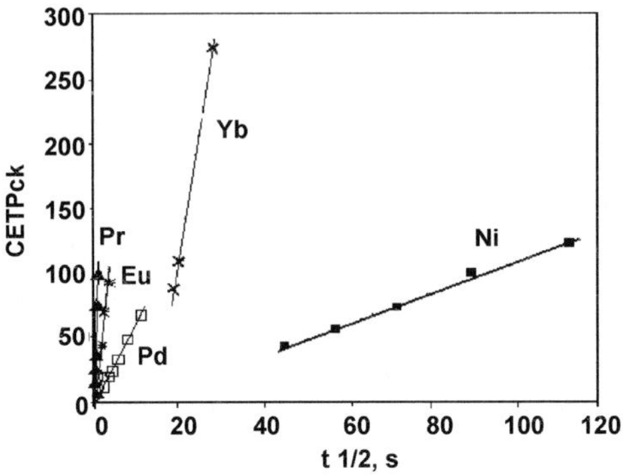

Fig. 9.6. $CETP_{ck}$ *vs*. $t_{1/2}$ determined in Triton X-100 or Brij 35 micelles. Pr, Eu, and Yb in heptane–H_2O liquid phase system with 0.1 *M* Cyanex 272; $\varphi_v = 0.18$; flow rate, 1 ml/min. Pd in heptane–H_2O liquid-phase system, $\varphi_v = 0.2$, flow rate 2 ml/min, [TOPO] = 0.1–0.5 *M*, pH = 3. Ni in $CHCl_3$–H_2O liquid-phase system, $\varphi_v = 0.4$, flow rate 0.4 ml/min, pH 5.8–6.2, [HPMBP] = 0.0045–0.01 *M* (adapted from Ref. [4]).

The soundness of the $CETP_{ck}$–$t_{1/2}$ relationship was further established by examining the CPC behaviour of lanthanides(III) with Cyanex 272 [13,23] and Ni(II) with HPMBP and HDSO [24,25]. In all cases, the $CETP_{ck}$ values correlated well with the $t_{1/2}$ values measured in the $CHCl_3$–H_2O phase system using a microporous Teflon phase separator. A good correlation was also obtained with the $CETP_{ck}$ values and the $t_{1/2}$ values measured in Brij 35 micelles, as shown in Fig. 9.6. It should be stressed that reactions with half-lives of even about one second can adversely affect the CPC efficiencies. The rapid extraction and back-extraction reactions observed in single-stage equilibrations may still be slow enough to reduce the efficiencies of multi-stage separations [4,14].

The group headed by Spivakov [5,26] also investigated the influence of kinetics on the separation and preconcentration of some elements by CCC. The extraction and back-extraction mechanisms of Fe(III), Nb(V) and Ta(V) are different in the DEHPA–*n*-decane–HCl system. As shown in Fig. 9.1, a part of Fe(III) eluted from the column after a column volume of mobile phase, while the other part of Fe(III) was eluted slowly only when a high concentration of HCl (5.0 *M*) was used.

The kinetics of cation-exchange reagents (anionic such as DEHPA or neutral such as the organophosphorus extracting reagents DTPMDP [26] used) are different, and the latter is kinetically more favourable. With similar dynamic distribution ratios, the peak-width of Eu(III) and

DEHPA is very broad in contrast to the corresponding Eu(III)-DTPMDP peak. The value $K_D = 1.0$ is the optimal value to obtain a well-resolved separation, but Eu(III) with a K_D^{dyn} value of 1.2 in the DEHPA-based phase system was poorly retained in the column. It was eluted immediately after the dead volume in the form of a broad peak and this hindered its isocratic separation from other elements, whereas it was retained in the separation column after a volume of V_M had been pumped through the column of CCC [5,26]. The previously mentioned higher efficiency could be attributed to the greater total mass-transfer rate of the DTPMDP-based system compared to that of DEHPA-based system.

The shapes of the elution curves of Eu(III) are not influenced by the rotation speed and the pumping speed of the mobile phase [5,26]. Therefore, the kinetic properties may have a greater effect on the chromatographic process than the operational parameters of CCC.

Fedotov et al. [15] established the empirical relationship expressed by Eq. (9.28) for the total mass-transfer rates (k) ratio. This ratio is related to the batch distribution ratios (K_D^{bat}s) of solvent extraction systems and the half-widths ($W_{1/2}$) of chromatograms in the separation of inorganic substances:

$$\frac{\dfrac{K_{D(1)}^{bat} V_{s(1)}}{W_{1/2(1)}}}{\dfrac{K_{D(2)}^{bat} V_{s(2)}}{W_{1/2(2)}}} \approx \frac{k_{(1)}}{k_{(2)}} \tag{9.28}$$

where $W_{1/2}$ is the width of the chromatographic peak at half height, and subscripts (1) and (2) relate to two different extraction systems. Thus, having the batch extraction data (partition coefficients and mass transfer coefficients) for two systems and the results of a chromatographic experiment for one of them, $W_{1/2}$ of the other chromatographic peak can be estimated. The authors pointed out that this equation is only valid for partition coefficient (distribution constant) values in the range of 1–10 and for the reversible processes of extraction and back-extraction.

In our theoretical research [6,7,17], the height equivalent to a theoretical plate (HETP) was used to characterize the separation efficiency of inorganic separation. Among the factors affecting the separation efficiency of CCC, the vortex diffusion could be neglected for there is no solid support in CCC, and the extra column diffusion is also negligible. Three factors are discussed: (1) the molecule vertical diffusion; (2) mass-transfer diffusion in the two phases; and (3) kinetic effects owing to the existence of chemical reactions.

The concentration gradient in the column of CCC will lead to the molecule diffusion, which is proportional to the retention time. The contribution to HETP caused by the molecular diffusion is expressed by H_1:

$$H_1 = \frac{C_{m,1} D_m + C_{s,1} D_s \varphi_v K_D^{dyn}}{u} \tag{9.29}$$

where the suffixes m and s denote the mobile phase and stationary phase, respectively. The quantities $C_{m,1}$ and $C_{s,1}$ are the characteristic coefficients of the metal ion and its extraction complex related to the molecular diffusion in the aqueous and organic phase. D_m is the diffusion coefficient of the metal ion in the mobile phase, and D_s is the diffusion coefficient of the extraction complex in the stationary phase. Equation (9.29) shows that H_1 is determined by the design and operational parameters, e.g., the diameter of the column, flow rate, and the hydrodynamic properties of the liquid system used. The value of H_1 increases with increase of the distribution coefficient, K_D^{dyn}, of the inorganic compounds.

The contribution to the HETP resulting from mass-transfer diffusion can be divided into two parts: (1) the contribution of the chemical reaction kinetics, and (2) the contribution of physical mass transfer between phases. It should be noted that the chemical forms of the species distributing in the two phases are different in inorganic separation. The HETP contribution of the physical mass transfer, H_2 can be expressed by Eq. (9.30):

$$H_2 = u \left(\frac{C_{m,2} d_m^2}{D_m} + \frac{C_{s,2} d_s^2}{D_s \varphi_v} \frac{1}{K_D^{dyn}} \right) \tag{9.30}$$

where $C_{m,2}$ and $C_{s,2}$ are the characteristic constants related to the apparatus parameters as well as the chemical forms of the distribution species in the two phases: d_m and d_s are the lengths of the fragments of two phases or the thickness of the dispersed droplets in the two phases; H_2 is mainly determined by instrumental parameters and the properties of the two phases used.

The contribution to the HETP caused by the kinetic effect is denoted H_3. It was deduced according to the mass-transfer model of Fig. 9.2 [7].

$$H_3 = \frac{2\mu\varphi_v k'_f}{(\varphi_v k'_f + k_r)^2} = \frac{2\mu\varphi_v K_D^{dyn}}{k_r (K_D^{dyn}\varphi_v + 1)^2} \tag{9.31}$$

It can be concluded from the expression for H_3 that the efficiency of an inorganic separation made by CCC is affected by the mechanism of the extraction reaction: H_3 will decrease with increase of k'_f and k_r. Consequently, the systems with the faster extraction rate will have the higher efficiency. The separation efficiency will be improved by promoting the extraction kinetics. Factors such as the temperature, the concentration of inorganic counter-anion, the concentration of ligand in the organic phase, and the pH are usually employed to raise the reaction rate.

The total HETP is expressed by:

$$H = H_1 + H_2 + H_3 = \frac{C_{m,1}D_m + C_{s,1}D_s\varphi_v K_D^{dyn}}{u} + u\left(\frac{C_{m,2}d_m^2}{D_m} + \frac{C_{s,2}d_s^2}{D_s\varphi_v}\cdot\frac{1}{K_D^{dyn}}\right) + \frac{2u\varphi_v K_D^{dyn}}{k_r (K_D^{dyn}\varphi_v + 1)^2} \tag{9.32}$$

The theoretical expression, Eq. (9.31), is consistent with the empirical expression, Eq. (9.28), and with the investigations by Freiser et al. [3,4,11,14]. According to Eq. (9.31), H_3 is proportional to $t_{1/2}$. Since the back-extraction rate constant (k_r) is inversely proportion to $t_{1/2}$, $\mathrm{CETP}_{ck} \equiv \mathrm{H}_3 \propto t_{1/2}$. Obviously, the mass-transfer rate (k) is inversely proportional to $t_{1/2}$, and $W_{1/2}$ is proportional to H. Thus, from Eq. (9.28), we have $H \propto W_{1/2} \propto 1/k \propto t_{1/2}$.

CCC provides useful clues on the chemical kinetics of extraction and back-extraction of a particular cation–ligand reaction. The dependence of CETP_{ck} on the concentrations of the species in the aqueous and organic phases could elucidate the mechanism of the dissociation reaction and its rate-limiting step [3]. The determined CETP is directly proportional to [TOPO] and inversely proportional to $[Cl^-]$[5]. This means that the kinetics of the back-extraction reaction is the major factor affecting the CPC efficiency for Pd(II) elution. The CETP_{ck} of $Ni(PMBP)_2$ and $Ni(DSO)_2$ is proportional to the pH value, indicating that the dissociation of these complexes is the predominant factor in determining the chromatographic efficiency [4].

9.3 METHODOLOGY

9.3.1 Experimental procedure

The standard procedure for inorganic separation is similar to that for organic separations. The CCC column or CPC partition cells are first filled with the liquid stationary phase. Next, the mobile phase is supplied at a proper flow-rate by a pressurized pump until the volume balance between the two phases is established, while the apparatus rotates at the desired speed. This is followed by the injection of the sample solution through the sample port. The effluent exiting from the end of the column is continuously monitored by an on-line detector or collected into test tubes and monitored by off-line methods.

The aqueous phase is usually used as the mobile phase for economic reasons and convenience of detection. Isocratic elution is the most common mode because other elution modes may produce variations of density, viscosity, and interfacial tension during the separation process. These variations may induce leaks of the stationary phase. The whole volume of mobile phase should be thoroughly equilibrated with the stationary phase. Also, the volume of the stationary phase should be lower than can be provided by the phase-pair of the least density difference. Constant *Sf* values could be obtained by establishing the initial phase equilibrium at a high flow-rate and a low revolution-speed and conducting the separation process at lower flow-rate and higher revolution-speed [27].

When CCC is used for enrichment (preconcentration), a large volume of sample solution will be involved. A non-equilibrium injection method [28] should be used in order to reduce the elution time and to achieve maximum efficiency of separation. In this case, the sample solution passes through the column full of stationary phase and displaces the part of organic solution to form the suitable phase equilibrium. The retention volume and resolution remained almost constant for erbium and holmium using the non-equilibrium injection method [28].

9.3.2 Elution mode

Three elution modes are employed in inorganic separations using CCC, namely, isocratic, step-elution, and gradient elution. The selection of the elution mode depends on the distribution ratios of components of interest in the analyte mixture as well as the kinetic character of the separation system. Values of K_D^{dyn} ranging from 0.1 to 10 are favourable for inorganic separations. Large distribution ratios result in long elution times and large elution volumes. This is unfavourable for both analysis and preparative purposes. If the systems do have very large distribution ratios, then the step-elution mode should be chosen. Preconcentration is preferable in the case of large volumes of diluted sample, but high distribution ratios, i.e., ratios larger than 100, are needed in this case. Isocratic elution could be selected for kinetically labile extraction systems. The systems with large distribution ratios or slow kinetic reactions should be paired with step-elution or gradient elution. Fedotov et al. [26] indicated that fast systems with high mass-transfer rates (about 10^{-3} cm/s) are suitable for isocratic separation as well as for separation by step-elution.

Step-elution is required for slower systems with mass-transfer rates around 10^{-4} cm/s. For kinetically very slow systems, with mass-transfer coefficients as low as 10^{-5} cm/s, high partition coefficients (10^2 and more) are necessary for quantitative extraction of the analytes, and step-elution should be used after preconcentration.

9.3.3 Two-phase system

The selection of the two-phase liquid system is perhaps the most important parameter with inorganic separations. A recent comprehensive review involving phase selection was compiled by Maryutina et al. [2]. Phase selection is complicated because the chromatographic behaviour of inorganic compounds depends on the type and concentration of the extractant(s) in the organic phase, the nature of the counter-ion, and concentration of complexing agents, the presence of other ligands, and the pH of the aqueous phase. Variations in the concentration of a constituent in the two phases will change the dynamic distribution ratio and/or selectivity and/or the mass-transfer rate of the analytes. The HETP depends on the K_D^{dyn} value, the properties of the two phases, and the mechanism of the separation. The efficiency can be improved significantly by adjusting the composition of the two phases.

The organic phase

Extractants: The CCC inorganic separations benefit from the comprehensive investigations and knowledge gained in solvent extraction. A number of effective and selective extractants can be used as separation reagents. The main factors for the selection of extractants are selectivity, loading capacity, and kinetic properties. The chemical stability and cost are always to be considered.

About twenty extractants have been used as chelating agents to separate inorganic compounds. Among them the most widely used are the organophosphorus extractants. Typical acidic organophosphorus extractants are DEHPA, EHPA and Cyanex 272. They have been used as additives to the stationary phase to separate lanthanoids for their good loading and stripping behaviour, low solubility in the aqueous phase, and relatively large separation factor [29]. Mutual separation of adjacent lanthanoids had been accomplished in CCC using acidic organophosphorus extractants. For example, the separation of the 14 stable lanthanoids was performed in one five-hour run [30] by applying an exponential gradient concentration of hydrochloric acid in the mobile phase and using DEHPA in heptane as stationary phase, as shown in Fig. 9.7.

It is difficult to elute completely the heavy lanthanoids [31], iron(III) [5], Nb(V) and Ta(V) [5] because of their low mass-transfer rate when acidic organophosphorus extractants are employed. Neutral organophosphorus compounds are known to be well distinguished from acidic organophosphorus compounds in their kinetic properties. The peak shape of lanthanoids was symmetrical when DHDECMP was used as the stationary phase [6]. It should be noted that the separation factor of

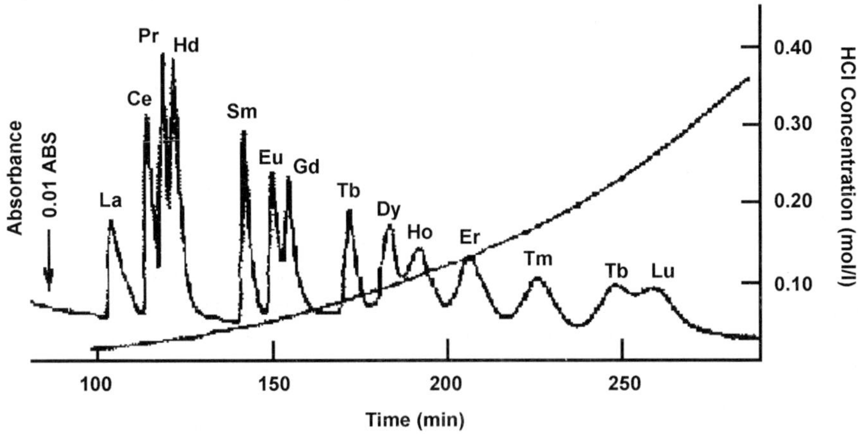

Fig. 9.7. Acid gradient separation of the 14 stable rare earth elements by HSCCC. Mobile phase: aqueous HCl solution with increasing concentration, stationary phase: heptane with 0.2 *M* DEHPA (compare with Fig. 9.3) (adapted from Ref. [30]).

neutral organophosphorus compounds is lower than that of acidic organophosphorus compounds. However, the separation factor could be improved by adding some polar ligand in the aqueous phase, e.g., NH_4SCN when TPMDPD was used in the stationary phase [5,32].

Alkali and alkaline earth elements were separated in the presence of crown ester DCH18C6 [33,34]. Cobalt dicarbolide (CD) [35,36] was used to separate Cs and Sr. Tetra-octylethylenediamine (TOEDA) [37] was employed to pre-concentrate and separate heavy elements such as Zr, Hf, Nb, Ta, Zn and Cd, and Y was successively isolated from the other heavy lanthanoids [38] using carboxylic acid with a large alkyl moiety. Heavy actinoids were separated from light lanthanoids by using TPMDPD [32], Ph2Bu2 [32] and LIX63 [39]. Americium was fractionated from lighter lanthanoids by gradient elution using LIX63 as a stationary phase [63]. The separation of Am from Eu with high recovery can be accomplished. Americium can be selectively retained on Cyanex 301 stationary phase [64]. Tertiary amines can be used to concentrate ultra-trace plutonium in highly concentrated HNO_3 phases [17].

Concentration effect: The concentration of the extractant in the organic solvent affects the elution curve shape and the dynamic coefficients. It has been observed that increases in the reagent concentration in the organic phase often leads to better separation [2]. A two-fold increase in the DEHPA concentration results in better selectivity of Sm extraction. The batch partition coefficient for Sm increases by a factor of 10, whereas the recovery of Nd remains practically the same, so favourable conditions are fulfilled for the separation of Nd and Sm ($\alpha > 10$). Similar effects of the reagent concentration on the elution curve shapes are observed with the use of reagents of other classes.

Diluents: The type of the diluent that is used to dissolve the extractant and make the stationary phase, has a great effect on the chromatographic process. The use of different solvents could alter the retention volume and separation efficiency.

When DEHPA was used to separate the rare earth elements (REE), a polar solvent was effective for both shortening the retention times and sharpening the peaks of the heavy REE^{3+} ions. Almost the whole series of REE ions will be mutually separated with DEHPA by adjusting the nature of solvent in the stationary phase, and the concentration of HCl in the mobile phase [40]. With regard to the diluent used, *n*-heptane was used for the separation of light REEs, La(III), Ce(III), Pr(III), Nd(III), and the more polar diluent, toluene, was suitable for the separation of the middle REEs Sm(III) and Eu(IIII) [41]. The most polar diluent, chloroform, was needed for the separation of the heavier REEs Er(III), Yb(III) and Lu(III) [40].

The number of carbon atoms in n-paraffin solvents can also affect the retention volume and separation efficiency [42]. When DEHPA was dissolved in the mixed solvent, CCl_4 and n-paraffin, to separate REEs, the K_D values decreased with increasing concentrations of CCl_4, in analogy with the TBP extraction system [12], and increased with the chain length of the n-paraffin. This result was attributed to variations in the activity of DEHPA coupled with the concentration of CCl_4 and chain length of the n-paraffin. The activity of DEHPA in the organic phase increases with the number of atoms of the n-paraffin since the association of polar DEHPA takes place through water molecules dissolved in the organic phase and the solubility of water in n-paraffin decreases when the alkyl-chain carbon number increases. The relative peak-width of Sm and Eu was reduced and the resolution increased by increasing the n-paraffin carbon number. The best separation efficiency was achieved using the long chain n-pentadecane. Pr, Sm, Eu, and Gd were successfully separated in the 30% DEHPA/15% CCl_4/55% n-pentadecane and 0.5 M HNO_3 two-phase liquid system [42]. Because the α values were not affected by the difference of the number of carbon, the improvement of the resolution between them is due to the reduction in the peak width (increased efficiency).

Aqueous phase

The composition of the aqueous phase also has a great effect on the partition coefficients of inorganic compounds and on the separation efficiency.

Acid concentration and pH: When an acidic extractant is present in the organic phase, the retention volume and the separation efficiency will depend considerably on the concentration of acid and the pH in the aqueous phase. For example, the chromatographic behaviour of the trivalent Eu ion is shown in Fig. 9.8. [5]. A planetary centrifuge with one-layer coil column was applied with a solvent system of DEHPA at different HCl concentrations in the aqueous phase. In all cases, the elution of Eu(III) begins immediately after a volume of V_m. The peak width increases with the reduction of HCl concentration.

The V_R values increase with pH, and the separation efficiencies of REEs were significantly improved [22,27]. The higher pH values lead to an excellent peak resolution and the Rs values reached almost 1 at pH 2.06, especially for the Er/Y pair with the low separation factor of 1.47. It

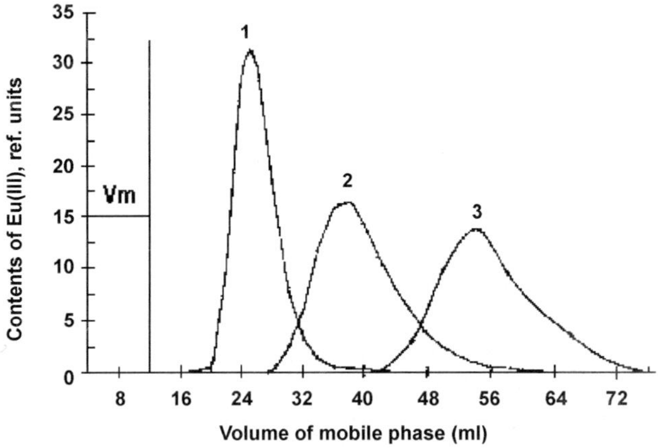

Fig. 9.8. Chromatographic behaviour of Eu(III) in 0.005 *M* TPMDPD chloroform/0.5 *M* NH_4SCN/HCl system. *Sf* = 0.5. Concentration of HCl in the mobile phase: (1) 1.0 *M*; (2) 0.6 *M*; (3) 0.4 *M*. *F* = 1 ml/min (adapted from Ref. [5]).

is important to control the pH of the mobile phase in order to optimize the distribution ratio of each component for adequate separation [27]. The light and middle lanthanoids have been successfully separated using EHPA and adjusting the mobile phase pH [22].

Salting-out effect: The distribution ratio and separation factor of REEs in tributyl phosphate (TBP) are small, but could be enhanced using the salting-out effect. Satisfactory separation of Pr and Eu was achieved by CPC in a 30% TBP/ *n*-dodecane extraction system containing 5 *M* $LiNO_3$ in the mobile phase [12].

Water-miscible organic solvent in the mobile phase: Several water-miscible organic solvents (ethylene glycol [EG], EtOH, DMSO, ACN, DMF, MeOH) were used as organic additives to the mobile phase to accelerate the extraction and increase the theoretical plate number. The elution behaviour of Sm and Gd was tested [22] by using 0.02 *M* $(EHPA)_2$ in kerosene as stationary phase, and 0.1 *M* Cl_2CHCOO(H, Na) in water at pH 1.90 as mobile phase. The chromatograms showed that the peak of Sm was followed and overlapped by the broad peak of Gd. By adding some ethylene glycol, EtOH and/or DMSO in the aqueous mobile phase, symmetrical chromatographic peaks were obtained and the *N* values for individual peaks, as well as the separation factors, increased markedly. The influence of the miscible organic additives may be related to a number of factors: (1) a lowering of the interfacial tension may cause a

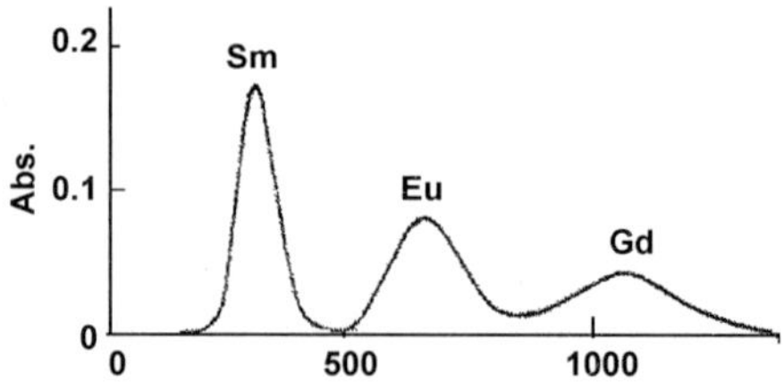

Fig. 9.9. Mutual separation of Sm, Eu and Gd, with a mobile phase containing 20% EG, pH 1.94. Stationary phase $CHCl_3$ with 0.1 *M* DEHPA (adapted from Ref. [43]).

reduction in droplet size; (2) an increase in distribution of EHPA into the aqueous phase accelerates the rate-determining step, such as the formation of the first complex $M(HA_2)^{2+}$; (3) a reduction in water activity reduces the extent of hydration of metal ion. These associated factors will enhance the extraction rate and improve peak efficiency and resolution [22].

When 20% ethylene glycol was added to the 0.08 *M* HCl mobile phase, well-resolved chromatograms were obtained for the separation of Er^{3+} and Tm^{3+} ions using a 0.1 *M* DEHPA-in-$CHCl_3$ stationary phase. A near-baseline separation of Tm^{3+} and Yb^{3+} ions was attained with 0.10 *M* HCl as the mobile phase containing 20% EG. However, the separation of Er^{3+} and Tm^{3+} deteriorated and became unsatisfactory [41]. The most difficult ions to separate are Sm, Eu and Gd. Employing the 20% ethylene glycol–water mixture as the mobile phase [43] did separate them, as shown in Fig. 9.9.

9.3.4 Operational parameters

Effect of rotational speed and flow rate

In the standard CCC procedure, the volume of stationary phase depends on the speed of revolution and flow rate. The V_s volume rises with increasing rotation speed and decreasing flow rate. Generally, the peak efficiency observed with classical (non-ionic) organic separations increases with the rotation speed and decreases with the flow rate of mobile phase. Therefore, the separation should be performed with the rotor spinning as fast as possible and the mobile-phase flow rate being as low as possible.

Several groups have examined the dependence of the separation efficiency on the speed of revolution and flow rate in inorganic separations. It was found that, at a fixed phase ratio *Sf*, the values of *N* and R_s were little affected by the revolution speed and flow rate [27] provided that the stationary-phase volume was really kept constant. At variable *Sf* values, a high stationary-phase volume results from the high speed of revolution and low flow rate. This contributes to the enhancement of the separation efficiency [27,42,44]. Another explanation is that the drop size of the mobile phase reduces in the increasing gravitational field, and the rate of extraction is enhanced [22].

When the analytes 3-picoline, (tetraheptylammonium)$_2$IrCl$_6$, and PdCl$_2$(TOPO)$_2$ were separated in CCC using the 1,2-dichloroethane–H_2O phase pair at a flow rate of 2 ml/min, Muralidharan and Freiser [10] discovered that the efficiency improved dramatically when the volume of stationary phase decreased. Separations were efficient when φ_v was below 1, and the best efficiencies were observed when φ_v was 0.3 or less. These results seem to contradict the preceding results [27,42,44]. Our theoretical research, Eq. (9.31), shows that H_3 is related to the K_D^{dyn} and φ_v values. The optimum value of φ_v is dependent on the K_D^{dyn} value. Figure 9.10 shows the calculated H_3 value versus φ_v for different values of K_D^{dyn}. The value of H_3 increases with φ_v up to a turn-point and then decreases. The turn-point of φ_v is $1/K_D^{dyn}$ derived from Eq. (9.31). In the cases with low K_D^{dyn} values, H_3 will generally

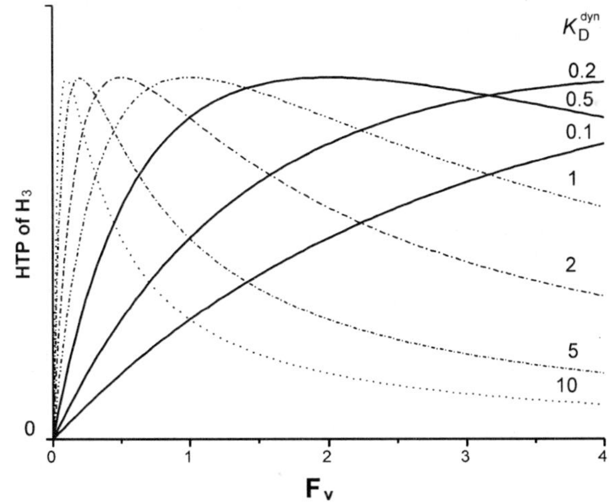

Fig. 9.10. Theoretical plate height contribution, H_3, *vs*. phase ratio, where F_v means φ_v for different ionic solute distribution coefficients, K_D^{dyn}.

TABLE 9.2

Reported working range of φ_v of which the separation efficiency is increased with φ_v

Phase-pair (stationary–mobile)	Metal ions	K_D^{dyn}	Turn-point of φ_v	Working range of φ_v	Ref.
0.02 *M* $(EHPA)_2$–0.1 *M*	Ho	2.7*	0.37	0.44–1.17	27
$Cl_2CHCOO(H,Na)$	Y	4.5*	0.22		
	Er	6.6*	0.16		
0.1 *M* Cyanex 272–H_2O	Gd	2.4	0.42	0.18–0.26	44
	Tb	4.4	0.23		
	Dy	7.2	0.14		

*Estimated from Fig. 7 of Ref. [27].

increase with the φ_v value and the separation efficiency will reduce with the increase of φ_v [10]. In contrast, the distribution constants, K_D^{dyn}, are high, and H_3 will decrease with k_v when the CCC process is performed at φ_v values larger than $1/K_D^{dyn}$. In this case, the separation efficiency and the resolution increased with φ_v [13,27,42] (the reported K_D^{dyn}, turn-points for φ_v, and the working ranges of φ_v are listed in Table 9.2).

Effect of temperature

The efficiency of organic separation increases with temperature, as was found with the binary system octanol–water and the separation of hydroquinone and pentachlorophenol [45]. Conducting the CCC process above ambient temperature is also advantageous for separation of inorganic compounds. Temperature changes induce numerous changes in physico-chemical properties in liquids. The density and, particularly, viscosity decrease with rise of temperature. The solvents' mutual solubility is generally very sensitive to temperature, increasing when the temperature increases [45]. Increasing temperature will promote the extraction reactions and thus improve the separation efficiency [7], and resolution [22,38,41].

When 0.02 *M* $(EHPA)_2$ in toluene–0.1 *M* $CHCl_2COO(H,Na)$ was used as solvent system to separate Ho, Y and Er, a rise in temperature improved the peak profile [38], and the *N* values for Y doubled, from 17 at 25°C to 30 at 40°C. The separation factors between Ho and Er were not very much changed, having similar values of 2.8 and 2.7 at 25°C and 40°C, respectively. The resolution was somewhat enhanced from 0.77 at 25°C to 0.83 at 40°C owing to the increase in *N*.

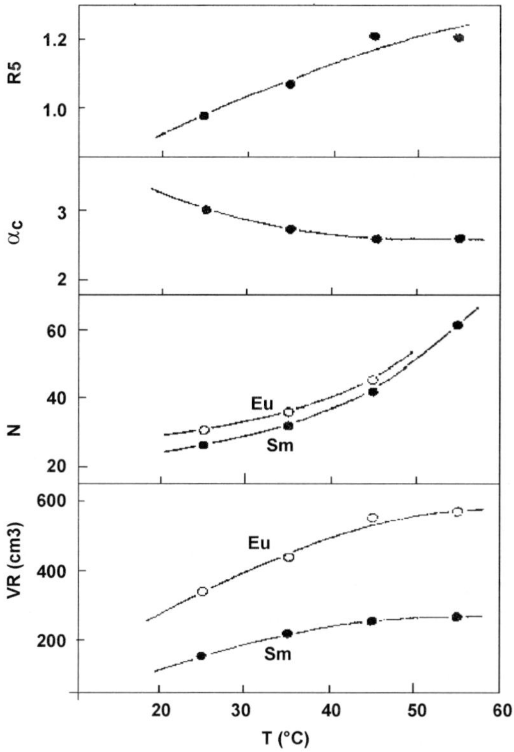

Fig. 9.11. Effect of temperature on the separation of Sm and Eu by CPC. CPC machine: Sanki LLN, 6 cartridges, 800 rpm: stationary phase, 0.02 *M* $(EHPA)_2$; mobile phase, 0.1 *M* $(H,Na)Cl_2CHCOO$ in 20% v/v EG–water, pH 2.04: samples, 10^{-3} *M* of each lanthanoid, 1 ml injected (adapted from Ref. [22]).

Takeo Araki et al. [41] reported that the *Rs* values for Sm^{3+}, Eu^{3+}, Gd^{3+}, Tb^{3+} and Dy^{3+} could be improved by raising the temperature. This result was obtained using a CPC machine and DEHPA in $CHCl_3$ as the stationary phase, and aqueous HCl solution as the mobile phase. Although Dy^{3+} and Ho^{3+} were found to be rather difficult to separate at 25°C, these two REE^{3+}s were partially separated, and the separation of Ho^{3+} and Er^{3+} became almost complete with 0.06 *M* HCl mobile phase by elevating the temperature to 45°C. The value of *N* increased considerably with temperature rises, especially for the components having higher t_R. The temperature rises did not significantly change the values of retention time (t_R).

The distribution coefficient also sometimes depends on temperature. The retention volumes of Sm and Eu increased with temperature and their V_R values at 55°C were about 1.7 times those at 25°C [22]. The separation factor reduced slightly with rising temperature, while the N values of Sm and Eu increased considerably. Therefore, the resolution of these two adjacent elements increased almost linearly with rising temperature, as seen in Fig. 9.11.

Effect of the channel number in CPC

Many inorganic separations were accomplished using CPC Sanki machines with cartridges (Model LLN) or disks (Model HPCPC) both containing channels or cells. The value of N will increase with the number of cartridges. The effect of the number of cells in CPC on the chromatographic behaviour of binary mixtures of $PrCl_3/NdCl_3$, and $CeCl_3/PrCl_3$ was examined with 0.1 M DEHPA n-heptane solution as the stationary phase and 0.07 M aqueous HCl solution as the mobile phase [40]. The results showed that the peak-resolution was improved from 0.37 for the three-cartridge (1200 channels or micro-cells) system to 0.62 for the six-cartridge (2400 micro cells) system. Of course, the retention volumes doubled. Increase of the number of micro cells is effective when the peak broadening is substantial, e.g., for REE ions with higher atomic numbers.

9.4 APPLICATIONS

The applications of CCC to inorganic compounds have so far involved the mutual separation of inorganic ions [29], analytical applications [46], purification of chemical reagents [47], and preconcentration of metal ions [36]. The majority of these researches are listed briefly in Table 9.3. To our knowledge, CCC is a useful method for pre-concentrating inorganic compounds for trace analysis. Since, as this book's title says, CCC is a support-free liquid stationary phase, it is free of contamination, and the loss of trace solute is completely eliminated. The sample-treatment is simple, for there are no problems of column clogging, and the great load capacity permits the injection of large volumes of the sample. Moreover, CCC can be coupled with high sensitivity inorganic measuring instruments such as ICP-AES, DCP-AES and ICP-MS.

TABLE 9.3

Applications of CCC in inorganic compound separation

Stationary phase	Mobile phase	Element	Apparatus	Ref.
		Extractant DEHPA		
0.5 *M* DEHPA in *n*-decane	Preconcentration: 0.1 *M* HCl Step elution: (1) 0.5 *M* HCl (2) 2 *M* HCl	 Ce Eu	Planetary centrifuge with a vertical column drum: R = 140 mm, r = 50 mm, i.d. = 0.75 mm, V_c = 20 ml	49
	Preconcentration: 0.5 *M* HCl Step elution: (1) 3 *M* HCl (2) 5 *M* HCl	Na, Cs, Sr, Mn, Fe Ce, Y Fe		
1 *M* DEHPA in *n*-decane	Preconcentration: 0.1 *M* HCl Step elution: (1) 0.5 *M* HCl (2) 3.0 *M* HCl	 Nd Sm	Planetary centrifuge with a vertical column drum: β = 0.35, V_c = 24 ml	26
0.1 *M* DEHPA in *n*-heptane	0.05 *M* HCl 0.04 *M* HCl 0.07 *M* HCl 0.14 *M* HCl	La, Ce, Pr La, Ce La, Pr, Nd Nd, Sm	CPC model LLN with cartridges	21, 40
0.1 *M* DEHPA in chloroform	0.1 *M* HCl	Er, Yb	Sanki CPC model MMF, 1200 and 2400 cells	40
0.1 *M* DEHPA in toluene	0.03 *M* HCl	Sm, Eu	Sanki CPC model LBP-V, 3 model 240W cartridges, 1200 cells, V_c = 63 ml	21
30% DEHPA / 15% CCl_4/55% *n*-pentadecane	0.5 *M* HNO_3	Pr, Sm, Eu, Gd	Sanki CPC Model-LLB; 6 rotors, 720 cells per rotor, V_c = 135 ml	42
0.1 *M* DEHPA in *n*-heptane	0.05 *M* HCl 0.06 *M* HCl 0.12 *M* HCl 0.15 *M* HCl	La, Ce, Pr Pr, Nd Nd, Sm Sm, Eu	Sanki CPC model NMF; cartridges of model 240W, 400×3 cells	56
0.5 *M* DEHPA in *n*-hexane	0.75 *M* HNO_3	Sm, Eu, Gd	Locular countercurrent chromatography. Two glass columns: 150×2 cm Cells: 2×130	57

TABLE 9.3 (continuation)

Stationary phase	Mobile phase	Element	Apparatus	Ref.
0.003 *M* DEHPA in *n*-heptane	Exponential gradient: 0–0.3 *M* HCl	La, Ce, Pr, Nd, Sm, Eu, Gd, Tb, Dy, Ho, Er, Tm, Yb, Lu	HSCCC. β = 0.5 to 0.75, *r* = 15mm, i.d. = 1.07 mm, V_c = 90 ml	30
0.02 *M* DEHPA in *n*-heptane	0.02 *M* HCl	La, Pr, Nd		
0.5 *M* DEHPA in *n*-decane	Preconcentration: 0.1 *M* HCl	Cr, Mn, Co, Ni, Cu, Al, Sr, Mo, Cd, Sn, Sb, Ba, Pb, Ti, Zr, B, V	Planetary centrifuge with a vertical column drum. *R* = 140 mm, *r* = 50 mm, i.d. = 0.75 mm, V_c = 20 ml	31
	Elution: 3.0 *M* HCl	REEs		
0.1 *M* DEHPA in chloroform	0.02 *M* HCl	Sm, Eu, Gd	Sanki CPC model LBP-V: 3 model 240W cartridges, 1200 cells, V_c = 63 ml, 800 rpm	41
	0.04 *M* HCl	Gd, Tb, Dy		
	0.03 *M* HCl	Eu, Gd, Tb, Dy		
	0.06 *M* HCl (*T*=45°C)	Dy, Ho, Er		
	0.08 *M* HCl in 20% v/v EG– water (*T*=45°C)	Ho, Er, Tm		
	0.1 *M* HCl	Er, Tm, Yb		
	0.1 *M* HCl in 20% v/v EG– water (*T*=45°C)	Er, Tm, Yb		
	0.12 and 0.14 *M*	Tm, Yb, Lu		
0.2 *M* DEHPA in *n*-heptane	7.0×10^{-3} *M* citric acid	Ni, Co, Mg, Cu	Shimadzu HSCCC-1A. *r* = 10 cm, β = 0.5–0.6, V_c = 300 ml	58
	0.059 *M* citric acid	Cu, Cd, Mn		
0.002 *M* DEHPA in *n*-heptane	0.14 *M* citric acid	Fe(II), Fe(III)		
0.102 *M* DEHPA in *n*-heptane	Displacement separation: (1) pH~6 deionized water with Co = Cu = Ni = 0.01 *M*; (2) pH = 5.53; (3) pH 3.5 deionized water	Co, Cu, Ni	CPHV 2000 from Société Francaise de Chromato Colonne. V_c = 153 ml Coil: 3×76 m×1.6 mm; *r* = 3.7 to 4.9 cm. *R* = 6 cm, β = 0.61–0.82	59

TABLE 9.3 (continuation)

Stationary phase	Mobile phase	Element	Apparatus	Ref.
		Extractant TOPO		
0.3 *M* DEHPA + 0.02 *M* TOPO in *n*-decane/MIBK (3:1)	Preconcentration: 1.0 *M* NH_4NO_3	Al, Ba, Cu, Co, Fe, Sr, Pb, Mn, Ti, Zr, Cr, V	Planetary centrifuge with a vertical column drum. *R* = 140 mm, *r* = 50 mm, i.d. = 0.75 mm, V_c = 20 ml	44
	Elution: 6.0 *M* HCl	REEs		
0.5 *M* TOPO in *n*-heptane	0.1 *M* HCl	Pd, Pt, Ir, Rh	Sanki CPC model SPL: 6 cartridges, 2400 total channels	60
0.1 *M* TOPO in MIBK	Preconcentration: 1.0 *M* NH_4NO_3	Al, Ba, Cu, Co, Fe, Sr, Pb, Mn, Ti, Zr, Cr, V	Planetary centrifuge with a vertical column drum. *R* = 140 mm, *r* = 50 mm, i.d. = 0.75 mm, V_c = 20 ml	44
	Elution: 6.0 *M* HCl	REEs		
		Extractant EHPA		
0.02 *M* EHPA in kerosene	20% EtOH in 0.1 *M* (H, Na)$Cl_2CO_2^-$, pH 1.90	Sm, Gd	Sanki CPC with 6 or 12 cartridges #30E, 54 cells per cartridge	22, 43
	20% EG in 0.1 *M* (H, Na)$Cl_2CO_2^-$, pH 2.5	La, Pr, Nd		
	20% EG in 0.1 *M* (H, Na)$Cl_2CO_2^-$, pH 1.91, 1.94	Sm, Eu, Gd		
	20% EG in 0.1 *M* (H, Na)$Cl_2CO_2^-$, pH 1.48	Gd, Tb, Dy		
	20% EG in 0.1 *M* (H, Na)$Cl_2CO_2^-$, pH 1.69	Ho, Er, Tm		
	20% EG in 0.1 *M* (H, Na)$Cl_2CO_2^-$, pH 1.35	Tm, Yb, Lu		
0.02 *M* $(EHPA)_2$ in toluene	0.01 *M* (H,Na)$CHCl_2$ COO, pH 2.92	La, Ce, Pr, Nd	Shimadzu HCC-1A: i.d. = 1.5 mm, *r* = 15 cm, V_c = 268 ml	50
	0.1 *M* (H,Na)$CHCl_2$ COO, pH 2.75	Sm, Eu, Gd		
	0.1 *M* (H,Na)$CHCl_2$ COO, pH 2.13	Gd, Tb, Dy, Ho, Y, Er		
	0.1 *M* (H,Na)$CHCl_2$ COO, pH 1.52	Tm, Yb, Lu		
	pH 2.99, 0.1 *M* (H,Na) $CHCl_2COO$, pH 2.68; pH 2.07; pH 1.54	La, Ce, Pr, Nd; Sm, Eu, Gd; Tb, Dy, Ho, Y, Er; Tm, Yb, Lu		

TABLE 9.3 (continuation)

Stationary phase	Mobile phase	Element	Apparatus	Ref.
0.02 *M* EHPA in toluene	0.1 *M* (H,Na)$CHCl_2$COO, pH 2.07, 40°C	Ho, Y, Er; Tb, Y, Tm; Dy, Ho, Y, Er	Sanki CPC model NMF with 6 cartridges, 2400 cells, V_c = 125 ml	38
0.02 *M* EHPA in toluene	0.1 *M* (H,Na)$CHCl_2$COO, pH 2.05, 40°C	Dy, Ho, Y, Er Ho, Y, Er	Shimadzu HCC-1A: i.d. = 1.5 mm, r = 15 cm, V_c = 265 ml	27
0.15 *M* EHPA in chloroform	Aqueous solution of HCl, pH 2.93	Nd, Pr, Ce, La	High-speed planet centrifuge. R = 192 mm, r = 100 mm, i.d. = 1.6 mm, V_c = 150 ml	61
		Other extractants*		
0.1 *M* Cyanex 272 in *n*-heptane	H_2O, pH = 2.1	Nd, Sm, Eu	Sanki CPC model SPL with 6 cartridges, 400 channels	10, 13
0.1 *M* Cyanex 272 in heptane	H_2O, pH gradient 2.5 to 1.4	La, Pr, Sm, Gd, Dy, Tm		
0.1 *M* Cyanex 272 in *n*-heptane	pH 1.36 to 2.8	La, Pr, Nd, Sm, Eu, Gd, Dy, Tb, Ho, Er, Tm, Yb		
0.2 *M* Ph2Bu2 in chloroform	3.0 *M* HNO_3	Ce(III), Am(III)	Planetary centrifuge with a vertical column drum. R = 140 mm	32
1 *M* Ph2Bu2 in chloroform	Preconcentration: 3.0 *M* HNO_3 Elution: 3.0 *M* NH_4NO_3	Al, Ba, Cu, Co, Fe, Sr, Pb, Mn, Ti, Zr, Cr, V REEs		3
5% DNTDC in MIBK	Preconcentration: 1 *M* HNO_3 Step elution: (1) 0.5 *M* NaCl in 0.5 *M* HCl (2) 0.5 *M* NaCl in 3 *M* HCl	Orthophosphate ion Pyrophosphate ion	Planetary centrifuge with a vertical column drum. R = 140 mm, r = 50 mm, i.d. = 0.75 mm, V_c = 20 ml	49
25% DHDECMP in cyclohexane	2.91 *M* HNO_3	14 REEs	GS-20A HSCCC, multilayer coil partition centrifuge. i.d. = 1.0 mm, V_c = 30.5 ml, β = 0.4 to 0.74	6

TABLE 9.3 (continuation)

Stationary phase	Mobile phase	Element	Apparatus	Ref.
0.005 *M* TPMDPD in chloroform	0.9 *M* HCl + 0.5 *M* NH_4SCN 2.0 *M* HCl	Eu, Fe(II) Fe(III)	Planetary centrifuge with a vertical column drum. *R* = 140 mm, *r* = 50 mm, i.d. = 1.5 mm, V_c = 24 ml	5
0.003 *M* TPMDPD in chloroform	1.0 *M* HCl + 0.5 *M* NH_4SCN	REE (10 mg)	Planetary centrifuge with a vertical column drum. *R* = 140 mm, *r* = 50 mm, i.d. = 1.5 mm, V_c = 20 ml	32
0.025 *M* HEDPA	Am(III)			
0.005 *M* TPMDPD in chloroform	1.0 *M* HCl +0.5 *M* NH_4SCN–0.03 *M* HEDPA 0.025 *M* HEDPA in water	Eu(III) Am(III)		
0.002, 0.0025 and 0.003 *M* TPMDPD in chloroform	0.5 *M* NH_4SCN–1 *M* HCl	Eu(III), Am(III)		
0.035 TDPPMB in chloroform	3.0 *M* HNO_3	Eu(III), Am(III)		
0.07 *M* Tol-triazine in chloroform	3.0 *M* HNO_3	Eu(III), Am(III), Ce(III)		
0.02 *M* Tol-triazine in chloroform	1.0 *M* HCl + 0.5 *M* NH_4SCN 0.025 *M* HEDPA	REE (20 mg) Am(III)		
3% DEHPA+0.05 *M* DCH18C6 in chloroform	10^{-3} *M* $Ba(NO_3)_2$ in 0.1 *M* HNO_3	Sr, Cs	Planetary centrifuge with a vertical column drum.	35, 36
0.1 *M* DCH18C6 in chloroform	5.0 *M* HNO_3 5.0 *M* HNO_3 0.5 *M* CH_3COOH	Rb, Ca Ba Sr		
0.03 *M* DCH18C6 in chloroform	0.005 *M* picrate acid 2.0 *M* HCl	Ca Rb		
DCH18C6/$CHCl_3$/1-BuOH	H_2O	Li^+, Na^+, K^+, Rb^+, Cs^+	CPC model L-90. 3 cartridges, 150 cells, V_c = 55.8 ml	19
0.01 *M* CD in nitrobenzene	Preconcentration: 5×10^{-4} *M* HNO_3 Elution: 5×10^{-3} *M* $Ba(NO_3)_2$ + 0.25% PEG 300 in 0.1 *M* HNO_3	 Cs, Sr	Planetary centrifuge with a vertical column drum.	35, 36

TABLE 9.3 (continuation)

Stationary phase	Mobile phase	Element	Apparatus	Ref.
0.1 *M* TOEDA in chloroform	Preconcentration: (1) 0.1 *M* HCl + 0.01 *M* $H_2C_2O_4$ (2) 0.1 *M* HCl +5% ascorbic acid			37
	Step elution: (1) 0.1 *M* HCl + 0.01 *M* $H_2C_2O_4$	Cu, Al, Co, Ni, Fe, alkali, alkaline earth, REEs		
	(2) 0.1 *M* HCl + 5% ascorbic acid	Fe		
	(3) 2 *M* HCl	Zr, Hf, Nb, Ta		
	(4) 1.0 *M* HNO_3	Zn, Cd		
30% TBP in *n*-dodecane	0.3 *M* HNO_3–0.05 *M* hydrazine	Nd(III), U(IV), U(VI)	Sanki CPC model LLN with 6 cartridges, 2400 cells	62
70%TBP in *n*-dodecane	(1, 2, 3) *M* $LiNO_3$–1 *M* HNO_3	Cu, Pr, Eu		
30% TBP 35% *n*-dodecane, 35% CCl_4	5.0 *M* $LiNO_3$ + 1.0 *M* HNO_3	Ce, Pr, Eu	Sanki CPC model-LLN. 6 cartridges, 2400 cells, V_c = 125 ml	12
25% TOA in toluene	Preconcentration: 8 *M* HNO_3 Step elution:		GS-20A HSCCC, a multilayer coil partition centrifuge. i.d. = 1.0 mm, V_c = 30.5 ml, β = 0.4 to 0.74	17
	(1) 10 *M* HCl	Th		
	(2) 3 *M* HNO_3	U		
	(3) 1 *M* HNO_3	Pu		
0.1 *M* PMBP in toluene	H_2O, pH = 2.32	Pr, Eu	Sanki CPC LLN	3
0.001 *M* PMBP in MIBK	3.0 *M* HNO_3	Hf	Planetary centrifuge with a vertical column drum. *R* = 140 mm, *r* = 50 mm i.d. = 1.5 mm, V_c = 20 ml	35
	0.1 *M* Na_2SO_4 in 3 *M* HNO_3	Zr		
0.25 *M* $(VA10)_2$ in kerosene	0.1 *M* $NaNO_3$ + 0.01 *M* (H,Na)$CHCl_2COO$	Y, Ho	Sanki CPC model NMF with 6 cartridges, 2400 cells, V_c = 125 ml	38
	0.1 *M* $NaNO_3$ + 0.01 *M* (H,Na)$CHCl_2COO$	Y (Dy, Ho, Er)		

*See extractant abbreviations listed in Table 9.4. (H,Na)$CHCl_2COO$ means that NaOH was added to the dichloroacetic acid solution to obtain the desired pH value.

TABLE 9.4

Abbreviations of reagents

ACN	Acetonitrile
BPHA	*N*-Benzoyl-*N*-phenylhydroxylamine
CD	Cobalt dicarbolide
Cyanex 272	Bis-(2,4,4-Trimethylpentyl)phosphoric acid.
Cyanex 301	Bis(2,4,4-trimethylpentyl)dithiophosphinic acid
DCH18C6	Dicyclohexano-18-crown-6
DEHPA	Di(2-Ethylhexyl)phosphoric acid
DHDECMP	Dihexyl(*N,N*-diethyl)methylenecarbamoyl phosphonate
DMF	Dimethylformamide
DMSO	Dimethyl sulfoxide
DNTDC	Dinonyltin dichloride
DPPMB	1,2,4-Tris (diphenylphosphinylmethyl)benzene.
EG	Ethylene glycol
EHPA	2-Ethylhexylphosphonic acid mono-2-ethylhexyl ester
HEDPA	Hydroxyethylidenediphosphonic acid
HMDTCA	*N,N*-Hexamethylenedithiocarbamic acid
LIX 63	5,8-Diethyl-7-hydroxydodecan-6-one oxime
PMBP	1-Phenyl-3-methyl-4-benzoyl-5-pyrazolone
Ph2Bu2	Diphenyl dibutylcarbamoylmethylphosphine oxide.
TBP.	tri-*n*-Butyl phosphate
TPMDPD	Tetraphenylmethylenediphosphine dioxide
TOA	tri-*n*-Octylamine
TOEDA	tetra-Octylethylenediamine
TOPO	tri-Octylphosphine oxide
Tol-triazine	2,4,6-tris(Ditolylphosphoryl)-1,3,5-triazine
VA10	Versatic 10, a bulky tertiary alkyl carboxylic acid

9.4.1 Preconcentration of metal ions

REEs in rocks were preconcentrated in CCC columns by using a 0.5 *M* solution of DEHPA in *n*-decane as stationary phase, and eluted with various concentrations of aqueous HCl solutions [31]. ICP-AES was used for the determination of REEs in the eluent. The data obtained were in good agreement with the certified values of reference rocks, except for heavy REEs. The quantitative recovery of both heavy and light REEs could be attained by using 0.03 *M* DEHPA + 0.02 *M* TOPO in the solvent mixture (3:1) of *n*-decane + MIBK as stationary phase, 1 *M* HCl aqueous solution as mobile phase, or 1.0 *M* Ph2Bu2 solution in chloroform as stationary phase, and 3 *M* aqueous HNO_3 as mobile phase [44].

High-speed CCC using EHPA in toluene as stationary phase was able to enrich selectively the adjacent REEs holmium and erbium. The sample volume could be as large as 2000 ml, and the peaks of holmium and erbium eluted using a pH 2 water phase were well resolved, even with the very low initial concentration of 2×10^{-7} *M* [28].

A 500 ml metal ion solution containing ppb to ppm levels of Ca, Cd, Co, Cu, Mg, Mn, Ni, Pb and Zn ions was continuously fed into a CCC column. The stationary phase (0.6 *M* DEHPA in heptane) volume was minimized and the concentrated metal ions were eluted with 10 ml of 1 *M* nitric acid and determined with ICP-AES. The recoveries of Ca, Cd, Mg, Mn, Pb and Zn ranged above 88% at a concentration of 10 ppb in the sample solution [48].

CCC can also be used as an enrichment separation method [3,17,26, 31,36,37,44,49]. The analyte could first be preconcentrated in the stationary phase, and then separated by isocratic or step-elution or gradient elution.

Enrichment is important not only for extracting metal ions from low concentrations, and preparation of super-pure reagents, but also for eliminating interferences during the determination of analyte. Nakamura et al. [50] accomplished enrichment separation of all REEs using 0.02 *M* $(EHPA)_2$ in toluene as stationary phase. In their experiment, 1000 ml of solution with concentrations of 5×10^{-7} *M* for each REE was pumped through the column and the enriched REEs were eluted by a stepwise pH gradient of mobile phase. The chromatogram in Fig. 9.12 shows that the REEs were well mutually separated.

9.4.2 Determination of plutonium in soil by ICP-MS after CCC pre-separation

Plutonium is one of the most toxic elements. It has been released into the environment by nuclear weapon tests, satellite and reactor accidents, as well as nuclear facilities. The quantities and the isotopic compositions of plutonium released into the environment are a concern for scientists in the environmental and radiochemistry fields. It is known that ^{239}Pu is the most abundant isotope since its half life is 24,110

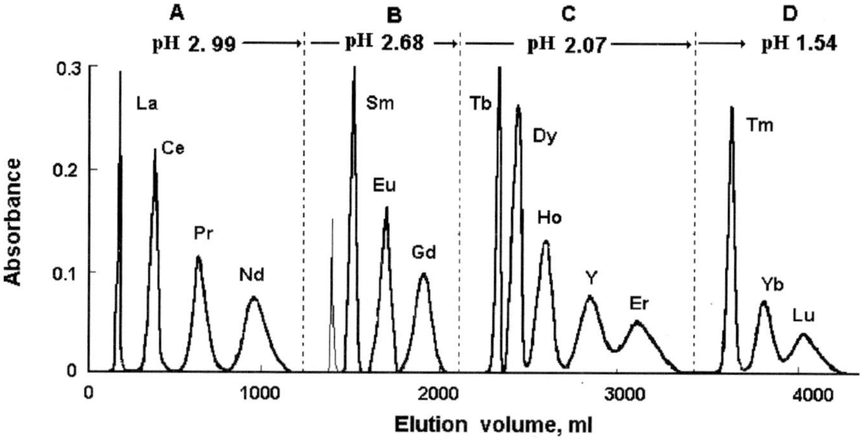

Fig. 9.12. Chromatographic separation of REEs including yttrium. Sample, 5×10^{-7} *M* of each rare earth; 1000 ml passed on the stationary phase, 0.02 *M* $(EHPA)_2$ in toluene. Next, elution with the aqueous mobile phase 10^{-3} *M* $(H,Na)CH_3COO$ adjusted at different pH: (A) pH 2.99; (B) pH 2.68; (C) pH 2.07; (D) pH 1.54 (adapted from Ref. [50]).

years (6,563 years for ^{240}Pu). The ratio $^{240}Pu/^{239}Pu$ can provide an opportunity to distinguish plutonium originating from different sources. The quantities of plutonium in environmental samples are minute. The mean concentration of ^{239}Pu in contaminated soils is estimated to be at a level of 10^{-13} g/g (0.1 ppt) [51]. It had been a challenge for radio-analytical chemists to perform the analysis of a large number of samples in the due time. The merits of inductively-coupled plasma mass spectrometry (ICP-MS), which has high sensitivity, high sample throughput, and the ability to measure isotopic ratios, make possible the routine analysis of ultra-trace plutonium [52–55]. When ICP-MS is used to determine ultra-traces of plutonium, the main interference is signal suppression caused by matrix elements, the polyatomic ion of $^{238}U^1H$, and the tailing of *m*/*z* 238. One should also be aware of the mass discrimination and memory effect existing in the determination process.

In our investigation [55], ultra-traces of plutonium in soil were pre-separated using CCC before the ICP-MS determination. The organic stationary phase was toluene containing 20% tri-*n*-octylamine. The matrix effect, polyatomic interference, was minimized by the pre-separation step. Most interfering matrix elements, thorium and uranium, were eluted by using 7.5 *M* HNO_3, 10 *M* HCl and 3 *M* HNO_3, respectively. The separation factors were 10^5 for thorium and 10^4 for uranium.

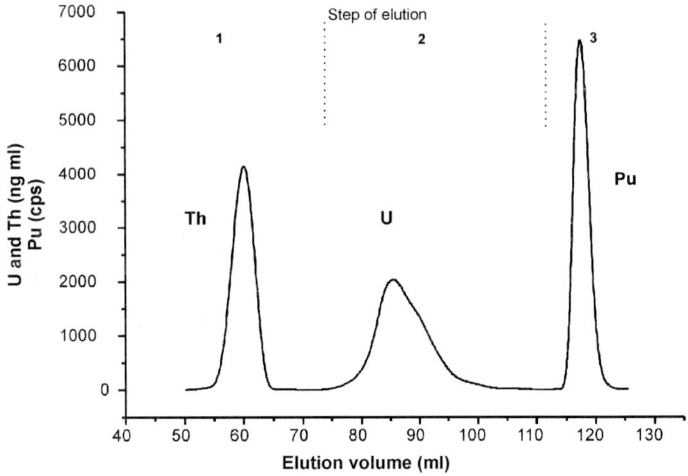

Fig. 9.13. Preconcentration of ultra-trace plutonium and step-elution of Th, U, Pu using CCC. GS-20A multilayer coil partition centrifuge machine, $0.4 < \beta < 0.74$; $V_c = 30.5$ ml; stationary phase; 20% tri-*n*-octylamine in toluene; $V_s = 15$ ml; $Sf = 0.50$. Mobile phase: (1) 10 *M* HCl; (2) 3 *M* HNO_3; (3) 0.025 *M* oxalic acid–0.15 *M* HNO_3. Sample solution, 50 ml of 7.5 *M* HNO_3 containing 20 μg Th, 10 μg U and 0.04 ng ^{242}Pu. Rotation speed, 1200 rpm; flow rate, 1.0 ml/min.

Plutonium was stripped using 5 ml 0.025 *M* oxalic acid–0.15 *M* HNO_3, as shown in Fig. 9.13. The plutonium was almost completely recovered. Aside from the high separation effect and recovery, it is convenient that the eluent, 0.025 *M* oxalic acid–0.15 *M* HNO_3, could be introduced directly into the plasma torch. The presence of oxalic acid causes a little signal suppression owing to the carbon deposits on the interface but, nevertheless, oxalic acid quickly eliminated all memory effects.

The isotope dilution (ID) method was also used to determine ultra-traces of plutonium: ^{242}Pu was used to spike the sample. The existence of mass discrimination was corrected by determining the ratio of the ^{239}Pu and ^{242}Pu isotopes. By applying ID-ICP-MS, plutonium 239 in the femtogram range was measured using a high-resolution ICP-MS. The limit of detection of ^{239}Pu was 5.9 fg/ml (0.0059 ppt or 25 attomoles). The practical detection limit was 25.5 fg/g for soil (0.025 ppt or 0.1 femtomole).

A spike-solution containing 0.04 ng ^{242}Pu was added to 50 g soil and the sample was heated at 500°C for 24 h to decompose the organic matter. The Pu was extracted by acid leaching using 150 ml 8 *M* HNO_3 + 5 ml H_2O_2. The residue was centrifuged. The supernatant was used directly for CCC separation. The eluent containing plutonium was used for ICP-MS determination. The concentration of ^{239}Pu determined was in the range $0.08–1.6\times10^{-12}$ g/g (0.08–1.6 ppt) in soils sampled in Xi'an and Gansu, China. The concentration of ^{239}Pu in the soil sampled in Xi'an almost corresponds to the common background level, while the concentration of ^{239}Pu in the soil sampled in Gansu is larger. Moreover, the ratio of $^{240}Pu/^{239}Pu$ is smaller than the mean value of the global fallout: this result means that there is a source of ^{239}Pu in Gansu.

9.5 SUMMARY

The main difference between the CCC separation of inorganic ionic compounds from organic non-ionic ones is the need of a chemical reaction to perform the CCC process. This difference has two results: (1) the retention volume deviates from the one predicted using the batch distribution ratios, and (2) the efficiency is much lower than that obtained with comparable organic separations, unless a kinetic labile system is used. The lower efficiency is a big disadvantage when some extraction systems are used in CCC. The present theoretical results are useful for the methodology of inorganic separations by CCC. The inherent relationship of the chromatographic behaviour with the chemical reactions during the CCC process can be used to study the kinetics as well as the mechanism of the reaction-complexation used. This might become another field for CCC studies. The methodology of inorganic separations by CCC will be improved with deeper understanding of the inorganic separation. This will help scientists to select the appropriate separation system and operating parameters to accomplish their separation purpose. CCC does have advantages for inorganic separation compared to other chromatographic methods, although it has not yet been widely used. Its unique capabilities, such as very large volumes of sample solution, no contamination during the separation process, no adsorption losses, and the ease of sample treatment are significant, especially for trace-metal analysis. In the present stage, CCC could be used as a relatively mature preconcentration and pre-separation method for inorganic analysis.

REFERENCES

1 W.D. Conway, *Countercurrent Chromatography, Apparatus, Theory, and Applications.* VCH Publishers, New York, 1990.

2 T.A. Maryutina, P.S. Fedotov and B.Ya. Spivakov, In: J.-M. Menet and D. Thiébaut (Eds.), *Countercurrent Chromatography*. Marcel Dekker, New York, 1999, pp. 171–221.
3 S. Muralidharan, G. Ma and H. Freiser, In: K.L. Nash and G.R. Choppin (Eds.), *Separations of f Elements*. Plenum Press, New York, 1995, pp. 177–197.
4 S. Muralidharan and H. Freiser, In: A.H. Bond, M.L. Dietz and R.D. Rogers (Eds.), *Metal-Ion Separation and Preconcentration*. American Chemical Society, 1999, pp. 347–389.
5 P.S. Fedotov, T.A. Maryutina, V.M. Pukhovskaya and B.Ya. Spivakov, *J. Liq. Chromatogr.,* 617 (1994) 3491–3506.
6 Y.R. Jin, L.Z. Zhang, S.J. Hang, L.X. Zhang, J.M. Zhang, G.Q. Zhou and H.B. Dong, *J. Chromatogr. A,* 888 (2000) 137–144.
7 Y.R. Jin, L.Z. Zhang, L.X. Zhang, L. Li and S.J. Hang, *J. Liq. Chromatogr. Relat. Technol.* 24 (2001) 1593–1605.
8 D.E. Schaufelberger, In: Y. Ito and W.D. Conway (Eds.), *High-Speed Countercurrent Chromatography*. Wiley, 1996, pp. 45–70.
9 N.E. Tayar, R.S. Tsai, P. Vallat, C. Altomare and B. Testa, *J. Chromatogr.,* 556 (1991) 181–194.
10 S. Muralidharan and H. Freiser, In: *Recent Prog. Actinides Sep. Chem. Proc. Workshop Actinides Solution Chem.,* WASC'94/1994, Pub. 1997, pp. 191–208.
11 S. Muralidharan and H. Freiser, In: A.P. Foucault (Ed.), *Centrifugal Partition Chromatography*, Chromatographic Science Series, Vol. 68. Marcel Dekker, New York, 1995, pp. 317–330.
12 H. Abe, S. Usuda, H. Takeishi and S. Tachimori, *J. Liq. Chromatogr.,* 16 (1993) 2661–2672.
13 S. Muralidharan, R. Cai and H. Freiser, *J. Liq. Chromatogr.,* 13 (1990) 3651–3672.
14 Y. Surakitbanharn, S. Muralidharan and H. Freiser, *Anal. Chem.*. 63 (1991) 2642–2645.
15 P.S. Fedotov, T.A. Maryutina, A.A. Pichugin and B.Ya. Spivakov, *Russian J. Inorg. Chem.,* 38 (1993) 1758–1764.
16 P.Z. Lu and Z.C. Tai, *Fundamentals of Chromatographic Theory*. Science Press, Beijing, China, 1989, pp. 13–40 (in Chinese).
17 Y.R. Jin, Ph.D. Thesis. Zhejiang University, Zhejiang. P.R. China, 2000 (in Chinese).
18 F. Wilkinson, *Chemical Kinetics and Reactions Mechanisms*. Van Nostrand Reinhold, New York, 1981.
19 T. Araki, Y. Kubo, T. Toda, M. Takata, T. Yamashita, W. Murayama and Y. Nunogaki, *Analyst,* 110 (1985) 913–916.
20 A. Berthod and D.W. Armstrong, *J. Liq. Chromatogr.,* 11 (1988) 567–583.
21 T. Araki, T. Okazawa, Y. Kubo, H. Ando and H. Asai, *J. Liq. Chromatogr.,*

11 (1988) 267–281.
22 K. Akiba, S. Sawai, S. Nakamura and W. Murayama, *J. Liq. Chromatogr.*, 11 (1988) 2517–2536.
23 G. Ma, H. Freiser and S. Muralidharan, *Anal. Chem..*, 69 (1997) 2835–2841.
24 F. Chen, H. Freiser and S. Muralidharan, *Langmuir,* 10 (1994) 2139–2144.
25 F. Chen, H. Ma, H. Freiser and S. Muralidharan, *Langmuir,* 11 (1995) 3235–3242.
26 P.S. Fedotov, T.A. Maryutina, O.N. Grebneva and B.Ya. Spivakov, In: H. Hiroyuki and H. Toshikiko (Eds.), *Int. Symp. Chromatogr. 35th Anniv. Res. Group Liq. Chromatogr.*, Japan, 1995, pp. 575–585.
27 K. Akiba, H. Hashimoto, S. Nakamura and Y. Saito, *J. Liq. Chromatogr.*, 18 (1995) 2723–2741.
28 K. Akiba, H. Hashimoto, S. Nakamura and Y. Saito, *J. Liq. Chromatogr. Relat. Technol.*, 20 (1997) 1995–2007.
29 K. Akiba, In: A.P. Foucault (Ed.), *Centrifugal Partition Chromatography*, Chromatographic Science Series, Vol. 68. Marcel Dekker, New York, 1995, pp. 219–239.
30 E. Kitazume, M. Bhatnagar, Y. Ito, *J. Chromatogr. A*, 538 (1991) 133–140.
31 V.M. Pukhovskaya, T.A. Maryutina, O.N. Grebneva, N.M. Kuz'min and B.Ya. Spivakov, *Spectrochim. Acta,* 48B (1993) 1365–1379.
32 B.F. Myasoedov, M.K. Chmutova and I.A. Lebedev, *Recent Prog. Actinides Sep. Chem. Proc. Workshop Actinides Solution Chem.*, WASC, 1994 (Pub. 1997), pp. 209–221.
33 B.S. Mohite and S.M. Khopkar, *Analyst,* 112 (1987) 191–194.
34 B.S. Mohite and S.M. Khopkar, *Talanta*, 32 (1985) 565–567.
35 Y.A. Zolotov, B.Ya Spivakov, T.A. Maryutina, V.L. Bashlov and I.V. Pavlenko, *Fresenius Z. Anal. Chem.* 335 (1989) 938–944.
36 B.Ya. Spivakov, T.A. Maryutina and Yu. A. Zolotov, *Process Metallurgy,* 7A (Solvent Extr., 1990, pt. A) 451–456. Chem. Abs., 117, 142436e.
37 P.S Fedotov, T.A. Maryutina, O.N. Grebneva, N.M. Kuz'min and B.Ya. Spivakov, *J. Anal. Chem..*, 52 (1997) 1034–1038.
38 S. Nakamura, M. Watanabe and K. Akiba, *J. Liq. Chromatogr.*, 18 (1995) 377–394.
39 H. Hoshi, S. Nakamura and K. Akiba, *J. Liq. Chromatogr. Relat. Technol.*, 22 (1999) 1319–1330.
40 T. Araki, T. Okazawa, Y. Kubo, H. Asai and H. Ando, *J. Liq. Chromatogr.*, 11 (1988) 2473–2485.
41 T. Araki, H. Asai, H. Ando, N. Tanaka, K. Kimata, K. Hosoya and H. Narita, *J. Liq. Chromatogr.*, 13 (1990) 3673–3687.
42 H. Abe, S. Usuda and S. Tachimori. *J. Liq. Chromatogr.*, 17 (1994) 1821–1835.
43 K. Akiba, Y. Ishibashi and S. Nakamura, In: T. Sekine (Ed.), *Proceedings of the International Solvent Extraction Conference, Japan 1990*. Elsevier,

1992, pp. 1677–1682.
44 V.M. Pukhovskaya, O.N. Grebneva, T.A. Maryutina, N.M. Kuz'min and B.Ya, Spivakov, *Spectrochim. Acta,* 50B (1995) 5–12.
45 A. Berthod and D.W. Armstrong, *J. Liq. Chromatogr.*, 11 (1988) 1457–1474.
46 E. Kitazume, In: Y. Ito and W.D. Conway (Eds.), *High-Speed Countercurrent Chromatography*. Wiley, 1996, pp. 415–443.
47 T.A. Marytina, B.Ya. Spivakov and P. Tschöpel, *Fresenius' J. Anal. Chem.*, 356 (1996) 430–434.
48 E. Kitazume, N. Sato and Y. Ito, *J. Liq. Chromatogr. Relat. Technol.*, 21 (1998) 251–261.
49 Yu. A. Zolotov, B.Ya. Spivakov, T.A. Maryutina, V.L. Bashlov and I.V. Pavlenko, *Fresenius' Z. Anal. Chem.,* 35 (1989) 938–944.
50 S. Nakamura, H. Hashimoto and K. Akiba, *J. Chromatogr. A*, 789 (1997) 381–387.
51 Yu. T. Chuburkov, V.P. Perelygin, I. Zvara, Z. Szeglowski, S.P. Shtanko, G.V. Buklanov, Yu. P. Kharitonov, A.G. Belov and T.P. Drobina, *Radiochim. Acta,* 68 (1995) 227–231.
52 A.E. Eroglu, C.W. McLeod, K.S. Leonard and D. McCubbin, *Spectrochim. Acta,* 53B (1998) 1221–1233.
53 W.C. Inkret, D.W. Efurd, G. Miller, D.J. Rokop and T.M. Benjamin, *Int. J. Mass Spectrom.,* 178 (1998) 113–120.
54 I. Rodushkin, P. Lindahl, E. Holm and P. Roos, *Nuclear Instruments and Methods in Physics Research, A,* 423 (1999) 472–479.
55 Y.R. Jin, L.Z. Zhang, S.J. Han, F.R. Zhu, J.F. Bai, G.Q. Zhou, F. Ma and L.X. Zhang, *Acta Chimica Sinica,* 58 (2000) 1291–1295 (in Chinese).
56 T. Araki, T. Okazawa, Y. Kubo, H. Ando and H. Asai, *J. Liq. Chromatogr.*, 11 (1988) 2487–2506.
57 T. Takeuchi and Y. Kabasawa, *J. Chromatogr.,* 538 (1991) 125–131.
58 E. Kitazume, N. Sato, Y. Saito and Y. Ito, *Anal. Chem.,* 65 (1993) 2225–2228.
59 K. Talabardon, M. Gagean, J.M. Mermet and A. Berthod, *J. Liq. Chromatogr. Relat. Technol.,* 21 (1998) 231–250.
60 Y. Surakitbanharn, S. Muralidharan and H. Freiser, *Solvent Extraction Ion Exch.,* 9 (1991) 45–59.
61 Z. Ma, L. Zhang and S. Han, *J. Chromatogr. A,* 766 (1997) 282–285.
62 S. Usuda, H. Abe, S. Tachimori, H. Takeishi and W. Murayama, In: T. Sekine (Ed.), *Solvent Extraction 1990, Part A*. Elsevier Science, Amsterdam, 1992, pp. 717–722.
63 H. Hoshi, A. Tsuyoshi and K. Akiba, *J. Radioanal. Nucl. Chem.*, 249 (2001) 547–550.
64 H. Hoshi, A. Tsuyoshi and K. Akiba, *Solvent Extr. Res. Dev., Jpn.*, 8 (2001) 159–164.

Chapter 10

The use of CCC in the pharmaceutical industry

Guy H. Harris

10.1 INTRODUCTION

The industrial use of countercurrent chromatography (CCC) has been almost entirely by natural products drug discovery programs in pharmaceutical research laboratories. The close collaboration of biologists, microbiologists, natural products chemists and medicinal chemists within these laboratories is a proven source of new drugs and lead structures for medicinal chemistry programs. The bioassay-guided approach to the isolation of natural products can present some challenging problems for which CCC has, many times, been the answer.

Despite the historical success of this effort, two major challenges currently face industrial natural products drug discovery efforts: (1) the adaptation of crude extracts to the high-throughput screening process used for pure compound collections; and (2) the perception that the bioassay-guided isolation process is too costly and time consuming. High throughput screening (HTS) technology utilizes small volumes and sensitive, fluorescence based detection methods. Natural product extracts can be rich in coloured, polyaromatic compounds which can interfere with these detection methods at screening concentrations. Extracts can also be rich in polyphenolic compounds and fatty acids which can interfere with older-types of receptor or enzyme based assays. Pre-fractionation, usually chromatographic, of crude extracts prior to high-throughput screening is emerging as a way to increase the "quality" of the crude extract by reducing assay interference caused by non-specific components. HTS requires large extract collections which

Comprehensive Analytical Chemistry XXXVIII
A. Berthod (Ed.)

are rapidly screened within a few days. The result is tens, or hundreds, of viable actives which much be examined by natural products chemists. To do this requires the adoption and development of generally applicable, parallel isolation methodologies. The nature of CCC makes it an ideal solution for these two problems.

The purpose of this chapter is to review these emerging uses of CCC for crude extract fractionation in the context of the challenges facing industrial natural products drug discovery programs. It is obvious from the literature that CCC is a very useful technique for natural products isolation. Gunawardana and McAlpine recently summarized the preparative applications of CCC and included a summary of the role of CCC in pharmaceutical natural products work [1]. From their summary, numerous other reviews of the application of CCC, and the numerous reports in the primary and patent literature, a legitimate question may be: "Is there a family of natural products that cannot be chromatographed and isolated using CCC?" The bioassay-guided isolation process practised in the pharmaceutical industry does present some unique problems and an attempt will be made to highlight their solution using CCC. Finally, large scale separations are primarily an industrial concern and the progress being made on the industrial scale-up of CCC will be surveyed.

10.2 THE USE OF CCC FOR NATURAL PRODUCT CRUDE EXTRACT FRACTIONATION

10.2.1 The problem

Screening natural sources for biological activity and the purification of natural products from these sources generally requires the preparation of an extract. The purpose of the extraction is to liberate the desired compounds from the whole organism matrix of biopolymers in a useful, concentrated form. Plant material, dry or fresh, is commonly extracted with alcohol or a sequential series of solvents of increasing polarity. Marine invertebrates are treated similarly. Extractions of microbial fermentations can be with either a miscible solvent, such as an alcohol or acetone, or an immiscible solvent such as ethyl acetate or methyl ethyl ketone.

The resulting crude extract is a mixture of tens to thousands of compounds, each ranging in quantity from trace to the major component of the mixture. Use of this crude extract requires that it be soluble in a more concentrated solution than the original extract and in a solvent suitable for further use or fractionation. For extract screening programs this is typically accomplished by utilizing the almost universal solvent DMSO. DMSO is not a useful solvent for further chromatographic fractionation however. Reverse phase stationary supports require the sample to be soluble in an aqueous dominated environment and normal phase supports require a non-polar, organic solvent. Both can be problematic for very crude extracts when applied on a general basis to a large number of samples. In contrast, the biphasic solvent system mixture used for CCC is very general and offers a wide range of solubility.

An ideal isolation scheme for a natural product involves an initial low resolution, easily scaled step combined with one or two complementary high resolution chromatographic separations. A natural product can rarely, if ever, be purified to homogeneity from a crude extract using a single step process. The purpose of the initial step of this process is to reduce the mass of the extract by removing bulk amounts of primary metabolites, media components, and other material while enriching the purity of the desired compound or activity. Sequential liquid-liquid partition is a useful initial step for immiscible fermentation extracts and plant and marine invertebrate whole organism extracts. Solid phase extraction using a neutral, hydrophobic, resins such as polystyrene-divinylbenzene or using ion exchange resins (i.e., Dianion HP-20 or Amberlite XAD) is a typical first step for microbial fermentation water soluble extracts.

10.2.2 CCC as a solution: methodology and solvent systems for crude extracts

The sequential liquid–liquid partition of the crude extract between water and a series of increasing polarity solvents mentioned above is a mild process resulting in nearly complete recovery of solutes. The process does not require pH adjustment and maintains the native ionic form of the solutes. The resulting fractions from the partition scheme generally exhibit better solubility than the crude extract and are several fold enriched by weight relative to a crude alcoholic extract. CCC is essentially a higher resolution form of this process and maintains all of the advantages.

Selection of a solvent system for a CCC separation has generally been specific to the particular problem. Numerous authors have reviewed methods for selecting solvent systems. If the target compound is known, partition coefficients can be experimentally determined for it in a series of solvent systems and this usually leads to the proper solvent system for separation. Similarly, bioactivity and sample weight distribution can be used for unknown compounds, mixtures or crude extracts. While these methods are effective, they are neither general nor high-throughput.

A single, isocratic solvent system effectively fractionates compounds only within a limited polarity range. Unfortunately, the polarity range which typically exists in a crude natural product extract is often broad. Step gradients have been used in a manner analogous to low pressure column chromatography to increase the fractionation range within a single run. Tian and coworkers have reported the use of stepwise elution HSCCC technique to purify tanshinone IIA, tanshinone I and cryptotanshinone, in a single run at over 95% purity, from *Salvia miltiorrhiza* [2]. Gradient elution, a standard HPLC methodology, is limited in CCC to solvent systems in which the phase volume range is not significantly altered over the gradient range but have been successfully applied and are relatively common.

CCC is unique compared to other chromatographic techniques in that the components retained in the stationary phase as the separation progresses are easily accessible. Two operational techniques, dual-mode and simple extrusion of the stationary post-run, are quite effective at extending the fractionation range of a typical isocratic solvent system. Both exploit the separation of components in the stationary phase which has occurred during a run but which were not eluted in the mobile phase. In dual-mode CCC the separation is first developed with mobile phase (MP) and then the phases are reversed during the run and the column eluted with the former stationary phase (SP). Thiebault and coworkers have described a mathematical model of this process for hydrodynamic CCC instruments [3]. It is analogous to ascending and descending flow reversal in hydrostatic, centrifugal partition chromatography (CPC), instruments. A similar result can also be accomplished by stopping the column rotation and mobile phase flow after one or more column volumes of mobile phase has been eluted and simply extruding the retained stationary phase while collecting fractions. Conway has presented some equations to calculate K_D values for components retained in the extruded stationary phase [4].

The challenge for solvent system selection for the parallel processing of crude extracts is to maximize the weight purification and bioactivity distribution over the entire fractionation range for the majority of

TABLE 10.1

Solvent systems for standardized crude extract fractionation

Solvent system	CCC Mode	Application	Ref.
Hexane:EtOAc:MeOH:H_2O (1:3:3:3 v/v/v/v)	Dual-mode HSCCC	Fungal and bacterial crude extract standardized fractionation	7
Hexane:EtOAc:MeOH:H_2O (1:1:1:1 v/v/v/v)	Isocratic HSCCC with SP extrusion	Fungal crude extract standardized fractionation	Harris et al. (unpublished results)
Heptane:EtOAc:MeOH:H_2O (6:1:6:1 v/v/v/v)	Dual mode CPC	Plant crude ethanol extract pre-fractionation	5
CH_2Cl_2:MeOH:H_2O (5:6:4 v/v/v)	Isocratic HSCCC with SP extrusion	Plant crude methanol extract pre-fractionation	12
nBuOH:EtOAc:H_2O (3:2:5 v/v/v)	Dual mode CPC	Plant crude ethanol extract pre-fractionation	5
Polar method 3: SP: MeOH/H_2O (95/5 %v/v) MP_A: heptane : toluene (7:3 v/v) MP_B: toluene MP_C: EtOAc MP_D: MeOH	Gradient elution HSCCC	Plant crude extract pre-fractionation	6

samples of interest. Several attempts have been made to determine general solvent systems for combination with the above methodology for standardized crude extract fractionation. Most of these have relied on a set of test natural products or crude extracts and the generality of the solvent system judged by the resolution of these compounds or crude mixtures. The solvent systems derived from these studies are summarized in Table 10.1 and the results further described below.

Ingkaninan and coworkers [5] surveyed a series of eight isocratic and four gradient solvent systems, based upon various mixtures of heptane, EtOAc, BuOH, MeOH, and H_2O, for their ability to resolve crude ethanol extracts of *Tabernaemontana pachysiphon* leaves, *Aconitum xhenryi* roots, *Tabernaemontanan pandacaqui* and *Catharanthus roseus* cell suspension cultures. The separation ability of each solvent was rated based upon the distribution of extract components observed by thin layer chromatographic analysis of the resulting fractions from centrifugal partition chromatography (CPC) separations. From these experiments, it was found that BuOH: EtOAc:H_2O (3:2:5 v/v/v) gave good separation of polar compounds and heptane:EtOAc:MeOH:H_2O (6:1:6:1 v/v/v/v) gave good separation of non-polar compounds.

Several gradient solvent systems for the pre-fractionation of crude plant extracts have been described by workers at Pfizer [6]. Up to four MP components, ranging in polarity from hexane to MeOH, were employed with an aqueous SP. A set of plant terpenes, alkaloids, and phenolic compounds were used to indicate the polarity range and separation ability of the solvent systems. Separations were carried out using a P.C. Inc. coil planet centrifuge equipped with either a 85 ml or 340 ml Tripple® coil. The authors concluded that the gradient systems described performed well on crude plant extracts however no single system worked equally well for both polar and non-polar components. The utility of the method was illustrated by the isolation of oregonin, **1**, from the SDH active extract of *Alnus incana* in a single step.

1

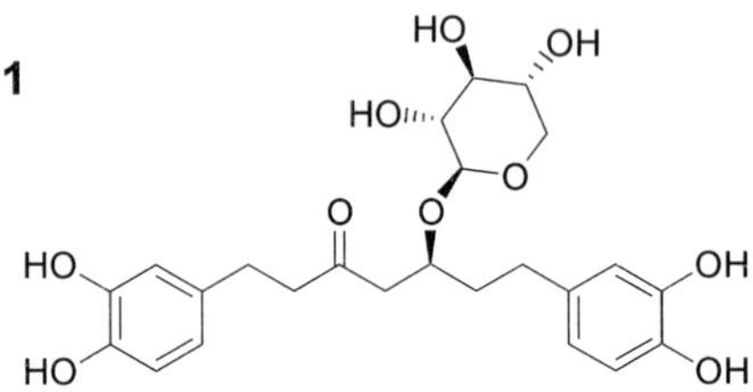

10.2.3 CCC as a general first isolation step for natural product extracts—examples

A standardized dual-mode HSCCC separation for use as a general first isolation step for microbial crude extracts has been described by the MDS Panlabs (now New Chemical Entities, Inc.) natural products group [7]. A P.C. Inc. coil planet centrifuge was used with a solvent system of hexane: EtOAc:MeOH:H_2O (1:3:3:3 v/v/v/v). A photodiode array detector was employed to provide a chromatogram and UV spectrum of the active fractions. Initially, the upper, less polar, phase was used as the MP (normal phase chromatography) eluting with approximately one column volume (360 ml) of solvent. The MP and flow direction were reversed (reverse phase chromatography) and the elution continued for an additional 360 ml. The system was typically used to fractionate a 400 mg sample of crude extract in a single run and a total of 80 chromatographic fractions, consisting of 9 ml each, were collected. The fractions were combined for a next isolation step based upon bioassay results. The next and final isolation step employed in many of the examples detailed below was preparative reverse phase HPLC. Natural products can rarely, if ever, be purified in a single chromatographic step. This two step combination of CCC, especially in normal phase mode, followed by reverse phase HPLC is a very effective and quite general combination for a surprisingly large number of compounds.

This standardized system has been applied to the identification of *in vitro* inhibitors of nitric oxide synthase, p56[lck] tyrosine kinase, and CD45 tyrosine phosphatase and the results are summarized in Table 10.2. Immediately obvious from this summary is the diversity of structural types for which the method has been successful. The retention time/volume in the standardized system coupled with the ultraviolet spectrum obtained using the diode array detector was used as a standard

TABLE 10.2

Structures and retention time in the standardized MDS Panlabs CCC system for reported compounds

Compound	Elution time	Structure	Ref.
Saquamycin A1 (**2**)	45–63	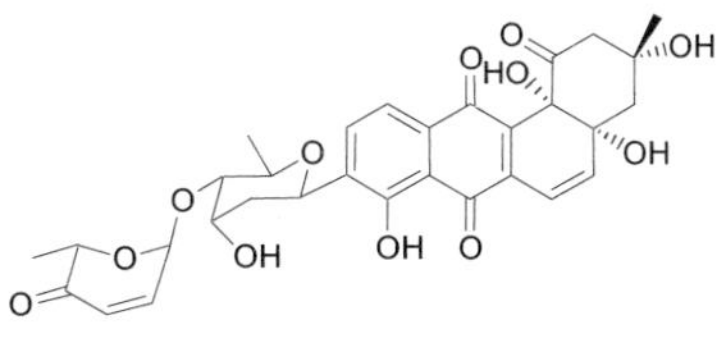	8
A7884 (**3**)	45–63	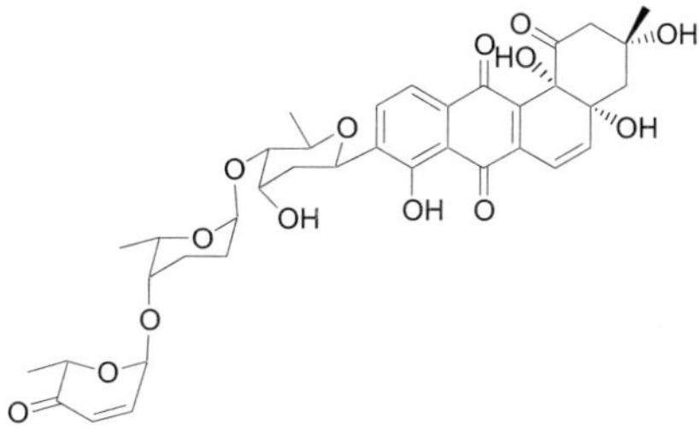	8

TABLE 10.2 (continued)

Compound	Elution time	Structure	Ref.
Vineomycin C (**4**)	45–63	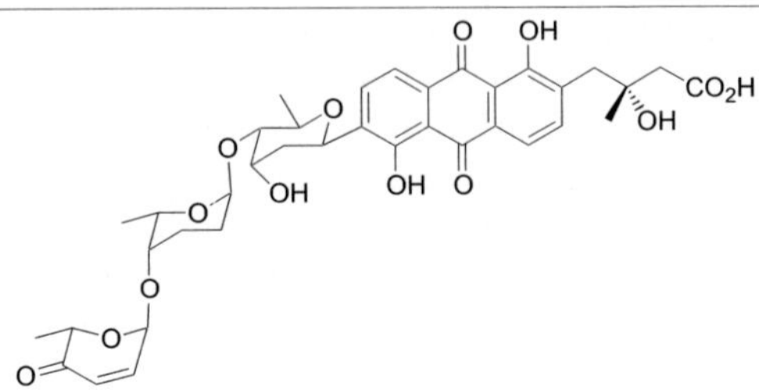	8
Geldanamycin (**5**)	57–76	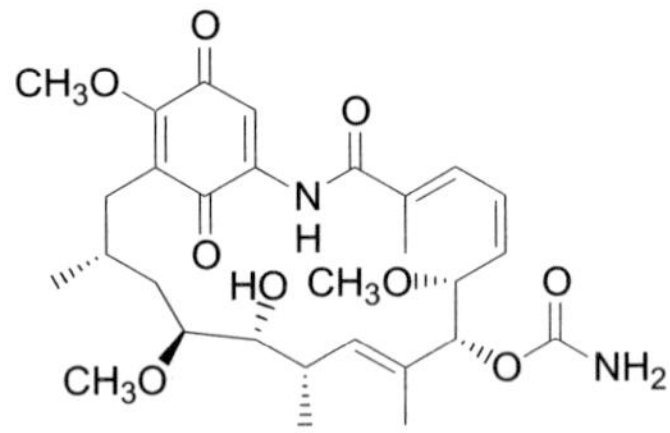	7
Phomacin A (**6**)	60–69	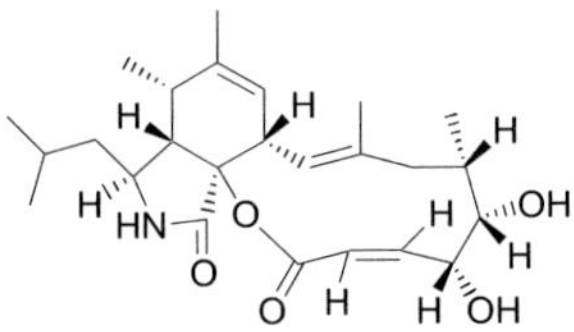	9
Rabelmycin (**7**)	69–84	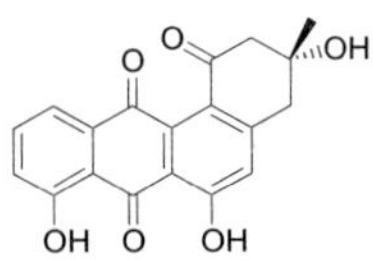	8
Pulchellalactam (**8**)	72–75	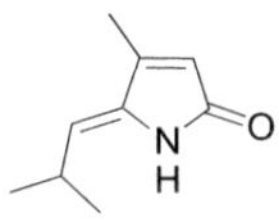	10

TABLE 10.2 (continued)

Compound	Elution time	Structure	Ref.
Xanthomegnin (**9**)	72–99	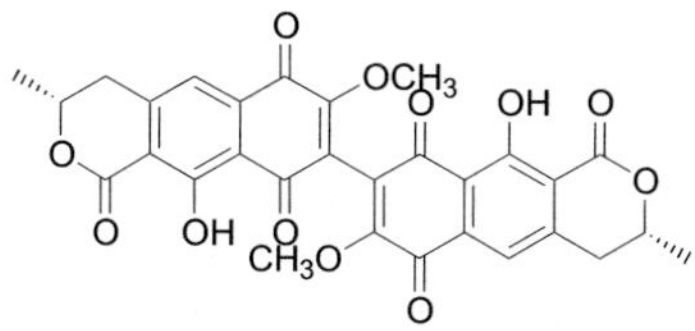	8
Phomacin C (**10**)	108–120	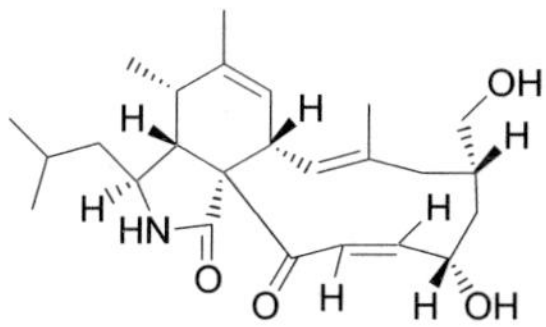	9
Elaiophylin (**11**)	151–171	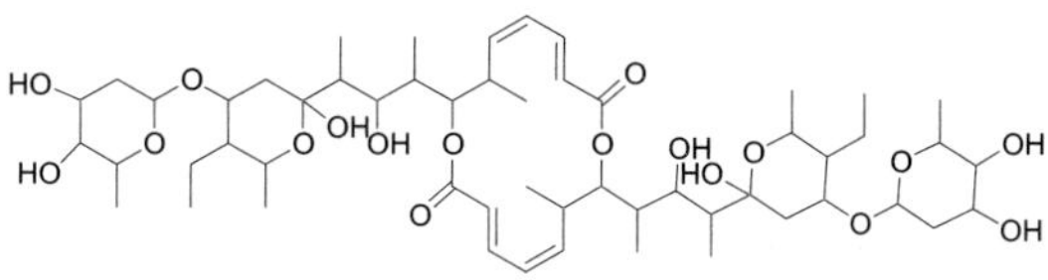	7
Dihydrocarolic acid (**12**)	156–162	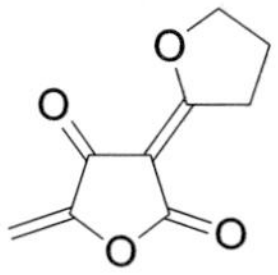	11
Penetricin D (**13**)	192–204	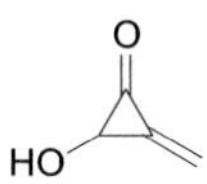	11
Phomacin B (**14**)	219–231	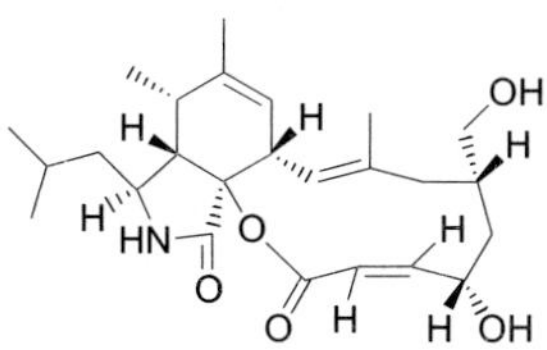	9

dereplication protocol. We have applied a similar system to the isolation and identification of novel broad spectrum antifungal agents. Production of multiple antifungal agents by a single microorganism is common and therefore the antifungal spectrum observed for the crude extract may be a composite of several antifungal agents present in the extract. A standardized CCC separation was used to do a "biological separation" of the antifungal activities allowing the determination of their individual antifungal spectrum prior to complete purification. A solvent system of hexane:EtOAc:MeOH:H_2O (1:1:1:1) was empirically found to distribute the antifungal activity and weight across the entire fractionation range.

Two sets of fungal extracts were processed in parallel using this system. The first consisted of 17 fungal culture extracts with broad spectrum antifungal activity versus *Candida albicans*, *Aspergillus fumigatus* and *Cryptococcus neoformans*. The extracts were subjected to CCC fractionation using the above solvent system in a PharmaTech CCC1000 equipped with three coils of 120 ml total volume. The separation was developed with one column volume of MP (upper layer) and then the rotation and MP flow stopped and the column contents extruded collecting fractions. The antifungal spectrum of each fraction was determined and those with activity of interest were further characterized using LC-MS to identify known antifungal compounds or by complete isolation and structure elucidation for unknowns. The comparative antifungal profile, versus *Candida albicans,* for all 17 extracts as well as the compound identification results is shown in Fig. 10.1. The antifungal compounds identified and the purity of the CCC fractions containing them is shown in Table 10.3. The average stationary phase fraction (*Sf*) for the set was 0.75–0.80.

A second set of 16 fungal extracts with broad spectrum antifungal activity was processed using this system except that a Conway CentriChrom DP1000 coil planet centrifuge equipped with 2–350 ml coils was used. This instrument allowed two simultaneous separations to be performed, doubling throughput. As for the first set, the antifungal activity was located by assaying the fractions against *Candida albicans* and the results shown in Fig. 10.2. The average *Sf* for this set was 0.87 and ranged from 0.74–0.92. The average purification of the antifungal activity was 16-fold and ranged from 1.1-fold to 96-fold.

Examination of Figs. 10.1 and 10.2 shows a general and even distribution of antifungal activity over the entire fractionation range. As expected, the more non-polar compounds, such as emestrin and cytochalasin A, elute early and the relatively polar lipopeptides typified by

TABLE 10.3

Structures and weight/purification factors for antifungal set 1

Extract	Taxonomy	Active component	Structure	Weight/ purification
F852044	*Crytosporiopsis* sp.	Echino-candin family	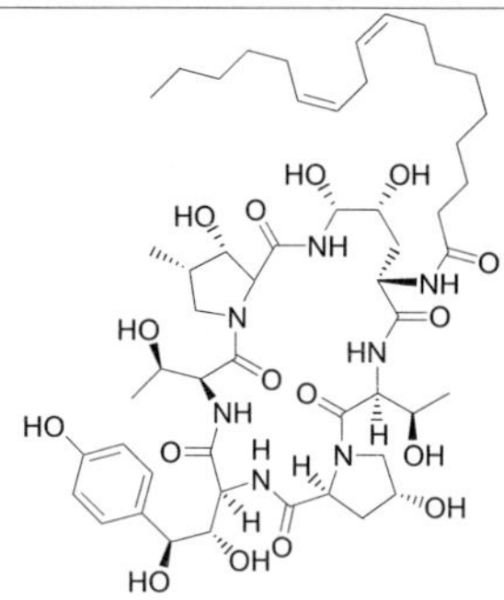	11 mg/8.9×
F852312	sterile mycelium			10 mg/8.0×
F852557	*Polychema* sp.			96 mg/2.4×
F852569	unidentified			71 mg/2.6×
F852574	*Aspergillus* sp.			30 mg/6.5×
F852019	unidentified	versiol / calbistrin mixture	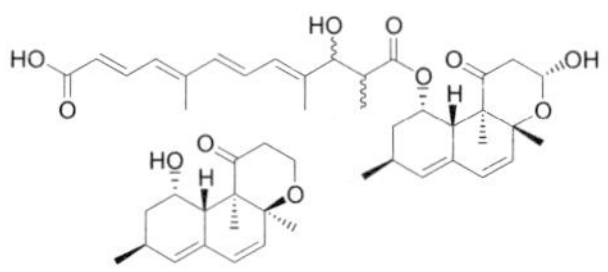	Fr 5–12; 12.3 mg/12× Fr 13–15; 5.2 mg/27× Fr 16–20; 8.9 mg/16×
F852613	sterile mycelium			Fr 4–17; 53.7 mg/1.6×
F851599	*Hypomyces polyporinus*	myco-parasitic acid B (**17**)	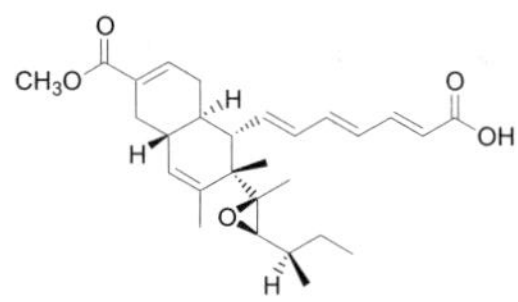	97 mg/1.8×
F851708	unidentified	cytochalasin A (**18**)	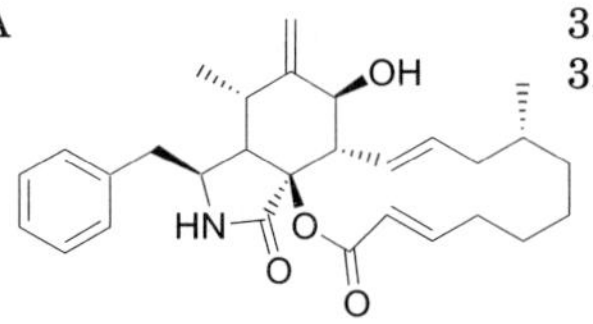	3.0 mg/14×
F851956	unidentified			3.0 mg/34×
F851972	*Nectria* sp.	monorden (**19**)	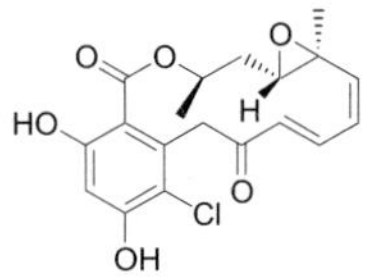	6.0 mg/45×

TABLE 10.3 (continuation)

Extract	Taxonomy	Active component	Structure	Weight/ purification
F852574*	*Aspergillus* sp.	emestrin (**20**)	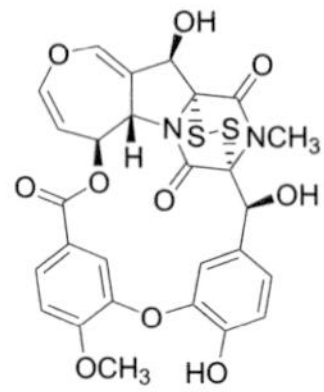	3.2 mg/61×
F852190	Xylariaceae	not identified		26 mg/6.3×
F851745	*Acremonium* sp.	not identified		78 mg/1.3×
F852183	*Fusarium* sp.	1) not identified 2) not identified		12 mg/53× 13 mg/50×
F852364	*Penicillium* sp.	1) not identified 2) not identified 3) not identified		102 mg/2.1× 0.7 mg/314× 28 mg/7.8×
F851569	unidentified	1) not identified 2) not identified		Fr 5–11; 6.9 mg/28× Fr 12–17; 2.8 mg/68 x
F851527	sterile mycelium			Fr 4–11; 5.9 mg/9.5× Fr 12–17; 2.0 mg/28×

echinocandin B was found late in the SP extrusion portion of the separation. As a first isolation step the separation provided satisfactory weight purification in most samples. Almost all samples were sufficiently pure to give useful ESI-LC/MS data. The reproducible elution of

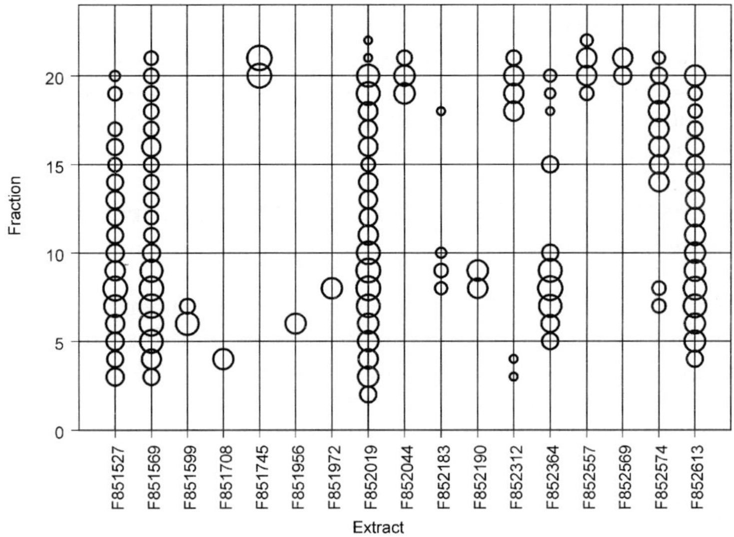

Fig. 10.1. Distribution of anti-*Candida* activity for Set #1 of 17 antifungal extracts fractionated by CCC. The size of the circle represents the zone diameter of antifungal activity.

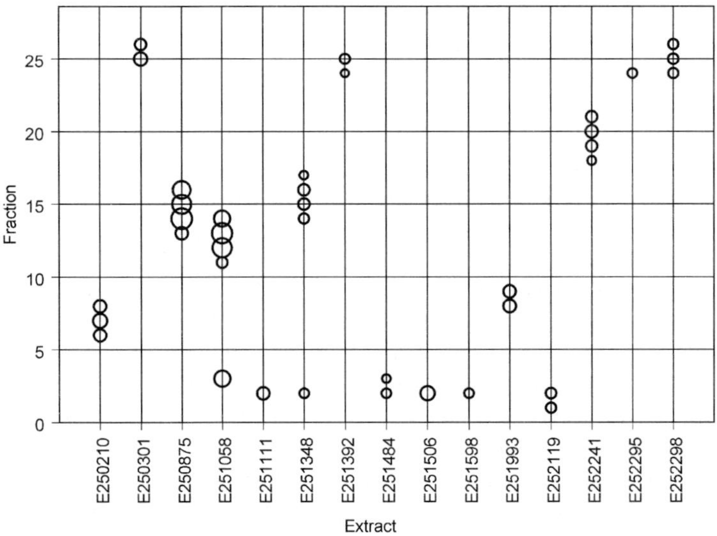

Fig. 10.2. Distribution of anti-*Candida* activity for Set #2 of 15 antifungal extracts. The size of the circle represents the zone diameter of antifungal activity.

cytochalasin A in two extracts, the versiol/calbistrin A mixture in two extracts, and the echinocandin family in five extracts provided useful sample classification information. Finally, the extracts of Set 1 had multiple antifungal activities which were revealed using the CCC fractionation.

10.2.4 Application of CCC to crude extract pre-fractionation for bioassay—examples

The CCC methodology described above has also been used for pre-fractionation of crude extracts prior to bioassay. For this use, the goal is a standardized system which reproducibly segregates compounds, such as fatty acids and polyphenolics, which interfere with bioassays and which provides some dereplication information.

We have described the pre-fractionation of a set of 64 crude methanol plant extracts prior to broad-based screening in approximately 25 bioassays [12]. The purpose of this experiment was two-fold: (1) to determine whether CCC was sufficiently robust for large numbers of diverse and randomly selected crude extracts; and (2) to determine if pre-fractionation was advantageous for broad bioassay screening. A solvent system of CH_2Cl_2:MeOH:H_2O was chosen because of the historically successful use of this solvent system for plant extracts. A ratio of 5:6:4 v/v/v was chosen, based upon a set of standard plant natural products, to favour the fractionation of the more polar components of the extracts. A 500 mg sample of each methanol extract was fractionated on a P.C. Inc. instrument with a coil volume of 300 ml. The lower phase (the chlorinated organic solvent phase) was chosen as the mobile phase to concentrate chlorophyll and fatty acids in the first one or two fractions, localizing these potential assay interferences. The separation was developed with one column volume of MP, until a component with a $K_D = 1$ eluted, and then the column rotation and MP flow was stopped, and the column contents were then extruded collecting fractions as for the MP.

The CCC performance was generally excellent and robust over the entire sample set. The *Sf* ranged from 0.29 to 0.82, with an average of 0.69, and no solubility problems were encountered. The measured sample recovery averaged 88% which was probably an underestimate due to the use of robotic fraction weighing equipment and the fact that no attempt was made to rigorously dry the feed sample. The overall weight distribution was skewed toward the extreme non-polar and polar ends of the separation. The polar fractions were dominated by polyphenolic

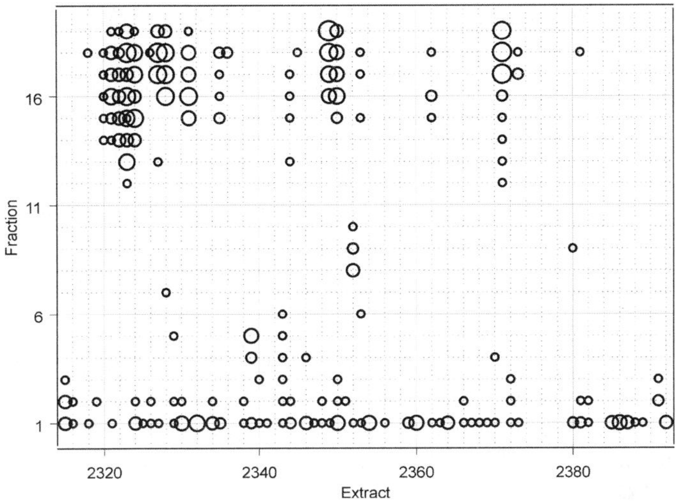

Fig. 10.3. Distribution of primary hits for the CCC fractions from 64 plant extracts tested in 26 bioassays. The size of the circle represents the number of assays in which the fraction tested positive.

compounds and carbohydrates and the non-polar fractions mainly contained fatty acids and other lipids. A screening sample concentration of 10 mg/ml was targeted which generally represented a concentration over simply screening the crude extract or simple partition of the crude extract.

The fractions were screened in 20 bioassays of diverse types. The distribution of the confirmed actives from all 64 samples and 19 fractions was examined as a measure of the usefulness of this pre-fractionation. Active fractions generally correlated well with the weight distribution and were skewed to the extremely polar and non-polar fractions (Fig. 10.3). Comparison of the activity distribution over several assays was also useful. For example, assays known to be effected by polyphenolic compounds exhibited the same pattern of extracts/active CCC fractions.

The solvent system of heptane:EtOAc:MeOH:H_2O (6:1:6:1 v/v/v/v) described by Ingkaninan et al. [13,14] was used to pre-fractionate seven plant EtOH extracts prior to screening of the fractions for activity in two receptor binding assays, adenosine A_1 and opiate, AChE inhibition, and anti-asthmatic activity. The reproducible pre-fractionation step was useful to localize fatty acids, linoleic acid in this example, in fractions

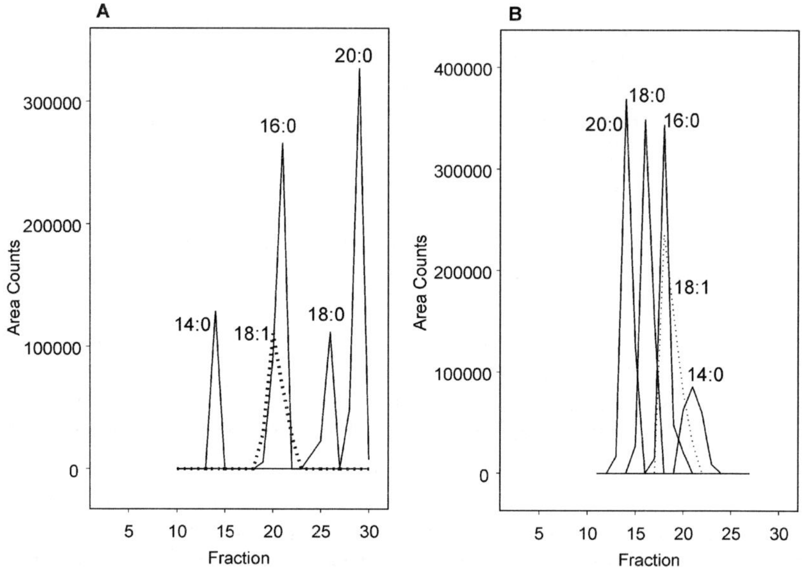

Fig. 10.4. (a) CCC separation of fatty acid mixture using heptane:MeOH:H_2O (100:95:5 v/v/v) containing 0.1% TFA, aqueous mobile phase. (b) CCC separation of fatty acid mixture using nonaqueous solvent system of hexane:CH_2Cl_2:ACN (50:15:35 %v/v/v), upper (hexane rich) mobile phase.

9–21 which yielded false positives in the receptor binding assays. In this study two common zones of inhibition to most of the extracts were observed, fractions 9–21 corresponding to fatty acids, and a polar zone, fractions 33–36. A second solvent system of EtOAc:MeOH:H_2O (43:22:35 %v/v/v), without heptane, was used to further chemically and biologically resolve the polar fraction.

Fatty acids are known to have some bioactivity through low micromolar ligands for many nuclear receptors. Fractionation of an extract in a standardized CCC system has been useful in our laboratory to rapidly distinguish extracts containing fatty acids as the sole active component from those containing "real" activity. We have utilized two solvent systems for this purpose, heptane:MeOH:H_2O (100:95:5 v/v/v) containing 0.1% TFA and the non-aqueous system of hexane:CH_2Cl_2:CH_3CN (50:15:35 v/v/v). Separation of a set of saturated fatty acids, C14, C16, C18, C20, and oleic acid (C18:1) in each of the two solvent systems is shown in Fig. 10.4. The fatty acid separations were performed using a Conway Centrichrome DP-1000 coil planet centrifuge equipped with two 165 ml bicoils. Baseline resolution of the even chain saturated fatty acids was easily achieved using these systems. The non-aqueous solvent system generally exhibited better solubility for crude extracts but retention of individual fatty acids was easier to control with the heptane:MeOH:H_2O system by simply adjusting the percentage of MeOH in the aqueous portion.

10.2.5 Application of CCC to the deconvolution of combinatorial libraries

A patent application from Zeneca Limited, UK, claimed the utility of CCC for the deconvolution of chemical libraries [15]. In the first of two examples, CCC was used to fractionate a sym-triazine nucleus based library using three amines in each of positions 2, 4, and 6 to provide a library of three mixtures, each containing nine components. The library (20 mg) was purified on a 60 ml coil using a solvent system of isohexane: EtOAc: MeOH:H_2O:triethylamine (2:1:2:1:0.6 v/v/v/v/v) with the lower aqueous phase mobile. Components were well distributed among all fractions as determined from HPLC analysis. The second example was a cysteine template based library containing 10 products. The mixture (60 mg) was purified, again on a 60 ml coil, using a solvent system of isohexane:EtOAc:MeOH:H_2O (1:1:1:1 v/v/v/v) and 13 fractions were collected. Again, the components were well distributed in most of the fractions. It was further claimed that fractions could be selected with "drug-like" ClogP values.

10.3 THE USE OF CCC FOR BIOASSAY-GUIDED ISOLATION OF NATURAL PRODUCTS

10.3.1 Bioassay-guided isolation of natural products using CCC

Modern high speed CCC is an established technique in most academic and industrial natural products groups. This is evidenced by numerous reviews of the application of CCC to the isolation of natural products. The Hostettman group has extensively reviewed the application of various CCC techniques to the isolation of plant natural products[16]. Zhang and co-workers (Chapter 8 and [17]) have reviewed the application of CCC to the fractionation of Chinese medicinal herbs. Fregeau and

Rinehart [18] reviewed the application of CCC to the isolation of marine natural products. McAlpine and his group at Abbott Laboratories was a pioneer in the application of various CCC techniques to the bioassay-guided isolation of fermentation products and their work has been summarized [19]. Winterhalter and coworkers have described several examples of the use of CCC for the isolation of labile, water soluble aroma precursors from wine and other foodstuffs [20,21].

Most natural products isolation performed by the pharmaceutical industry is bioassay-guided. Collections of plant, microorganism, marine invertebrate, etc. crude extracts are subjected to bioassay. Upon detection of a desired activity in one of these extracts, the problem for the natural products chemist becomes the isolation and identification of the responsible compound(s) using the bioassay as a guide. The success or failure of a chromatographic step in this process is judged by the weight purification achieved and the recovery of applied biological activity. It is a biological activity directed, not a chemically directed, activity.

Biological assays can be exquisitely sensitive and point to very minor, chemically unstable, natural products which would not otherwise be isolated. Most industrial natural products chemists seem to encounter CCC for the first time when they face one of these unstable, difficult to isolate compounds. The isolation of aroma precursors from wine and other foodstuffs are an example of the application of CCC the isolation of unstable, labile natural products. CCC is useful for other types of natural product isolation problems as well. A given solvent system can usually be used in either reverse phase of normal phase mode. This allows a normal phase separation of compounds which are not otherwise separable in this mode. The squalene synthase inhibitor and arthrinoside isolations described below are examples of this application. Compounds with tautomeric or chelating functional groups can bind to solid stationary phases resulting in poor peak shape or irreversible adsorption of material. CCC was used to solve this problem in the isolation of australifungin.

10.3.2 Application of CCC to bioassay-guided isolation problems—examples

Isolation of tricarboxylic acid based squalene synthase inhibitors

A family of novel picomolar inhibitors of squalene synthase, the zaragozic acids (squalestatins), were independently isolated from various fungal sources by researchers at Merck and Glaxo [22,23]. This family of

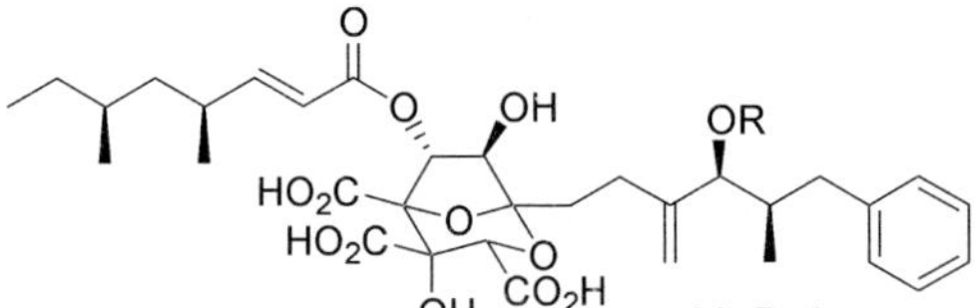

21 R=Ac
23 R=H

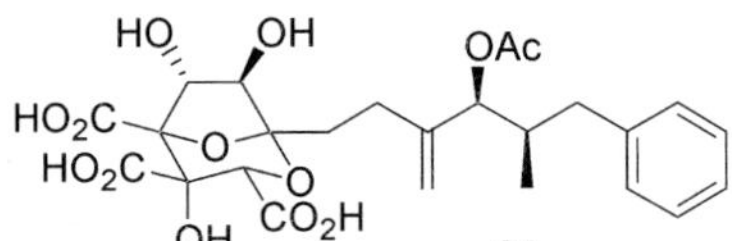

24

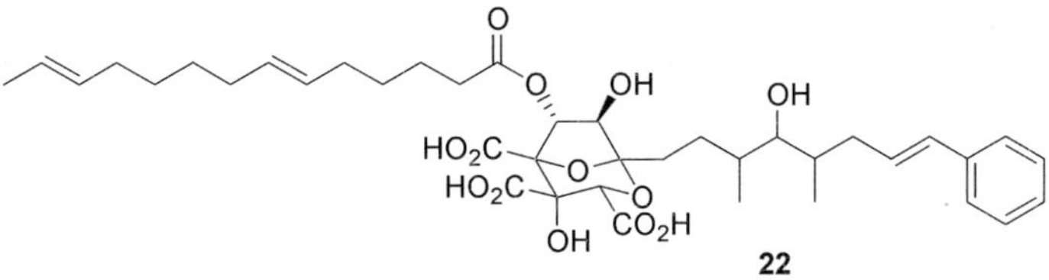

22

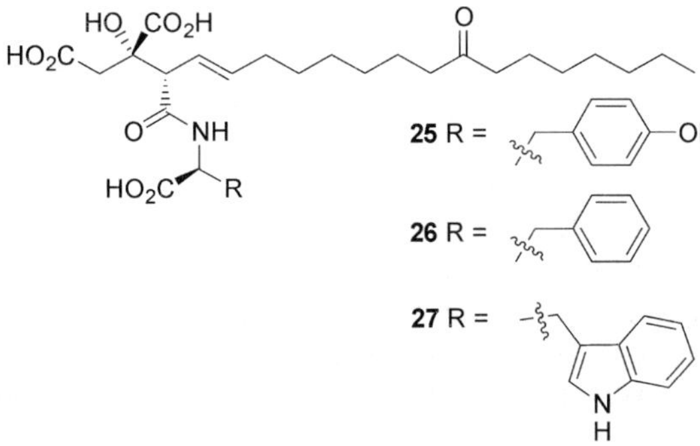

metabolites, typified by zaragozic acid A (squalestatin I), **21**, is characterized by a bicyclic, tricarboxylic acid core with two lipophilic side chains. A series of amino acid containing acyclic alkyl citrates, the viridiofungins, were also discovered by this screening program for squalene synthase inhibitors [24]. The triacid moiety of the zaragozic acids and alkyl citrates can be exploited for their isolation through the use of a selective ion exchange step but the triacid moiety also limits the use of classical normal phase silica gel based chromatographic methods. Thus the only high resolution chromatography available for separation of individual members of either family of compounds is reverse phase HPLC.

CCC was used for the original bioassay-guided isolation of the squalestatins by workers at Glaxo [23]. These workers employed a solvent

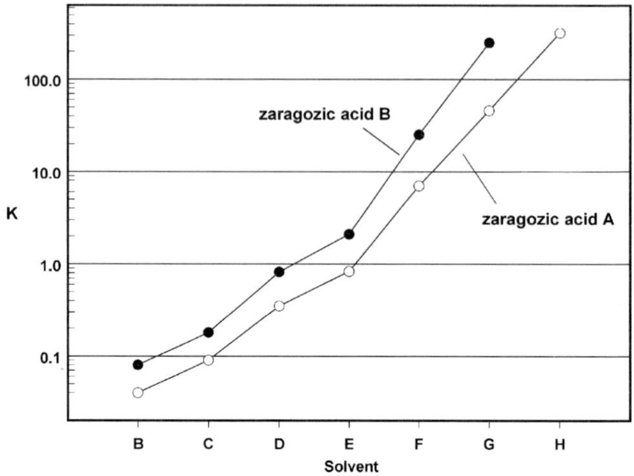

Fig. 10.5. Partition coefficients for a zaragozic acids A and B in a series of hexane: EtOAc: MeOH: H_2O solvent systems containing 0.1% H_3PO_4.

system of hexane:EtOAc:MeOH:0.01N H_2SO_4 (5:6:5:6 v/v/v/v) to resolve a crude fungal EtOAc extract into three zones of squalene synthase inhibitory activity corresponding to **21**, desacetyl squalestatin I, **23**, and a deacylated squalestatin I, **24**, in order of elution. The zones containing **23** and **24** were further purified using the same solvent system in phase ratios of (2:3:2:3 v/v/v/v) and (1:6:1:6 v/v/v/v) respectively.

We also exploited CCC as an alternative, high resolution separation for individual members of this group of compounds. To develop CCC separations for individual zaragozic acids, partition coefficients for two zaragozic acids, A and B (**22**), were measured in a series of hexane: EtOAc: MeOH:H_2O + 0.1% H_3PO_4 solvent systems based upon the work of Oka et al. [25]. A plot of K_D versus the approximate polarity of the solvent system is shown in Fig. 10.5. CCC separations were carried out using a P.C. Inc. coil planet centrifuge equipped with a 300 ml, 1.68 mm i.d. coil, using a phase ratio of 4.5:5:4.5:5 v/v/v/v derived from the data in Fig. 10.5. Excellent resolution of zaragozic acids A and B was obtained in either normal phase or reverse phase modes (Fig. 10.6A). Baseline resolution in normal phase mode was maintained up to loadings of 300 mg of each component (Fig. 10.6B).

Isolation of the viridiofungin family of amino acid containing alkyl citrates from the fungal fermentation was similar to that of the zara-

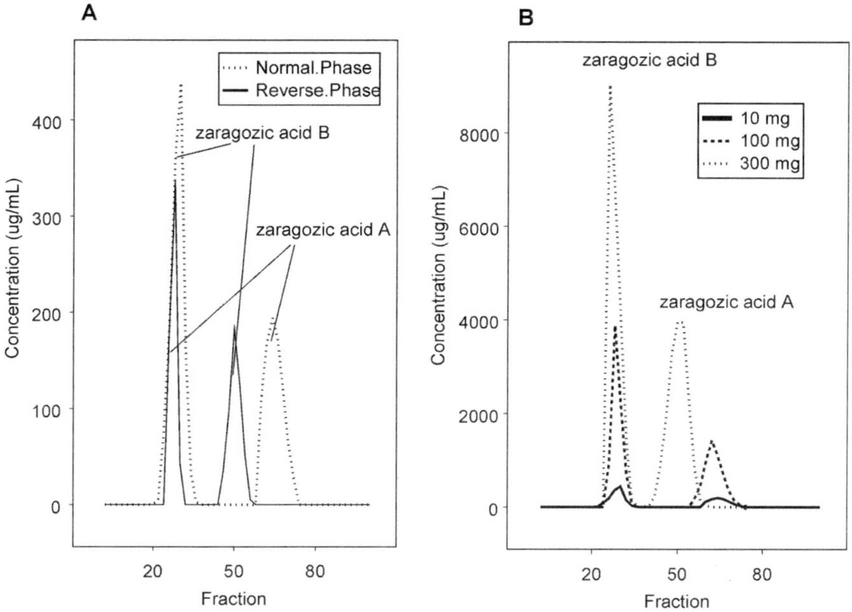

Fig. 10.6. (A) Normal phase and reverse phase CCC separation of zaragozic acids A and B. (B) Effect of loading on normal phase CCC separation of zaragozic acids A and B.

gozic acids except that numerous, related minor components were co-produced with the three major alkyl citrates. The presence of these minor components made obtaining highly purified material, even for the major components, somewhat challenging. Purification of viridiofungins A, B, and C (**25**, **26**, and **27**) consisted of an initial, selective ion exchange capture of tricarboxylic acids from the crude extract, followed by normal phase CCC (Fig. 10.7), and preparative reverse phase HPLC. A 950 mg sample of the crude tricarboxylic acid fraction, obtained by ion exchange, was separated using a solvent system of hexane:EtOAc: MeOH:H_2O containing 0.1% H_3PO_4 (1:1:1:1 v/v/v/v) on a P.C. Inc. MLCCC containing a 385 ml coil. Viridiofungin B (95 mg) eluted at K_D = 0.48, viridiofungin C (30 mg) eluted at K_D = 0.30, and viridiofungin A (558 mg) was recovered from the extruded stationary phase postrun. Just as for the zaragozic acids, the normal phase CCC step provided a separation dimension not obtainable with other chromatographic techniques. Resolution of viridiofungins B and C using reverse phase at preparative scale loading is difficult but was straightforward using CCC.

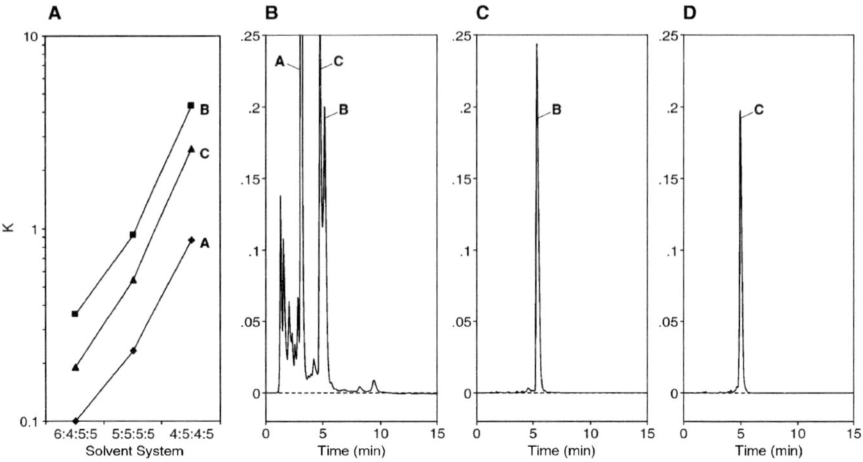

Fig. 10.7. CCC separation of viridiofungins A, B and C. (A) partition coefficients in a series of hexane:EtOAc:MeOH:H_2O solvent systems containing 0.1% H_3PO_4. Analytical RP HPLC analysis of: (B) crude viridiofungin fraction (CCC feed sample), (C) viridiofungin B fraction from CCC, (D) viridiofungin C sample from CCC.

Australifungin

The bioassay-guided isolation of the novel ceramide synthase inhibitors, australifungin, **28**, and australifunginol, **29**, from fermentation extracts of *Sporormiella australis* presented both stability and chromatographic challenges [26,27]. The ceramide synthase inhibition activity of crude culture extracts was unstable to extremes of pH and gave broad elution zones on solid chromatographic supports, particularly preparative reverse phase HPLC. These problems result from the combination of beta-ketoaldehyde and alpha-diketone functional groups of australifungin. These functional groups exhibit extensive keto-enol tautomerism, are somewhat reactive, and possess chelating properties, all of

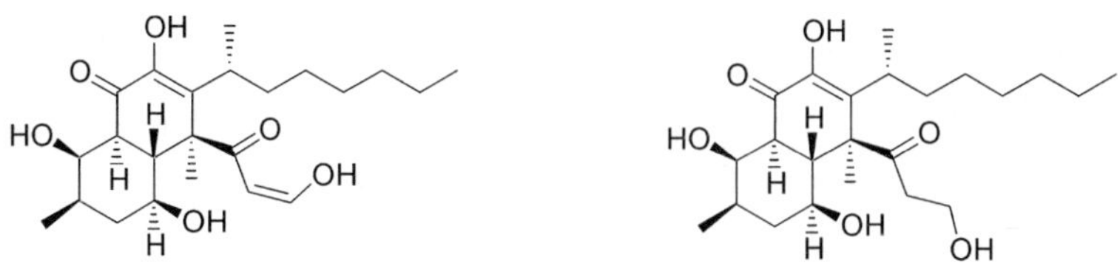

28 **29**

which are problematic for chromatographic separations. CCC was used to solve all of these chromatographic problems.

A two-step isolation process for australifungin and australifunginol was developed consisting of EtOAc extraction of solid fermentations of *Sporormiella australis*. The EtOAc extract, after partition between hexane and MeOH to remove non-polar material, was subjected to low pressure silica gel column chromatography under slightly acidic conditions to separate australifungin from australifunginol. Australifunginol, lacking the beta-ketoaldehyde, was then easily purified using preparative reverse phase HPLC. Analytical reverse phase HPLC of australifungin however required high carbon load C18 packings, elevated temperature (55°C), and buffering at pH 6.8 for good chromatographic peak shape. These conditions were not scalable and resulted in severe peak fronting and poor resolution (Fig. 10.8A). In contrast, CCC separation of the crude australifungin obtained after silica gel chromatography using a solvent system of hexane:EtOAc: MeOH:25 m*M* K_2HPO_4 pH 6.8 (7:3:5:5 v/v/v/v) resulted in excellent peak shape at room temperature.

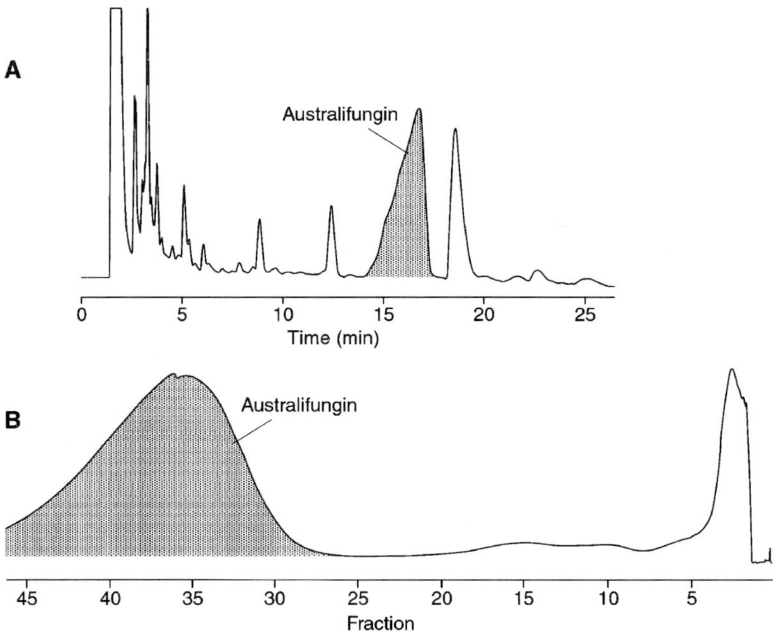

Fig. 10.8. (A) Semi-preparative reverse phase HPLC separation of australifungin rich cut from silica gel. (B) CCC separation of australifungin rich cut from silica gel.

The separation was scalable to several hundred milligrams and yielded pure australifungin (Fig. 10.8B).

Arthrinosides

A family of glycosylated acyclic polyketides were discovered as antifungal agents from culture extracts of an *Arthrinium* sp. The initial sample of arthrinoside A, **30**, was obtained in two bioassay-guided isolation steps, normal phase chromatography on diol silica followed by semi-preparative RP HPLC. It was evident from the initial isolation of arthrinoside A that several additional arthrinosides were present in the fungal extracts but neither chromatographic step provided further resolution of the mixture. The diacid character of the arthrinosides was successfully exploited for pH zone-refining CCC [28]. A pH zone-refining CCC separation of the crude arthrinoside mixture obtained after the diol silica gel step is shown in Fig. 10.9. The pH of effluent was monitored online and the fractions were assayed for anti-*Candida* activity. Three distinct pH plateaus were observed, implying three different major components, which also corresponded to antifungal activity zones. The middle zone contained **30** and an unrelated compound. The first and last zones contained two new members of the family, arthrinosides B (**31**) and C (**32**), respectively. The new compounds were essentially pure and the samples were used directly for structure elucidation.

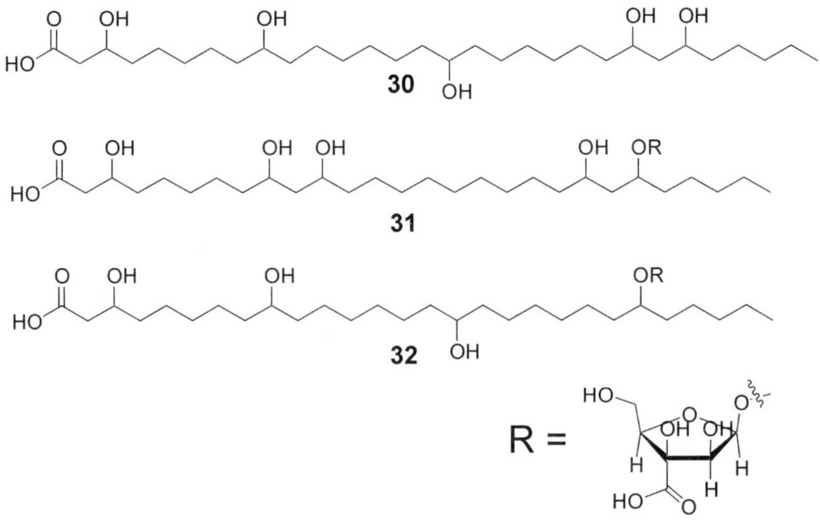

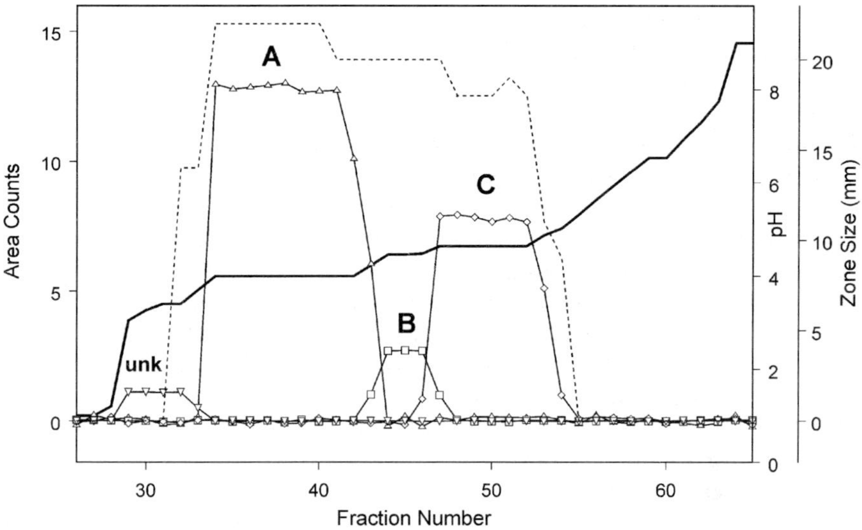

Fig. 10.9. pH Zone-refining CCC separation of arthrinosides A, B, and C.

10.4 INDUSTRIAL SCALEUP OF CCC

As has been previously stated, CCC is fundamentally a preparative chromatographic tool [29]. Most of the literature applications of CCC have been to prepare a research quantity of a compound, ranging from a few milligrams to several grams, for further use such as biological evaluation or structure elucidation. These separations may only be repeated a few times. They are generally performed in a research laboratory environment, and the cost to prepare the product is usually not a factor. In contrast, industrial separations are a routine process which is regularly repeated, generally performed in a pilot plant or process scale environment, and must be cost effective.

Scale does not change the separation principle for CCC. The CCC process can be effectively described using standard chromatographic theory and the separations can be well modeled [30]. This predictability, as well as solute recovery and lack of a solid support, has made CCC an attractive process to apply to industrial separations [31]. But the true industrial scale-up use of CCC has been limited by the lack of commercially available reliable instrumentation.

A UK government funded research program (BBSRC/DTI LINK) of academic (Brunel University, University of Wales Swansea, University College of London) and industrial interests (Glaxo Wellcome, Shell Research, AstraZeneca Pharmaceuticals, Zeneca Agrochemicals, Romulus Technology, AECS) recognized this problem and is beginning to address the engineering, hydrodynamic, and chromatographic variables involved in the scale up of CCC. As part of this effort, the predictable scale-up of a CCC separation from analytical to process scale of test mixture of benzyl alcohol and 2-phenyl-ethanol has been demonstrated by Sutherland et al. [32]. They found that scale-up on a single machine could be predicted from the partition coefficients of the samples to be separated, derived from an analytical scale separation, and knowledge of the stationary phase retention at scale-up flow rates and coil rotation speeds. A maximum sample throughput of 2.72 kg/day was predicted for this ideal example.

Other workers have described a low-speed rotary countercurrent chromatograph using a convoluted helical tube for industrial scale separations [33]. The instrument described utilized convoluted tubing to enable higher stationary phase retention and enhanced mixing of the phases when compared with standard wall tubing. A 10-litre volume coil was mounted in a rotary frame which allowed the flow tubes to rotate at 0–100 rpm without twisting. The simple design and low flow rates allow unattended operation of the instrument over extended periods of time. As an example, 150 g of crude extract of tea leaves were separated using a solvent system of hexane:EtOAc:nBuOH:water:acetic acid (0.5:1:2:6: 0.2 v/v/v/v/v), lower aqueous phase mobile, at a flow rate of 5 ml/min and coil rotation of 21 rpm. Resolution of the polyphenolic compounds ECG, epigallocatechin gallate (EGCG, which comprised 30% of the sample by weight), and epicatechin, as well as caffeine, from the extract was demonstrated. EGCG, 40.05 g, was recovered at a purity of 92.7% in 82.6% yield after 72 h.

The scale-up potential of CCC has been compared with pilot and process scale HPLC for industrial purification by several workers. Graham and coworkers [34] at AstraZeneca reported an economic evaluation of process scale HPLC versus HSCCC for the separation of a pharmaceutical active, the structure of which was not disclosed. The existing process was a RPLC separation on a 150 mm diameter C18 Kromasil column. A HSCCC separation was developed for the target compound using the same material quality targets as for the RPLC process. CCC separations were carried out on a J-type machine with a coil volume of 928 ml using a solvent system of nBuOH:CH_3CN:H_2O: glacial acetic acid containing NaCl. A comparison of the monthly throughput and capital cost of HPLC and CCC was roughly neutral. The CCC separation used failed to yield material of sufficient purity to meet quality targets however and stationary phase leakage during the large scale separations was a problem. Total solvent consumption was approximately 75% less for CCC.

Sutherland and coworkers [35] compared the scale-up of a typical HPLC separation with that of a CCC separation. Several CCC separations were presented. Analytical HPLC and small scale (10 mg) CCC separations of an engine oil anti-wear agent were compared and the CCC separation was scalable to a 1 g loading. These workers noted that scale-up of the HPLC separation in loading ranges of 10 mg to 1 g would require three different instruments whereas one typical lab scale CCC could be used for the entire loading range. It was also noted that preparative HPLC uses 5 to 80-fold more solvent than CCC for the equivalent separation.

10.5 FUTURE DIRECTIONS

CCC has been very successfully used for the bioassay guided isolation of natural products by pharmaceutical industry drug discovery groups. The number and size of these groups worldwide is declining however. Natural product extracts are steadily being replaced as a lead source, not by combinatorial chemistry as once thought, but by the rapid, high throughput screening of pure compound collections. Without these pockets of interest the application of CCC seems destined for decline within the industry. On the positive side, the application of CCC in the flavour and foodstuff industry is expanding.

The timing of this decline is unfortunate for both the continuing development of CCC and for natural products lead discovery. CCC is uniquely suited to the chromatographic pre-fractionation of crude extracts for screening in bioassays and the development of general CCC methodology is an area of interest of several research groups. As shown above, CCC can have a significant impact on the speed of natural products isolation. The recent development of CCC instrumentation capable of the parallel processing of multiple samples is applicable to both of these problems. Further work on generalized solvent systems and modes of CCC operation to expand the fractionation range of a single separation is necessary.

A true industrial scale CCC separation has yet to be reported. However, interest in this aspect of CCC is increasing. The recent interest in the design and engineering of process scale instruments is promising. Several engineering groups are studying the fluid dynamics of the various types of CCC instruments, a prerequisite for the scale-up of CCC. Finally, the full potential of the pH-zone-refining technique developed by Ito and coworkers [36] has yet to be realized and could significantly impact process scale CCC separations.

ACKNOWLEDGEMENTS

The author is indebted to his coworkers at Merck for their microbiology, bioassay, and mass spectral data support in generating some of the data reported here. Dr. John Thompson, Dr. Jan Onishi, Mr. Mark Rosenbach, Ms. Sarah Dreikorn, Ms. Angela Basilio, and Dr. Francisca Vicente provided antifungal assays. Dr. Gerald Bills and Mr. Jon Polishook provided fungal taxonomy. Dr. Anne Dombrowski and her group provided fermentations of the fungal cultures. Ms. Debbie Zink and Dr. Zhiquang Guan provided LCMS and ESI-FT-MS. Finally, my lab associates Ms. Wendy Clapp-Shapiro, Mr. Ping Dai, Ms. Jean Armbruster and Ms. Tamara Connelly who performed many of the CCC separations described here and were willing to step out of the box to become one of those pockets of industrial CCC.

REFERENCES

1 G. Gunawardana and J. McAlpine, in J.M. Menet and D. Thiebaut (Eds.), *Countercurrent Chromatography*. Marcel Dekker, New York, 1999, pp. 249-271.
2 G. Tian, Y. Zhang, T. Zhang, F. Yang and Y. Ito, *J. Chrom. A*, 904 (2000) 107.
3 M. Agnely and D. Thiebaut, *J. Chromatogr. A*, 790 (1997) 17.
4 W.D. Conway, *Countercurrent Chromatography: Apparatus, Theory and Applications*. VCH, New York, 1989, pp. 349–351.
5 K. Ingkaninan, A.C.J. Hermans-Lokkerbol and R. Verpoorte, *J. Liq. Chromatogr. Relat. Technol.*, 22 (1999) 885.
6 T. Burzycki, G. Schulte and W.D. Conway, presented at the 37th Annual Meeting of the American Society of Pharmacognosy, Santa Cruz, California, July 23–27, 1997.
7 K.A. Alvi, In: S.J. Cutler and H.G. Culter (Eds.), *Bioactive Natural Products*. CRC Press, Boca Raton, FL, 2000, pp. 185–195.

8 K.A. Alvi, D.D. Baker, V. Stienecker, M. Hosken and B.G. Nair, *J. of Antibiot.*, 53 (2000) 496.
9 K.A. Alvi, B.G. Nair, H. Pu, R. Ursino, C. Gallo and U. Mocek, *J. Org. Chem.*, 62 (1997) 2148.
10 K.A. Alvi, A. Casey and B.G. Nair, *J. Antibiot.*, 51 (1998) 515.
11 K.A. Alvi, B.G. Nair, J. Rabenstein, G. Davis and D.D. Baker, *J. Antibiot.*, 53 (2000) 110.
12 J.A. Armbruster, R.P. Borris, Q. Jimenez, N. Zamora, G. Tamayo-Castillo and G.H. Harris, *J. Liq. Chromatogr. Relat. Technol.,* 24 (2001) 1827.
13 K. Ingkanian, A.C. Hoek, S. Balconi and R. Verpoorte, *J. Liq. Chromatogr. Relat. Technol.,* 23 (2001) 2195.
14 K. Ingkanian, J.K. von Frijtage Drabbe Kunzel, A.P. IJzerman and R. Verpoorte, *J. Natural Prod.*, 62 (1999) 912.
15 R.E. Shute and B.G. Main, PCT Int. Appl. WO 9624842 A1 19960815.
16 M. Maillard, A. Marston and K. Hostettmann, In: Y. Ito and W.D. Conway (Eds.), *High-Speed Countercurrent Chromatography*. Wiley, New York, 1996, pp. 179–223.
17 T.Y. Zhang, In: Y. Ito and W.D. Conway (Eds.), *High-Speed Countercurrent Chromatography*. Wiley, New York, 1996, pp. 225–263.
18 N.L. Fregeau and K.L. Rhinehart, In: Y. Ito and W.D. Conway (Eds.), *High-Speed Countercurrent Chromatography*. Wiley, New York, 1996, pp. 265–300.
19 J.B. McAlpine and J.E. Hochlowski, In: G.H. Wagman and R. Cooper (Eds.), *Natural Product Isolation*. Elsevier, Amsterdam, 1989, pp. 1–54.
20 P. Winterhalter, H. Knapp and M. Straubinger, In: R. Teranishi, E.L. Wick and I. Hornstein (Eds.), *Flavor Chemistry: Thirty Years of Progress*. Kluwer Academic/Plenum, New York, 1999, pp. 255–264.
21 A. Degenhardt, P. Winterhalter and E. Chou, *Chem. Innovation*, 30 (2000) 25.
22 J.D. Bergstrom, M.M. Kurtz, D.J. Rew, A.M. Amend, J.D. Karkas, R.G. Bostedor, V.S. Bansal, C. Dufresne, F.L. VanMiddlesworth, O.D. Hensens, J.M. Liesch, D.L. Zink, K.E. Wilson, J. Onishi, J.A. Milligan, G. Bills, L. Kaplan, M.N. Omstead, R.G. Jenkins, L. Huang, M.S. Meinz, L. Quinn, R.W. Burg, Y.L. Kong, S. Mochales, I. Martin, F. Pelaez, M.T. Diez and A.W. Alberts, *Proc. Nat. Acad. Sci., USA*, 90 (1991) 80.
23 M.J. Dawson, J.E. Farthing, P.S. Marshall, R.F. Middleton, M.J. O'Neill, A. Shuttleworth, C. Stylli, R.M. Tait, P.M. Taylor, H.G. Wildman, A.D. Buss, D. Langley and M.V. Hayes, *J. Antibiot.*, 45 (1992) 639.
24 G.H. Harris, E.T.T. Jones, M.S. Meinz, M.N. Omstead, G.L. Helms, G.F. Bills, D. Zink and K.E. Wilson, *Tetr. Lett.*, 34 (1993) 5235.
25 F. Oka, H. Oka and Y. Ito, *J. Chromatog. A*, 538 (1991) 99.
26 O.D. Hensens, G.L. Helms, E.T.T. Jones and G.H. Harris, *J. Org. Chem.*, 60 (1995) 1772.

27 S.M. Mandala, R.A. Thornton, B.R. Frommer, J.E. Curotto, W. Rozdilsky, M.B. Kurtz, R.A. Giacobbe, G.F. Bills, M.A. Cabello, I. Martin, F. Pelaez and G.H. Harris, *J. of Antibiot.*, 48 (1995) 349.

28 G.H. Harris and P. Dai, *J. Liq. Chromatogr. Relat. Technol.,* 24 (2001) 1775.

29 A. Berthod and B. Billardello, *Advances in Chromatography, Vol. 40.* Marcel Dekker, New York, 2000, p. 503.

30 I.A Sutherland, A.J. Booth, L. Brown, B. Kemp, H. Kidwell, D. Games, A.S. Graham, G.G. Guillon, D. Hawes, M. Hayes, L. Janaway, G.J. Lye, P. Massey, C. Preston, P. Schering, T. Shoulder, C. Strawson and P. Wood, *J. Liq. Chromatogr. Relat. Technol.,* 24 (2001) 1533.

31 I.A. Sutherland, L. Brown, A.S. Graham, G.G. Guillon, D. Hawes, L. Janaway, R. Whiteside and P. Wood, *J. Chromatog. Sci.,* 39 (2001) 21.

32 I.A. Sutherland, L. Brown, S. Forbes, G. Games, D. Hawes, K. Hostettmann, E.H. Mckerrell, A. Marston, D. Wheatly and P. Wood, *J. Liq. Chromatogr. Relat. Technol.,* 21 (1998) 279.

33 Q. Du, P. Wu and Y. Ito, *Anal. Chem.*, 72 (2000) 3363.

34 A.S. Graham, I.F. McConvey and P. Shering, *J. Liq. Chromatogr. Relat. Technol.,* 24 (2001) 1811.

35 I.A. Sutherland, L. Brown, S. Forbes, D. Games, D. Hawes, K. Hostettmann, E.H. McKerrell, A. Marston, D. Wheatley and P. Wood, *J. Liq. Chromatogr. Relat. Technol.,* 21 (1998) 279.

36 Y. Ito and Y. Ma, *J. Chromatog. A*, 753 (1996) 1.

Chapter 11

Enantioseparations in countercurrent chromatography and centrifugal partition chromatography

Laureano Oliveros, Cristina Minguillón, Pilar Franco and Alain P. Foucault

11.1 INTRODUCTION

The interest in countercurrent chromatography and centrifugal partition chromatography (CCC&CPC) in the field of chiral separations lies in their preparative character. The production of enantiomerically pure compounds by separation of racemic mixtures of enantiomers, or the purification of enantiomerically enriched mixtures, which result from enantioselective syntheses, is now an important challenge, and these techniques are of value in this domain. Nevertheless, a total of eighteen papers in eighteen years—including those dealing with the "CPC chiral reactor", which is not exactly chiral chromatography—constitutes a modest output compared with the literature for other chiral separation techniques [1–18]. Moreover, only three or four efficient chiral selectors are found in these eighteen papers, as compared to hundreds for HPLC, GC or CE!

Countercurrent chromatography and centrifugal partition chromatography (CCC&CPC), although powerful preparative techniques because of their high capacity, low cost of liquid stationary phases (i.e., solvent mixtures) and low solvent consumption (a tenth of that for HPLC), show rather poor efficiency (1000 theoretical plates is often the maximum). For this reason, the enantioselectivity factor (isomer k' ratio) induced by the chiral selector has to be high enough, i.e., greater than 1.4. The major difficulty is in finding chiral selectors that are highly selective, as well as a combination of solvents that does not destroy this

Comprehensive Analytical Chemistry XXXVIII
A. Berthod (Ed.)

selectivity and retains the capacity to elute the chiral isomers of interest. The chiral selectors used in CCC generally came from other separation techniques. We shall highlight the origins of the use of the chiral selectors which have proved most efficient in chiral CCC, and which are described below.

- N-Dodecanoyl-L-proline-3,5-dimethylanilide was introduced by Oliveros et al. [6], who were among the first to synthesize a chiral L-proline-bonded silica for HPLC in 1979. Oliveros's group extrapolated certain data of Pirkle and Murray [19] and introduced a simple method of synthesizing N-dodecanoyl-L-proline-3,5-dimethylanilide. Some studies are under way in order to improve this chiral selector (CS).
- Sulfated β-cyclodextrins were introduced by Breinholt and his co-workers after remarkable results were obtained in capillary electrophoresis (a resolution factor greater than 100 or, alternatively, a world-record 10 s separation time) and then had to be applied to preparative needs [14,15].
- Vancomycin was introduced by Duret, Foucault, and Margraff [16]. Margraff, a specialist in the purification of natural products, had been impressed by the Armstrong group's experiments on HPLC with vancomycin and other chiral macrocyclic antibiotics grafted to silica.
- Quinine and quinidine carbamate derivatives have been used extensively by Lindner and his co-workers for the resolution of chiral acidic compounds [17,20–28]. These chiral selectors (CSs) proved to be highly enantioselective when used in HPLC, CE, and extraction experiments, among other separation techniques. Their ability to show very high separation factors in solution (extraction experiments) made them very suitable as potential chiral selectors for CCC.

In all cases the idea is the same, i.e., to borrow a chiral selector from different separation techniques that displayed high selectivity and then to determine whether it would remain robust for chiral CCC. We will describe here a few relevant examples.

11.2 CHIRAL SEPARATIONS

The direct separation of enantiomers is always based on the differences in stability of the diastereomeric complexes (A_-/A_+–CS) formed between each enantiomer (A_-/A_+) and the chiral selector involved (CS).

$$A^- + CS \overset{K_-}{\rightleftharpoons} A^- - CS$$

$$A^+ + CS \overset{K_+}{\leftrightharpoons} A^+ - CS$$

The difference in stability for these two equilibria can be expressed as

$$\Delta\Delta G^o = RT \ln(K_-/K_+); \quad (\text{with } K_- > K_+)$$

In HPLC, assuming that the complexes refer to the amount of analyte in the stationary phase, and the free enantiomers to the amount of analyte in the mobile phase[1]; the ratio of association constants expresses the enantioseparation factor, α. These equilibria also take place in CCC, but in the stationary phase where the CS is placed. However, a non-enantioselective partition phenomenon occurs simultaneously between the two phases. Thus, the solvent system controls not only the retention of the analyte but also its possibility of reaching the selector.

In this context, several considerations have to be taken into account when one is trying to make chiral separations in CCC. First, the selector should stay in the stationary phase. Leaks of selector in the mobile phase, owing to partition phenomena, have to be avoided to ensure the maximum stability for the stationary phase and reproducibility of the results. Next, the partition of the analyte into both phases has to be controlled to provide an efficient interaction with the CS within a reasonable retention time. Moreover, additional factors interfering with the enantioselective process have to be considered. For example, the loadability of the system depends on the amount of chiral selector in the stationary phase. Therefore, the stationary phase should contain a good solvent for the CS. Unfortunately, it is likely that such a good CS solvent will interact through hydrogen bonds or dipolar interactions with the CS. By doing so, it will compete with the analyte and prevent the association of both species, thus reducing the enantioselectivity of the process.

Enantioselectivity changes are observed when switching techniques. Reductions or increases in enantioselectivity were observed in CCC with some chiral selectors compared with the enantioselectivity obtained in HPLC. Such changes can be attributed to the different natures of the

1 This assumption is easily acceptable for low-molecular-weight chiral selectors. In cases using proteins as the chiral stationary phase, a non-enantioselective distribution between stationary and mobile phases can be considered [29].

stationary phases, where the enantioselective process occurs. The CS is dissolved in the liquid stationary phase in CCC; it is bonded onto the solid stationary phase in HPLC. The mobile phase is also important. Even if the compositions of both phases are closely related, in CCC the stationary and the mobile phase are saturated one with the other. Another factor that can account for these changes in enantioselectivity is that in CCC the CS is motion-free in the liquid stationary phase. Up to now the possible role of the group that fixes the CS to the stationary phase in HPLC (steric hindrance, introduction of an additional interacting group) has been scarcely evaluated [30].

11.3 CHIRAL CCC: THE PIONEERS

To our knowledge, the first attempt to resolve a pair of enantiomers using CCC was made in 1982 in Switzerland by Hostettmann and Prelog and their co-workers [1]. Following the good results obtained by liquid–liquid partition [2], these researchers used (*R*,*R*)-di-5-nonyl tartrate as a chiral selector to partly resolve (±)-norephedrine as its hexafluorophosphate salt. The 1,2-dichloroethane–water system was used, with the CS in the organic mobile phase and sodium hexafluorophosphate in the aqueous stationary phase. Although no baseline separation was achieved, the researchers showed that CCC could produce practically pure enantiomers. The instrument was a rotation locular countercurrent chromatograph (RLCCC), which showed poor efficiency and was extremely slow (four days for a run). It would be interesting to repeat these experiments with a modern instrument.

The second attempt, in Japan in 1984, was a transposition to CCC of ligand exchange chromatography, a technique quite popular at the time. Taking their inspiration from Davankov [31], Takeuchi et al. [3] synthesized N-*n*-dodecyl-L-proline (C_{12}-PRO), which remains in the organic phase of the *n*-butanol–water system. By adding Cu(II) ions, which were extracted in the organic phase as complexes with C_{12}-PRO, they partly resolved several D,L-amino acids by “ligand exchange CCC”. Baseline resolution was achieved for the D,L-isoleucine, but the separation time (2.5 days) was excessive, owing to the use of a prototype of a droplet countercurrent chromatograph (DCCC) made by connecting 400 pieces of Teflon tubing of two sizes alternately in series, which gave a total volume of around 2 litres. As for the first attempt, it would be interesting to reinvestigate “ligand exchange CCC” using recent instruments.

The third and last attempt in the pioneering research concerns the resolution of various substituted bicyclo[2.2.1]hept-5-ene-2-carboxylic acids. This took place in 1986 in the USA [4]. Oya and Snyder adopted CCC after several attempts with more traditional methods such as fractional crystallization or esterification to form diastereoisomers. (–)-(R)-2-aminobutanol proved to be a powerful chiral selector for the above-mentioned compounds in conjunction with chloroform–methanol–aqueous phosphate buffer, where it stays in the upper aqueous phase. Resolution was complete using a DCCC instrument with a volume of around 280 ml and a run-time of about 2.3 days.

11.4 N-DODECANOYL-L-PROLINE-3,5-DIMETHYLANILIDE

11.4.1 Following the Pirkle path

The 3,5-dimethylanilide group is considered an efficient π-donor structure in π-donor-HPLC chiral stationary phases having an amino acid as chiral moiety [5]. The enantioselectivity of such chiral stationary phases can be particularly remarkable for certain analytes when this amino acid is L-proline [19]. From these observations we have developed a simple method for synthesizing N-dodecanoyl-L-proline-3,5-dimethylanilide, a chiral π-donor molecule which is very useful in charge-transfer complexation (Fig. 11.1) [6]. This π-donor selector allowed the first speedy (80 min) and complete resolution of two neutral DNB-amino-acid derivatives, namely N-(3,5-dinitrobenzoyl)-*tert.*-butylvalinamide and N-(3,5-dinitrobenzoyl)-*tert.*-butyl-leucinamide, using the heptane–ethyl acetate–methanol–water 3:1:3:1 biphasic system and a CPC apparatus with a column volume of 240 ml [6].

In an attempt to improve this selector the analogous derivative of 4(R)-hydroxy-L-proline and the derivatives of both the preceding amino acids containing a 3,5-dimethylbenzoyl group instead of the N-dodecanoyl chain, have been prepared and tested in the separation of DNB-(±)-leucine [7]. This study showed some of the drawbacks of chiral CCC. Unlike N-dodecanoyl-L-proline-3,5-dimethylanilide, the new selectors were not soluble in the conditions used for the former. They could only be tested using a highly polar solvent system [methyl isobutyl ketone (MIBK)– phosphate buffer, pH 8] (Fig. 11.2). In such conditions the enantioselectivity of the original N-dodecanoyl selector for the derivatized leucine was reduced. However, N-dimethylbenzoyl-4(R)-hydroxy-

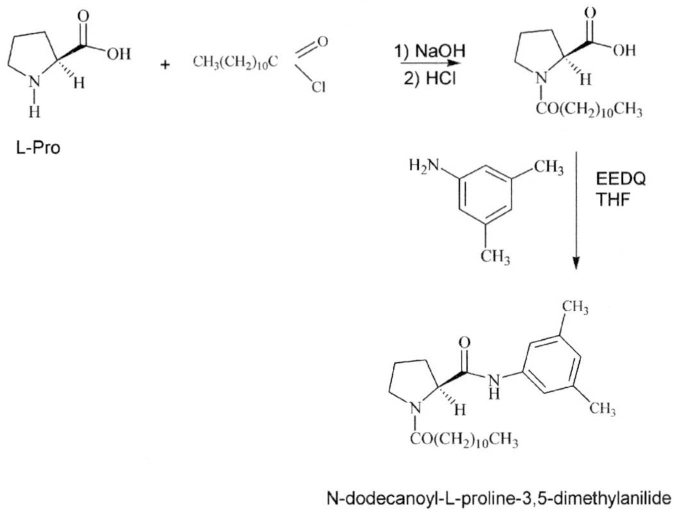

Fig. 11.1. L. Oliveros and co-workers' synthesis of N-dodecanoyl-L-proline-3,5-dimethylanilide. EEDQ, 2-ethoxy-1-ethoxycarbonyl-1,2-dihydroquinoline; THF, tetrahydrofuran. Room temperature.

L-proline-3,5-dimethylanilide was found to be more enantioselective than the original N-dodecanoyl selector in the resolution of the leucine derivative. It should be noted that the analyte was not resolved under similar conditions (mobile phase: heptane–MIBK) by HPLC on a chiral stationary phase bonded with the same selector.

N-dodecanoyl-L-proline-3,5-dimethylanilide was also used in an authoritative way by Ma et al. to perform complete analytical studies and to introduce the pH-zone-refining mode of CCC for chiral discrimination [8–11]. These authors verified that enantioselectivity was increased by: (1) augmenting the net amount or concentration of the chiral selector in the organic stationary phase, and (2) adjusting the hydrophobicity of the solvent system so that the mean distribution coefficients for the racemates fell between 0.6 and 0.8 (except for the pH-zone-refining mode). They were able to fractionate up to 1 g of DNB-amino acids using a CCC apparatus with a column volume of 330 ml in the classical elution mode, and up to 2 g in the pH-zone-refining mode described later. Figure 11.3 shows a good separation of four DNB–amino-acid racemates using the π-donor selector. The instrument, a prototype of a semi-analytical CCC, equipped with a set of three column holders for a total volume of 180 ml, was manufactured at the

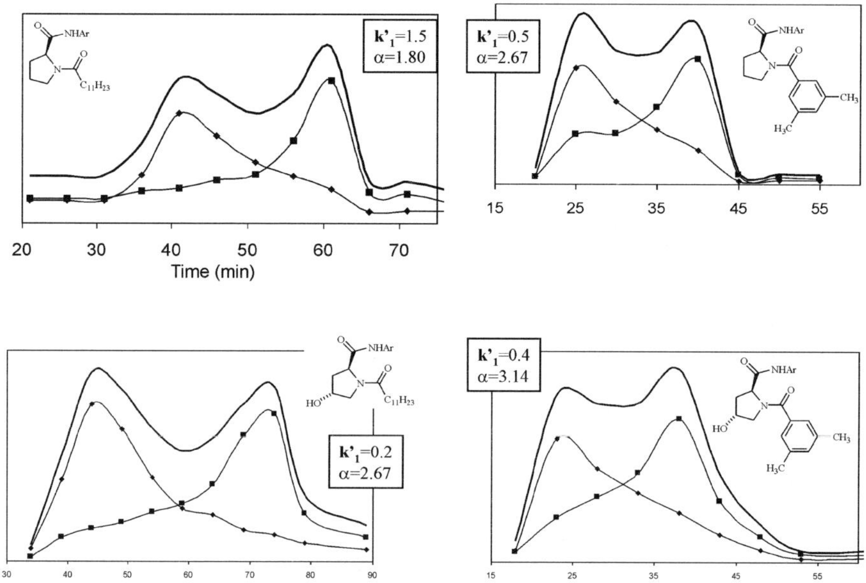

Fig. 11.2. Separation of 150 mg of DNB–Leu by CPC using the chiral selectors indicated (10 m*M*). The amount of each enantiomer eluted was determined by chiral HPLC. Both elution profiles are shown. Ar, 3,5-dimethylphenyl; solvent system, phosphate buffer 0.2 *M* (pH 8)–MIBK (1:1); flow, 3 ml/min; ω, 1000 rpm. (MIBK; methyl isobutyl ketone).

NIH machine shop (a similar apparatus is available from Pharma-Tech Research). One drawback is the rather long time required for experiments, which indicates the need for faster apparatus.

11.4.2 Chiral separation by pH-zone refining

Ito's group's introduction of the pH-zone refining [12] mode in CCC, as a variant of displacement chromatography [8], was highly successful. This approach applies to the purification of compounds whose electrical charge is pH-dependent (HA/A^- or BH^+/B pairs). One characteristic of pH-zone refining is that the separated products elute from the column in contiguous blocks arranged according to their pK_a values and partition coefficients (see Chapter 7). Another characteristic of this mode of CCC is its powerful preparative-scale capability. This has been dramatically demonstrated by Renault et al. [13] who injected 7 g of a crude mixture of *Catharanthus roseus* alkaloids onto a 250 ml CPC column, using HCl as retainer in the aqueous stationary phase and triethylamine as

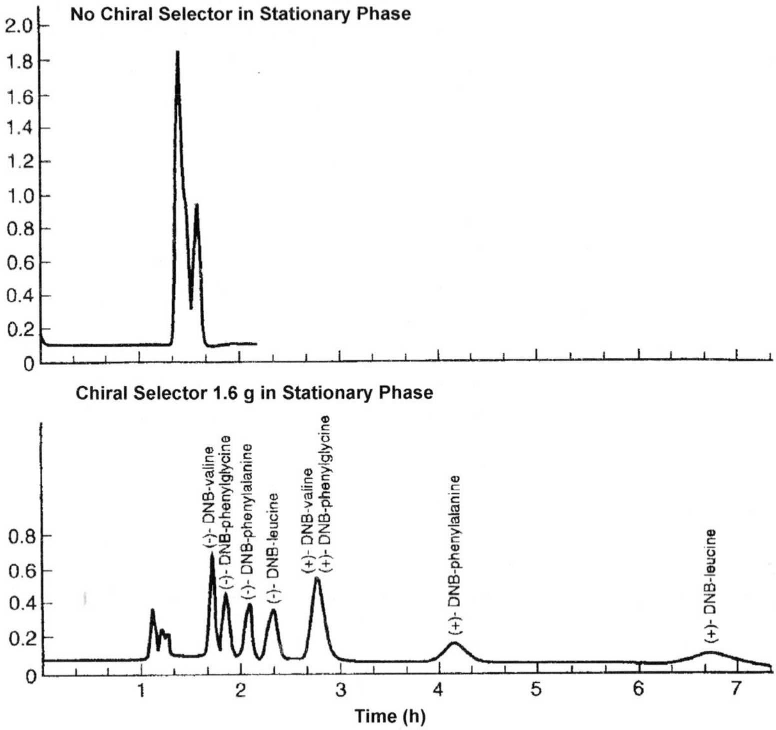

Fig. 11.3. Separation of four DNB–amino-acid racemates by CCC using N-dodecanoyl-L-proline-3,5-dimethylanilide as chiral selector. Solvent system, hexane–ethyl acetate–methanol–10 m*M* HCl, 8:2:5:5. CCC column volume, 180 ml; flow rate, 1 ml/min; rotation, 1000 rpm. Detection; UV 280 nm; samples, 10 mg each of (±) DNB-amino acid. Reproduced, with permission, from Ref. [10].

displacer in the organic mobile phase. The four major indole alkaloids were isolated from this crude mixture in one run.

In the absence of a complexing agent in the stationary phase, the separation of the analytes and their elution order depend on the pH value of their respective zones. The pH of a zone is expressed by the following equation:

$$\mathrm{pH} = \mathrm{p}K_{\mathrm{a}} + \log\,(K_{\mathrm{D}}/K_{\mathrm{R}} - 1) \qquad (11.1)$$

where K_{D} is the distribution ratio of the neutral analyte, and K_{R} the distribution ratio for the retainer acid or base (see Chapter 7, Eq. (7.8)).

For chiral applications, a chiral selector (CS) that should not partition in the mobile phase is added to the stationary phase. The CS complexes the analyte, with complex-formation constants, K_+ and K_- ($K_\pm$ in the formula). Consequently, Eq. (11.1) must be modified to:

$$pH_\pm = pK_a + \log (K_D/K_R (1 + [CS]_s K_\pm) - 1) \tag{11.2}$$

where $K_\pm$ are the complex-formation constants of the + or – isomer and $[CS]_s$ is the concentration of the chiral selector in the stationary phase. Since both the pK_a and K_D values of the two enantiomers are identical, and K_R is governed by the ratio of the retainer and displacer, the separation of the two enantiomers is governed by the difference between K_+ and K_-, and the concentration of CS in the stationary phase. Ma et al. [8] showed the capabilities of chiral separation through pH-zone refining by using the chiral selector introduced by Oliveros and his group [6] to resolve 2 g of DNB-(±)-leucine in about 3 h on a 330 ml CCC column, using trifluoroacetic acid as retainer in the organic stationary phase and ammonia as displacer in the aqueous mobile phase (Fig. 11.4). Chiral pH-zone refining for the resolution of ionizable racemates is probably the most promising preparative technique with CCC.

11.5 SULFATED β-CYCLODEXTRIN

The following example shows clearly that CCC can be complementary to other separation techniques. It indicates how the thinking of scientists not initially involved in CCC can contribute to the technique when the scaling-up of analytical results is required.

Breinholt et al. [14,15] noted that when, "using sulfated β-cyclodextrin (S-β-CD) as the chiral selector, resolution factors (*Rs*) greater than 100, or, alternatively, a world-record (at the time) 10 s separation time, were obtained for the enantiomeric pair of 7-*des*-methyl-ormeloxifene (7-DMO) (1 and 2), of which 1 is a major metabolite *in vivo* of (–)-(3R, 4R)-levormeloxifene (3), a partial oestrogen receptor agonist" (Fig. 11.5). However, this was with capillary electrophoresis (CE)! The authors then added: "an inherent limitation of CE is sample capacity, and the technique lacks in its present technical shape a direct preparative counterpart. ... A fundamental feature of CE (a separation mechanism governed by solution phenomena involving no solid support) is shared by countercurrent chromatography (CCC). ... The true strength

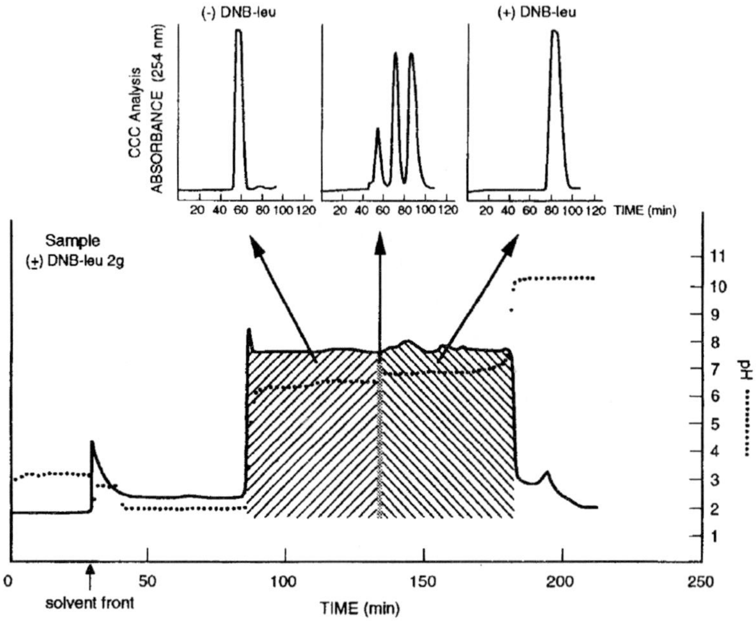

Fig. 11.4. Separation of DNB-(±)-leucine by pH-zone-refining CCC. Solvent system; methyl *tert.*-butyl ether–water; stationary phase, upper organic containing trifluoroacetic acid (40 m*M*) and N-dodecanoyl-L-proline-3,5-dimethyl-anilide (40 m*M*); mobile phase, lower aqueous containing ammonia (20 m*M*); CCC column volume, 330 ml; flow rate, 3.3 ml/min; rotation, 800 rpm; detection monitoring pH and UV at 280 nm. Sample, 2 g of DNB-(±)-leucine. Note that the analysis of the fraction from the mixing zone (middle chromatogram) shows three peaks (from left to right) corresponding to DNB-(–)-Leu, impurity, and DNB-(+)-Leu. Reproduced, with permission, from Ref. [8].

of CCC lies in its preparative applications, and we envision that the speed of CE, utilized for fast screening of several chiral selectors in order to identify particularly efficient ones, might be well complemented by the preparative aspect of CCC." [15].

Moreover, as sulfated β-cyclodextrin with a degree of substitution of 7–11 (Fig. 11.5) is a commercial product (Aldrich 38,915-3), other scientists can easily test this powerful chiral selector. In fact, this is the first example in which the CS is in the aqueous phase, a property shared with vancomycin (described later in this review). Fig. 11.6 shows the influence of solvent composition on the CCC resolution. The formation of a 7-DMO/*S*-β-CD inclusion complex is believed to rely at least partially on hydrophobic interactions, and it is expected that such a complex

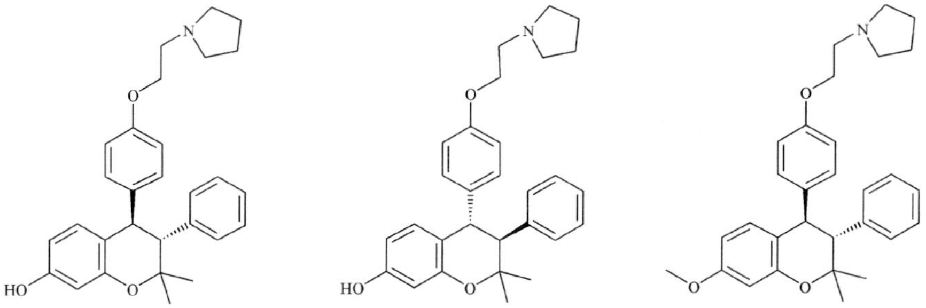

1 (-)-(3*R*,4*R*) **2** (+)-(3*S*,4*S*) **3** (-)-(3*R*,4*R*)

4 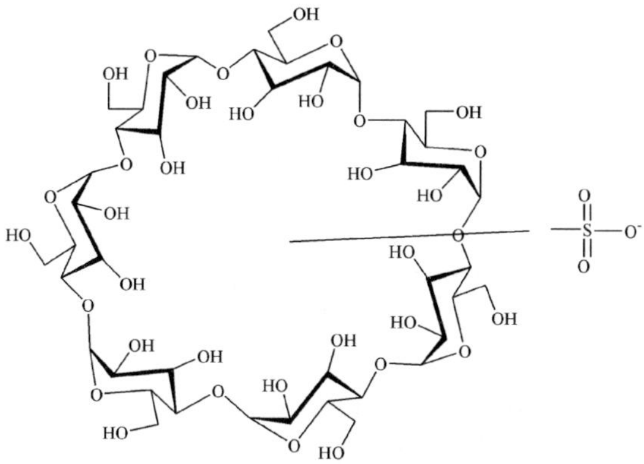 7-11

Fig. 11.5. (**1**) and (**2**): (–)-(3*R*,4*R*) and (+)-(3*S*,4*S*)-7-*des*-methylormeloxifene: (**3**); (–)-(3*R*,4*R*)-levormeloxifene: (**4**); sulfated β-cyclodextrin (degree of substitution, 7–11).

could be stabilized if the polarity of the aqueous phase was increased, i.e., from A to C in Fig. 11.6. It is to be hoped that this innovative use of a sulfated β-cyclodextrin will enlarge the CCC field for chiral preparative applications.

11.6 BOVINE SERUM ALBUMIN (BSA)

To our knowledge, Stewart and Doherty first published on chiral separation on an immobilized BSA stationary phase in 1973 [32]. A significant improvement in column performance and ease of operation was obtained through the introduction of a silica support for the immobilized albumin [33]. However, the low capacity and low stability of these

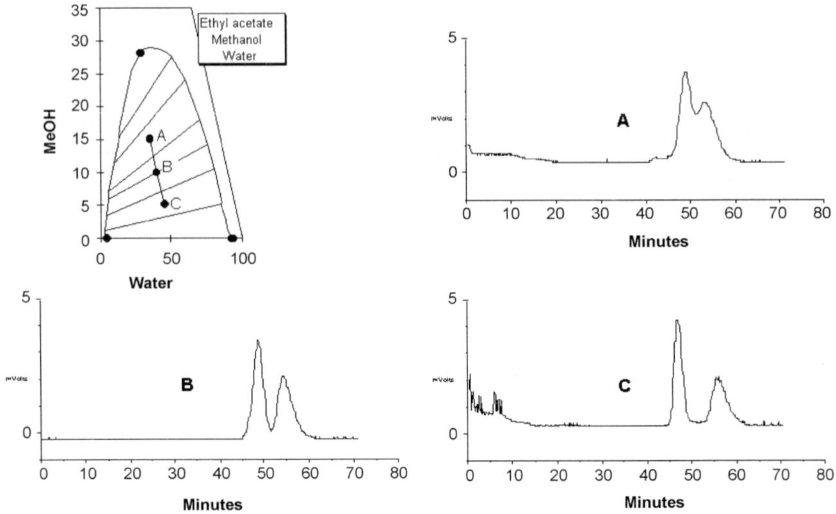

Fig. 11.6. Resolution of 7-*des*-methylormeloxifene using sulphated β-cyclodextrin as chiral selector. The solvent systems are ethyl acetate–methanol–aqueous triethylammonium acetate 10:3:7 (A) 10:2:8 (B) and 10:1:9 (C), containing 2% (w/v) of *S*-β-CD in the aqueous stationary phase. The first peak represents the (+)-enantiomer of 7-DMO.

chromatographic systems makes them mainly useful for analytical purposes.

Partition in an aqueous two-phase system for direct chiral resolution using BSA in one phase has been tried in two ways: countercurrent distribution, and CCC. The phase system, consisting of two immiscible aqueous phases, is obtained by mixing two polymers, or one polymer and a salt, with water [34]. Under suitable conditions, a protein can be confined almost completely to one of the phases, while low-molecular-weight compounds partition more equally between the phases.

BSA has been used in countercurrent distribution to resolve D,L-tryptophan [35] and ofloxacin [36], and in CCC to resolve commercial D,L-kynurenine [37]. However, the resolution is poor in each case. It is said that liquid–liquid partition is easy to scale-up and that the aqueous polymer–phase system should be of interest for large-scale separation of enantiomers: however, a significant increase in efficiency is required to accomplish this. This is not easy, since the polymer-phase systems are rather viscous and mass transfer during the chromatographic process is mediocre. Another factor making scale-up difficult is the molecular weight of BSA (6.6×10^4). Regardless of the stoichiometry of the complex, liquid–liquid partitions involving a complexation equilibrium commonly generate Langmuir isotherms, whose plateau is governed by the concentration of the complexing molecule (here BSA). Given its molecular weight of 6.6×10^4, a significant mass of BSA in one phase does not give a large concentration. That is why CCC and CPC may be of interest for interaction studies between free protein and other molecules rather than for their preparative capabilities, in this specific case.

11.7 VANCOMYCIN

Macrocyclic antibiotics have recently been introduced by Armstrong et al. as powerful and 'general' chiral selectors in liquid chromatography (LC) [38], thin-layer chromatography (TLC) [39], and capillary electrophoresis (CE) [40]. Native or derivatized vancomycin, ristocetin A, teicoplanin, rifamycin B and thiostrepton have been used in both the normal and reversed-phase modes in LC, with acetonitrile–1% triethylammonium acetate buffer, or 2-propanol–hexane [38] as mobile phases. A tremendous variety of racemates has been resolved on these chiral stationary phases, such as coumachlor and warfarin, devrinol, 5-methyl-5-phenylhydantoin, and various native or derivatized amino acids, with separation factors (α) as high as 2.42 (coumachlor, pH 3.6). Vancomycin has been used as a chiral additive in a water–acetonitrile mixture to resolve 6-carbamyl-quinolyl-N-protected amino acids, racemic drugs, and dansyl amino acids by TLC on diphenyl-type stationary phases [39].

Vancomycin is soluble in water (> 140 mg/ml) and dimethyl sulfoxide, somewhat soluble in methanol, and insoluble in higher alcohols and other less-polar organic solvents. Thus, a typical biphasic system which is suitable for chiral CCC consists of a water-rich stationary phase containing vancomycin and a less-polar mobile phase in which the racemic compounds stay preferentially, so that their partition ratio is in the 0.5–1 range, favouring rapid recovery.

Duret et al. [16] completely resolved racemic dansylnorleucine (DNS-nLeu) using a CCC apparatus (13 ml column) and a CPC apparatus (90 ml column). In both cases, the biphasic system was a mixture of toluene and an aqueous solution of 140 mg/ml vancomycin, adjusted to pH 4.7. The dual mode was performed since the D-enantiomer is strongly retained in the vancomycin aqueous solution. The D,L-DNS-nLeu is injected with the vancomycin aqueous solution as the stationary phase (tail to head for CCC, or ascending mode for CPC). When the L-enantiomer has been eluted, the mode is switched (head to tail, descending mode) and the D-enantiomer is eluted in the vancomycin aqueous mobile phase. Fifty milligrams of racemate mixture were resolved in 50 min using CPC. Despite this success, the conclusions of Duret et al., given below, are pessimistic.

"(1) The experimental conditions where vancomycin in solution will effectively recognize an enantiomer are extremely narrow:

- an aromatic solvent is needed to make a biphasic system with water containing vancomycin;
- addition of a non-aromatic solvent destroys chiral recognition;
- the pH range is narrow;
- the concentration of vancomycin must be high to get a high selectivity.

"(2) The high molecular weight of vancomycin (MW ≈ 1400) is a major drawback when considering preparative applications. Since two molecules of vancomycin are needed to complex one molecule of enantiomer, and since the Langmuir shape of the isotherm suggests that it could be a supra-molecular organization involving four to six additional molecules of vancomycin, we must conclude that even with large amounts of vancomycin in solution, injected quantities of racemate will remain small, even if the selectivity is high. We feel now that an ideal chiral selector for CCC must have a low molecular weight, in order to get a high molarity in a solvent by dissolving a reasonable mass of it. Since chiral chromatography involves the formation of a complex between an enantiomer and the chiral selector, it is probable that we will always encounter Langmuir shaped isotherms, and one way to extend the linear zone of these isotherms is to use higher molarities of chiral selector, the limit being that this chiral selector will be one of the solvents making the biphasic system [16]".

11.8 CINCHONA ALKALOID DERIVATIVES

Lämmerhofer et al. described the ability of cinchona derivatives, especially quinine (QN) and quinidine (QD), as efficient HPLC chiral selectors in anion-exchange mode [20,21]. Their high potential for forming very stable complexes with chiral acidic compounds has allowed their efficient use as chiral selectors also in CE [22,23], capillary electrochromatography (CEC) [24], supercritical fluid chromatography (SFC) and extraction experiments [25,26]. Exhaustive investigations have been performed in order to enhance the chiral recognition abilities of this type of selectors and to understand the influence of every substituent and functionality [21,27]. On one hand, the possibility of protonating the quinine–quinidine system makes it especially suitable for the resolution of acidic compounds. On the other hand, the derivatization with a carbamate group bearing bulky substituents (*tert.*-butyl or adamantyl groups), in the O-9-position, leads to selectors with higher enantioselective capability. Selectivity factors of about 30 were obtained at 25°C in HPLC for the resolution of DNB-(±)–leucine (DNB-Leu) with some of these carbamate derivatives [28]. Furthermore, this remarkable stereoselectivity was also displayed in solution: QN and QD derivatives are able to produce an enrichment of a liquid phase when extracted with an immiscible one containing them [24]. These experimental data give promising signs for thinking about an interesting application for CCC.

For the preceding purpose, modified selectors were designed, bearing an adamantylcarbamate function to enhance its chiral recognition ability and a lipophilic octadecylthio moiety to avoid its distribution into the mobile phase [17,41]. Ternary and quaternary solvent systems were evaluated for use in the CPC runs for the resolution of several amino acid derivatives. Selectivity factors ranging from 2 to 2.5 were obtained for the resolution of DNB–Leu or N-(3,5-dinitrobenzyloxycarbonyl) neopentyl-glycine (DNZ-NPG), using ammonium acetate 0.1 *M* (pH = 6–*tert.*-amyl alcohol–methanol–heptane biphasic systems. The scaling-up of these separations was performed up to 50 and 100 mg injected under these conditions (Fig. 11.7). Although no complete baseline separation was obtained, there was only a small mixed zone, and essentially pure enantiomers were eluted around the maxima of the chromatographic peaks. Changes to higher pH values and some other types of secondary and ternary solvent systems (buffer–MIBK and buffer–acetone–MIBK) are also under investigation (Fig. 11.8). Further optimizations are under way.

11.9 CHIRAL CCC REACTOR

Bruening, a pioneer in CPC, first mentioned the chiral CPC reactor in June 1990 at the San Mateo CPC symposium. He was working with Derguini in Nakanishi's laboratory at Columbia University in New York [42]. His talk was entitled, "The CPC reactor, a new tool for continuous organic synthesis" [43]. The abstract noted that, "the concepts of two-

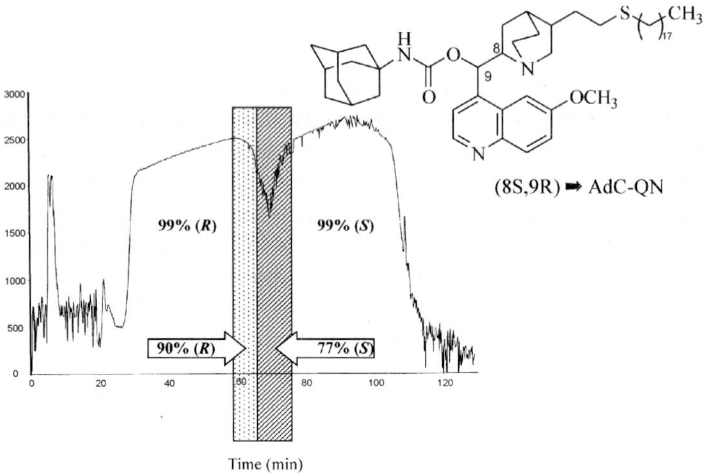

Fig. 11.7. Separation of 50 mg of DNZ-NPG by chiral-CPC using the quinine derivative shown (AdC-QN) (10.6 m*M*). Solvent system; ammonium acetate buffer, 0.25 *M* (pH 7.0)–*tert*-amyl alcohol –heptane–methanol (10:3:7:1); flow, 3 ml/min; ω, 1000 rpm.

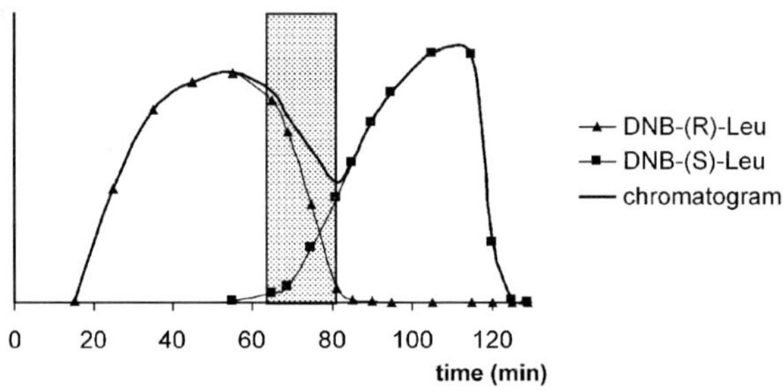

Fig. 11.8. Resolution of 150 mg of DNB-Leu. The area out of the shadowed region showed an ee > 90%. Solvent system: ammonium acetate buffer, 0.1 *M* (pH = 8)– MIBK (1:1)–CS–AdC-QN (10 m*M*); flow, 3 ml/min; ω, 1200 rpm.

phase reaction media and phase-transfer catalysis are well established in synthetic organic chemistry. Moreover, they reflect the reaction conditions found in biological systems, where the 'active sites', as in enzymes for example, are frequently lipophilic domains in a hydrophilic environment. A system where the catalytic component could be held stationary, while reactants are introduced at the same time as products are removed, would allow one to run certain chemical reactions continuously. Many attempts in this direction have failed, however, mainly because of the problems connected with the necessary mixing and separating (decanting) of the components in order to guarantee phase-transfer conditions. In CPC, thorough mixing of the two phases is produced at the inlet of each 'column', while decanting occurs concomitantly at the outlet. Thus, a single 'column' in a CPC cartridge could be seen as a micro reaction vessel for phase-transfer catalysis". The example was the enzymatic hydrolysis of D,L-*sec*-phenethyl acetate. However, Bruening never published these results. During the same symposium, Armstrong gave an example of an off-column enzyme reactor, followed by CPC separation of L-tryptophan from D-tryptophan-methyl ester [18]. As this was not directly related to the CCC reactor concept, five years passed before the first paper was published on the topic by Bousquet et al. [44]. Their example was the enantioselective hydrolysis of 2-cyanocyclopropyl-1,1-dicarboxylic acid dimethyl ester by esterase 30 000. The biphasic system was 1,1,1-trichloroethane–aqueous phosphate buffer (0.1 *M*, pH 7.5), with 7 g/l of the enzyme in the upper aqueous stationary phase. Hydrolysis of one isomer makes it go into the stationary phase, while the non-reacting isomer remains in the mobile phase, thus allowing 100% recovery. One drawback is the fact that the diester must remain 8 h in the column to achieve an enantiomeric purity of 80%. This paper failed to indicate the characteristics of the CCC instrument or provide details such as the temperature, flow rate, stationary phase hold-up, and elapsed time before the stationary phase needs to be regenerated. This French group has not published any subsequent papers on the subject.

A major contribution was made by the Kluyver Institute for Biotechnology in Delft (the Netherlands), headed by Luyben and van der Wielen. After fundamental research in CCC&CPC reported in the thesis [45] and corresponding papers [46–50] of van Buel et al., this group began studies of the CPC reactor for the production of chiral amino acids (corresponding to papers by den Hollander et al. [51,52]). The concept of a chromatographic reactor for preparative purposes has been studied in great detail [53,54], and a systematic approach to the potential of CPC to function as a chromatographic reactor was indispensable. A complete mathematical model was obtained by starting from van Buel's data [49] and expanding them with a reaction in the stationary phase depicted by μ_i (the stoichiometry coefficient of component, i) and R, the rate of the enzymatic conversion (mol s^{-1} m^{-3}). This leads to two differential equations which can be studied by numerical simulation.

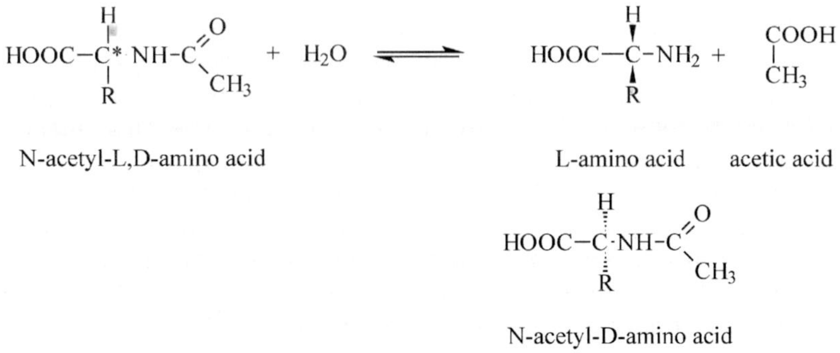

Fig. 11.9. Selective hydrolysis of N-acylated amino acids (R is the amino-acid side chain).

For this example, an aqueous two-phase system (ATPS), consisting of potassium phosphate (16%, w/w), PEG600 (16%, w/w) and water (68%, w/w) was used. Acylase was dissolved (0.2 g/l) in the stationary (PEG-rich) top phase, and the lower phase was used as the mobile phase. The substrate was injected (2 ml, 70 m*M*, 0.14 millimoles) onto the column, and the effluent profile was monitored with UV. The CPC column has 1068 individual cells for a total volume of 101 ml. The column was rotated at 800 rpm, and the flow-rate was 4 ml/min. With this set-up, the stationary-phase hold-up was 61% ($Sf = 0.61$). Figure 11.9 shows the enzymatic reaction in the stationary phase. The enzyme cleaves the acyl moiety of the L-enantiomer in a racemic mixture of an

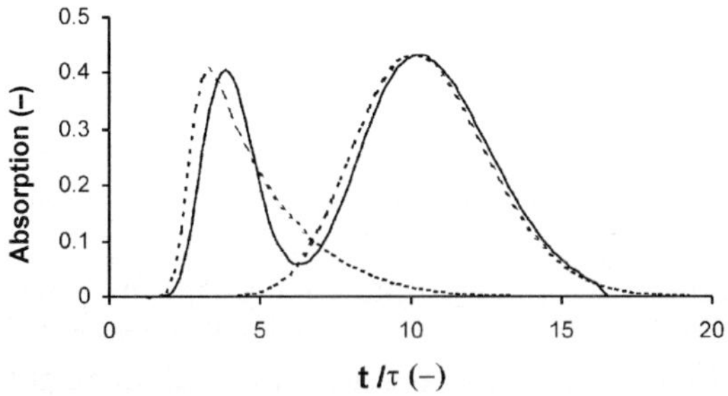

Fig. 11.10. Chiral CPC reactor, selective hydrolysis and chromatographic elution: solvent system, aqueous two-phase system PEG600 and potassium phosphate, containing acylase I (0.2 g/l); CPC column, 101 ml; flow-rate, 4 ml/min; rotation, 800 rpm. Injection of 0.14 millimoles (2 ml of 70 m*M*) N-acetyl-D,L-methionine. Calculated profiles of L-methionine and N-acetyl-D-methionine are also depicted (dashed lines). τ is the residence time for an unretained compound (V_M/F). Reproduced, with permission, from Ref. [52].

N-acylated amino acid. This hydrolysis is equilibrium-limited at higher initial concentrations. Chromatographic reaction increases the conversion. Figure 11.10 shows the effluent profile from the CPC column, indicating the simultaneous enzymatic hydrolysis in the stationary phase and the separation of the L-amino acid and the unconverted N-acylated amino acid. The predicted effluent profiles, based on the mathematical model, are also shown. This laboratory is considering larger-scale production, and further improvements in the overall efficiency of the process can be expected owing to a better design of the CPC cells for adaptation to these rather viscous ATPS.

11.10 CONCLUSION

Improvements of CCC and CPC apparatuses, and the arrival of new companies with analytical and preparative instruments, have provided scientists with new tools for resolving their purification problems. Chiral separations are still challenges for CCC and CPC, owing to the difficulty of finding chiral selectors which are highly efficient in liquid phases. The most promising way to perform chiral preparative purifications by CCC or CPC is probably the pH-zone-refining mode introduced by Ito, which will soon give us new applications, once the choice of suitable chiral selectors has been solved.

REFERENCES

1 B. Domon, K. Hostettmann, K. Kovacevic and V. Prelog, *J. Chromatogr.*, 250 (1982) 149.

2 V. Prelog, Z. Stojanac and K. Kovacevic, *Helv. Chim. Acta*, 65 (1982) 377.

3 T. Takeuchi, R. Horikawa and T. Tanimura, *J. Chromatogr.*, 284 (1984) 285.

4 S. Oya and J.K. Snyder, *J. Chromatogr.*, 370 (1986) 333.

5 L. Oliveros, C. Minguillón, B. Demazières and P.-L. Desbène, *J. Chromatogr.*, 589 (1992) 53.

6 L. Oliveros, P. Franco Puértolas, C. Minguillón, E. Camacho-Frias, A. Foucault and F. Le Goffic, *J. Liq. Chromatogr.*, 17 (1994) 2301.

7 C. Santano, C. Minguillón, Separación de enantiomeros por cromatografía de reparto centrífugo (CPC). Ventajas e inconvenientes. Presented at the XXX Scientific Meeting of the Group of Chromatography and related Techniques, Valencia (Spain) April 2001.

8 Y. Ma, Y. Ito and A. Foucault, *J. Chromatogr. A*, 704 (1995) 75.
9 Y. Ma and Y. Ito, *Anal. Chem.*, 67 (1995) 3069.
10 Y. Ma and Y. Ito, *Anal. Chem.*, 68 (1996) 1207.
11 Y. Ma, Y. Ito and A. Berthod, *J. Liq. Chromatogr.*, 22 (1999) 2945.
12 A. Weisz, A.L. Scher, K. Shinomiya, H.M. Fales and Y. Ito, *J. Am. Chem. Soc.*, 116 (1994) 704.
13 J.H. Renault, J.M. Nuzillard, G. Le Crouérour, P. Thépenier, M. Zèches-Hanrot and L. Le Men-Olivier, *J. Chromatogr. A*, 849 (1999) 421.
14 J. Breinholt and A.R. Varming, Poster presented at the 11th International Symposium on Chiral Discrimination, Chicago, IL, USA, July 25–28, 1999.
15 J. Breinholt, S.V. Lehmann and A.R. Varming, *Chirality*, 11 (1999) 768.
16 P. Duret, A. Foucault and R. Margraff, *J. Liq. Chromatogr.*, 23 (2000) 295.
17 P. Franco, W.R. Oberleitner, N.M. Maier, W. Lindner and C. Minguillón, Enantiomer separation by Centrifugal Partition Chromatography (CPC) using cinchona alkaloid derivatives as chiral selectors. Poster presented at the ISCD'2000 (Chirality 2000), Chamonix (France) September 2000.
18 D.W. Armstrong, R. Manges and I.W. Wainer, *J. Liq. Chromatogr.*, 13 (1990) 3571.
19 W.H. Pirkle and P.G. Murray, *J. Chromatogr.*, 641 (1993) 11.
20 M. Lämmerhofer and W. Lindner, *J. Chromatogr. A*, 741 (1996) 33.
21 M. Lämmerhofer, N.M., Maier and W. Lindner, *Am. Lab.*, 30 (1998) 71.
22 M. Lämmerhofer, E. Zarbl and W. Lindner, *J. Chromatogr. A*, 892 (2000) 509.
23 P. Franco, P.M. Klaus, C. Minguillón and W. Lindner, *Chirality*, 13 (2001) 177.
24 E. Tobler, M. Lämmerhofer and W. Lindner, *J. Chromatogr. A*, 875 (2000) 341.
25 K.-H. Kellner, A. Blasch, H. Chmiel, M. Lämmerhofer and W. Lindner, *Chirality*, 9 (1997) 268.
26 E. Tobler, M. Lämmerhofer, W.R. Oberleitner, N.M. Maier and W. Lindner, *Chromatographia*, 51 (2000) 65.
27 N.M. Maier, L. Nicoletti, M. Lämmerhofer and W. Lindner, *Chirality*, 11 (1999) 522.
28 N.M. Maier, W.R. Oberleitner and W. Lindner, Improved enantioseparation of N-acylated amino acid derivatives with a novel generation of O_6-modified cinchona alkaloid carbamate type selectors. Presented at the 23rd Symposium on High Performance Liquid Phase Separation and Related Techniques (HPLC'99), Granada (Spain) May–June 1999.
29 T. Fornstedt, G. Götmar, M. Andersson and G. Guiochon, *J. Am. Chem. Soc.*, 121 (1999) 1164.
30 A. Berthod, D.W. Armstrong and C.D. Chang, *Talanta,* 40 (1993) 1367.
31 V.A. Davankov, *Adv. Chromatogr.*, 18 (1980) 139.
32 K.K. Stewart and R.F. Doherty, *Proc. Natl. Acad. Sci. USA*, 70 (1973) 2850.

33 S. Allenmark, B. Bomgren and H. Borén, *J. Chromatogr.*, 264 (1983) 63.
34 P.Å Albertsson (Ed), *Partition of Cells and Macromolecules.* Wiley, New York, 1971.
35 B. Ekberg, B. Sellergren and P.Å. Albertsson, *J. Chromatogr.*, 333 (1985) 211.
36 T. Arai, H. Kuroda, *Chromatographia*, 32 (1991) 56.
37 K. Shinomiya, K. Kabasawa and Y. Ito, *J. Liq. Chromatogr.*, 21 (1998) 135.
38 D.W. Armstrong, Y. Tang, S. Chen, Y. Zhou, C. Bagwill and J.R. Chen, *Anal. Chem.*, 66 (1994) 1473.
39 D.W. Armstrong and Y. Zhou, *J. Liq. Chromatogr.*, 17 (1994) 1695.
40 D.W. Armstrong, K. Rundlett and G.L. Reid III, *Anal. Chem.*, 66 (1994) 1690.
41 P. Franco, J. Blanc, W.R. Oberleitner, N.M. Maier, W. Lindner and C. Minguillón, *Anal Chem.*, 74 (2002) in press.
42 R.C. Bruening, F. Derguini and K. Nakanishi, *J. Chromatogr.*, 357 (1986) 340.
43 R.C. Bruening, *3rd International Colloquium on CPC*, San Mateo, CA, USA, June 21–22, 1990.
44 O.R. Bousquet, J. Braun and F. Le Goffic, *Tetrahedron. Lett.*, 36 (1995) 8195.
45 M.J. van Buel, Thesis, Delft University of Technology, The Netherlands, 1997.
46 M.J. van Buel, L.A.M. van der Wielen and K.Ch.A.M. Luyben, In: A.P. Foucault (Ed.), *Centrifugal Partition Chromatography*, Chromatographic Science Series, Vol. 68. Marcel Dekker, New York, 1994, Ch. 3, p. 51.
47 M.J. van Buel, L.A.M. van der Wielen and K.Ch.A.M. Luyben, *J. Chromatogr. A*, 773 (1997) 1.
48 M.J. van Buel, L.A.M. van der Wielen and K.Ch.A.M. Luyben, *J. Chromatogr. A*, 773 (1997) 13.
49 M.J. van Buel, L.A.M. van der Wielen and K.Ch.A.M. Luyben, *AIChE J.*, 43 (1997) 693.
50 M.J. van Buel, F.E.D. van Haldema, L.A.M. van der Wielen and K.Ch.A.M. Luyben, *AIChE J.*, 44 (1998) 1356.
51 J.L. den Hollander, B.I. Stribos, M.J. van Buel, K.Ch.A.M. Luyben and L.A.M. van der Wielen, *J. Chromatogr. B*, 711 (1998) 223.
52 J.L. den Hollander, Y.W. Wong, K.Ch.A.M. Luyben and L.A.M. van der Wielen, *Chem. Eng. Sci.*, 54 (1999) 3207.
53 M. Sardin, D. Schweich and J. Villermaux, In: G. Ganetsos and P.E. Barker (Eds.), *Preparative and Production Scale Chromatography.* Marcel Dekker, New York, 1993, pp. 477–521.
54 M. Mazzotti, B. Neri, D. Gelosa and M. Morbidelli, *Ind. Eng. Chem. Res.*, 36 (1997) 3163.

Chapter 12

Supercritical fluids as mobile phases in CCC

Tiing Yu and Yu Chuan Chen

12.1 INTRODUCTION

12.1.1 Definition of supercritical fluids

A pure substance usually exists in three thermodynamically stable phases, i.e., solid, liquid, and gas, as a function of temperature and pressure, as shown in Fig. 12.1. When the temperature increases the liquid–gas phase boundary reaches a point, the critical temperature, at which the phase boundary disappears and the two phases join to form a uniform phase. At or above the critical temperature, T_c, the liquid phase can no longer exist even under elevated pressures. The corresponding vapour pressure at the critical temperature is the critical pressure, P_c, of the substance. When the pressure sustained on the pure substance is well below the critical pressure, the substance behaves as a gas even if the temperature is above the critical temperature. In practice, supercritical fluids may be defined as fluids that are compressed to possess liquid-like solvation strength when the temperatures are above their critical temperatures.

Under high pressures, a supercritical fluid possesses a density much higher than a gas and comparable to a common liquid. For example, the density of carbon dioxide at its critical point (T_c = 31.3°C, P_c = 72.9 bar) is 0.47 g/ml. This liquid-like density of the supercritical fluid carbon dioxide ($SFCO_2$) enables the dissolution of many low-polarity compounds. The diffusion coefficients of solutes in supercritical fluids are almost one order of magnitude higher and the viscosities are one

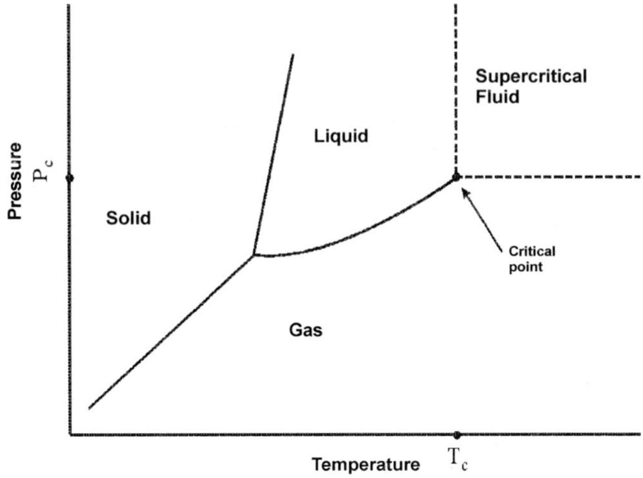

Fig. 12.1. Phase diagram of a substance. For CO_2, the most commonly used fluid, P_c = 73 bar and T_c = 31.3°C.

order of magnitude lower than those in liquid solvents. These two properties indicate an efficient solute mass transfer in the fluid. This is one of the reasons that helped to develop the applications of supercritical fluids in the area of chromatography and extraction.

12.1.2 Applications of supercritical fluid chromatography and extraction

Although a number of supercritical fluids have been investigated in various applications, we will mainly emphasize the widely used CO_2 in this chapter, due to its mild critical point and non-toxic nature. The first paper to demonstrate the use of supercritical fluid chromatography (SFC) appeared in 1962 [1]. Since then, SFC has been developing rapidly as an alternative technique to liquid and gas chromatography. Due to the intermediate properties of supercritical fluids between gases and liquids, SFC is a blend of some characteristics of both liquid and gas chromatography. The lower viscosity of supercritical fluids enables a higher flow rate of SFC over liquid chromatography (LC). However, the higher diffusivity of supercritical fluids usually results in greater band broadening of SFC over LC. One major advantage of SFC over gas chromatography (GC) is due to its ability to elute nonvolatile or thermal–labile compounds under low temperatures.

The SFC technique accommodates very well many LC and GC detectors. One example is the good compatibility of the flame ionization detector (FID) employed in SFC. This universal detector for organic compounds is known for its excellent signal response, linear dynamic range and ruggedness. In fact, the major SFC applications are for those nonvolatile or thermal labile compounds in which no suitable detector is available for LC separations. For example, a SFC machine equipped with a FID detector fits very well the hydrocarbon analysis of fossil fuels, especially for those compounds of higher molecular weight. In addition, SFC has been applied to a variety of areas, such as various agricultural products [2,3], natural products [4,5], synthetic polymers [6,7], environmental samples [8–10], and chiral separations [11,12]. A comprehensive review can be found in the literature [13].

In addition to SFC, supercritical fluid extraction (SFE) has developed very rapidly for solid-sample preparation in chemical analysis. Supercritical fluids, such as CO_2, become gases and can be released to the air without harm under ambient conditions after sample extraction. Accordingly, the extraction fluid separates from the analytes directly and instantly. The higher diffusivity of supercritical fluids compared with liquid solvents helps penetrate sample matrixes and accelerate the extraction rate considerably. SFE thus became an attractive alternative to conventional liquid extraction techniques, such as Soxhlet extraction. The applications of SFE cover very broad areas, such as extraction of pesticides and herbicides [14–16], extraction of chlorinated hydrocarbons [17–19] and polycyclic aromatic hydrocarbons [20–22] from various matrices. SFE has also been applied to natural product analyses and semi-preparations [23,24]. Direct extraction of analytes from aqueous samples was also reported [25–27]. With the addition of suitable chelating reagents [28–30] and surfactants [31–33], even metal ions and protein molecules could be readily extracted using CO_2.

12.1.3 Supercritical fluids as mobile phase in CCC

As is generally known, CCC has been recognized as a useful preparative separation technique. Since "conventional" two-phase liquid solvent systems have been extensively described in previous chapters, they will not be reviewed in this chapter. Instead, we will briefly specify the idea of replacing common liquids with $SFCO_2$. First, since CCC is basically a preparative technique, the sample collection and recovery after separation should be seriously considered. Additional separation or further concentration of analytes from the mobile phase seems an inevitable endeavour after sample collections. As already mentioned, one of the advantages of using $SFCO_2$ for sample preparations lies in the fact that $SFCO_2$ becomes a gas and separates instantly from the collected analytes when it comes back to ambient temperature and pressure. The second point is that it is interesting to know if the idea of a supercritical mobile fluid and a true liquid stationary phase would be feasible. We will call this technique SF/CCC to distinguish it from the conventional liquid–liquid CCC in the following.

12.2 EXPERIMENTAL APPARATUS AND PROCEDURES

Due to the critical temperature and pressure of CO_2 (31.3°C and 72.9 bar or 1050 p.s.i.) considerable modifications in apparatus and experimental procedures are required to enable the use of $SFCO_2$ in CCC. Accordingly, the complete experimental set-up and the experimental procedures will be described.

12.2.1 Experimental set-ups

Rotary seals are not compatible with the high-pressures needed to work with supercritical CO_2. A CCC machine with continuous tubing, a rotor and anti-twisting function is preferred. We started CCC without any experience. The slow speed centrifuge with glass columns shown in Fig. 12.2 was originally designed by Ito and Bhatnagar [34]. It seemed a reasonable choice and could readily be fabricated in our machine shop with stainless steel tubing. Several years later, we tried to run supercritical fluids on a multiplayer high-speed coil planet centrifuge (Type J machine). The column-connecting tubings did not survive the anti-twisting operation. The appropriate choice of the centrifuge ensured the future success. First, we wound the separation column eccentrically on the column holder. Using a thin 1.78 cm (outer diameter) column holder, a polyether ether ketone (PEEK) tubing (3.18 mm o.d., 1.57 mm i.d.), a pressure-resistant and flexible tubing, the first column was prepared although this tubing was actually still too rigid to make small turns on the small column holder. A high rotation speed (for example, 450 rpm) was required to retain only a mediocre stationary-phase ratio (Sf ~ 30%). We found later that if the tubing was wound coaxially onto the column holder, a slower rotation speed (~80 rpm) helped to hold more

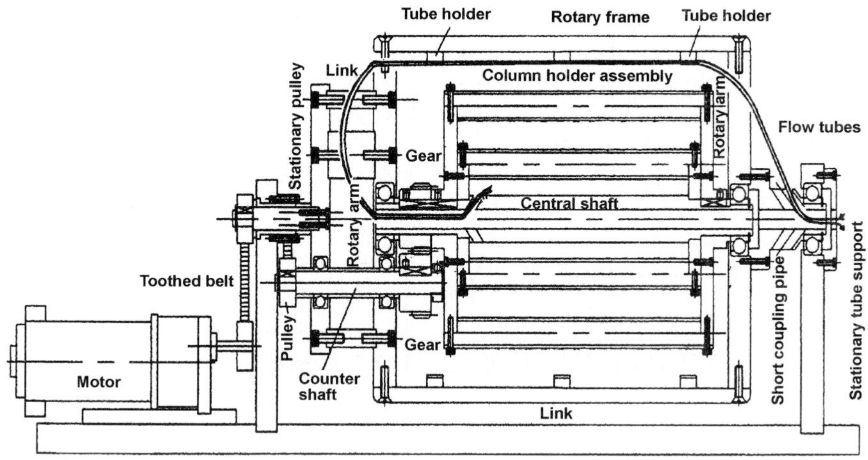

Fig. 12.2. The CCC centrifuge. The separation column (PEEK or stainless steel tubing) can be wound either eccentrically or coaxially onto the column holders (adapted from Ref. [34]).

stationary phase in the column. In addition, the PEEK tubing was replaced by metal tubing for some experiments. However, in all experiments a smaller-bore PEEK tubing (1.59 mm o.d., 0.25 mm i.d.) was used to connect from the pump to the column and from the column to the detector. It reduced the dead volume and provided a flexibility adequate to survive the anti-twisting operation.

As mentioned, the fundamental difference in constructing a SF/CCC apparatus instead of a conventional CCC is the high-pressure requirement. In addition to the modification of the centrifuge, we could borrow the ideas from SFC for the rest of the set-ups. A typical schematic diagram for the set-up is shown in Fig. 12.3. The apparatus layout was basically the same for most experiments and only slightly modified for some experiments. A syringe pump commonly used in SFC units was also employed to transport the high-pressure supercritical fluids. An UV-Vis detector with a high-pressure cell was used to monitor the effluent. Due to the low viscosity of supercritical fluids, a restrictor (essentially a small-bore tubing, e.g., a 30 cm × 20 μm fused silica or stainless tubing) connected to the system outlet was required to hold the desired pressure in the column. As with most conventional CCC, droplets of the stationary phase carried out occasionally by the mobile phase produced noise in the UV detector. A phase separator was placed before

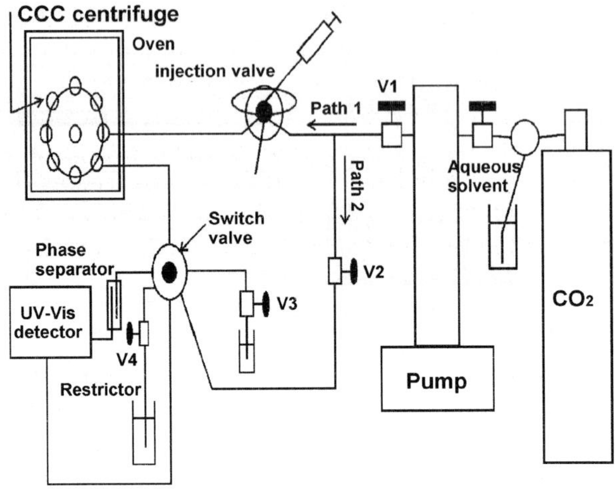

Fig. 12.3. The layout of experimental set-up for chromatography and extraction using $SFCO_2$.

the detector to ensure a stable baseline; this separator added some band broadening and inevitably degraded the separation efficiency. In contrast to the conventional CCC, the centrifuge was housed in an oven to control the temperature which was one of the important experimental parameters for SF/CCC.

12.2.2 Experimental procedures

The procedures described here will be slightly different from those published in the original papers. In fact, this reflects some improvements according to the experience gained. Therefore, we will illustrate the operations in the following according to the currently improved practice in our laboratory.

Phase retention
As shown in Chapter 2, a higher *Sf* value promotes the separation resolution. The column was filled with the desired stationary phase using the syringe pump. The pump was emptied and rinsed with some liquid CO_2 and then filled with CO_2 and compressed to the desired pressure. Since there was no need to monitor the eluent during this step, a flow path was set bypassing the detector. The centrifuge was turned on and set at the desired rotation speed while the oven was heated to the

experimental temperature. CO_2 in the liquid state was then pumped to the column via Path 1. With proper adjustment of the switch valve (a six-port valve), the $SFCO_2$ flow went directly from the CCC centrifuge to the waste through Valve V3 without passing through the phase separator and the detector. The stationary phase was gradually replaced by the CO_2 flow and finally the CO_2 bubbles appeared at the restrictor. The equilibrium was obtained after 10 more minutes. The centrifuge was stopped and the stationary phase was pushed out by N_2 flow and collected by a graduated cylinder.

Chromatography
Once the equilibrium was reached, Valve V3 was turned off. Valve V2 was opened to allow CO_2 flow following Path 2 (Fig. 12.3) to the phase separator, the detector and then to Valve V4 which was attached to a restrictor. The pressure built up rapidly with Valve V4 closed. Valve V2 was closed again, and the mobile phase flow was switched to pass the CCC machine, the phase separator, the detector, and the restrictor with Valve V4 open. The purpose of the process described here was two-fold. During the replacement of the stationary phase the fluid flowed without moving across the phase separator and the detector, thus preventing the stationary liquid from filling the phase separator. The phase separator and the detector were first filled with the mobile phase and pressurized before switching the flow to pass through the CCC machine. After this operation, the route switching of the mobile phase flow was very smooth. Otherwise, the large pressure difference (e.g., from 100 bar to 1 bar) between the CCC column and the downstream apparatuses would easily create a burst of stationary liquid to the phase separator and the detector.

Extraction
Although SFE has mainly been applied to solid-state samples, there are potential applications for liquid samples. We were also engaged in studies of SFE for aqueous samples using CCC. In fact, CCC has proved to be a very efficient machine in extracting analytes from liquid samples [35].

An aqueous sample with known volume was injected from the column outlet using a medical syringe (not shown in the set-up illustration). The column was then reconnected to the downstream apparatus. Usually the extraction effluent does not need to be monitored in a SFE course; however, we attached an UV-Vis detector to monitor the extraction process for obtaining analyte-recovery information. The addition of a detector to the end of the line imposed to balance the pressure drop. The phase separator and the detector had to be filled with the high-pressure CO_2 before the supercritical fluid passed the stationary phase (the extraction sample). Note that the liquid sample (the stationary phase) and the effluent ($SFCO_2$) were not pre-saturated with each other. Doing chromatographic runs, we could allow the $SFCO_2$ mobile phase to flow for the time necessary to reach equilibrium. However, during extraction, such a procedure would produce sample depletion and create serious errors in the result. Direct extraction with the unequilibrated $SFCO_2$ flow indeed pushed a considerable amount of the sample solution out of the column. This problem was resolved by introducing a simple modification: once the sample was loaded from the column outlet side, the column was charged with $SFCO_2$. The switch valve was set to break the connection between the column outlet and any openings. This was a static extraction at the desired temperature and pressure. After ~50 min, the switch valve was opened to connect to the phase separator and the dynamic extraction was done until its completion. Excellent extraction signals were usually obtained with none, or a trace amount, of liquid left in the phase separator. Under these experimental conditions, the density of the $SFCO_2$ mobile phase was much lower than that of the stationary phase. The lighter mobile phase was reported [34] passing the heavier stationary phase drop by drop towards the outlet while the heavier phase moved in the opposite direction. It seemed that once the unilateral phase distribution formed, the whole system also reached equilibrium. As a result, the aqueous stationary phase firmly remained in the column even without being mutually saturated with the mobile phase.

12.3 SEPARATION OF SOME SMALL ORGANIC COMPOUNDS

12.3.1 The first chromatograms

At first, the amount of stationary phase retained in the column, *Sf*, was optimized. The parameters to optimize include rotation speed, solvent composition, tubing diameters and temperature. Under various conditions, the *Sf* values never exceeded 10% using water as the stationary phase. Since water and $SFCO_2$ are only sparingly soluble in each other [36], we decided to add another component to increase their mutual solubility. Significant improvements were observed with the addition of methanol in water. As shown in Fig. 12.4, the phase retention was

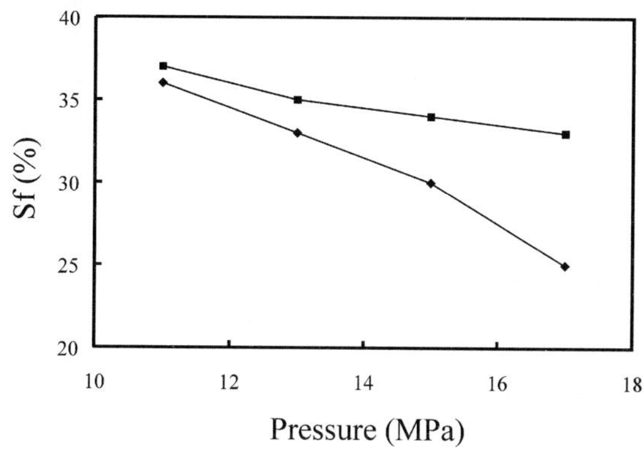

Fig. 12.4. Phase retention ratio (*Sf*) in two stationary phases under different pressures at 435 rpm and 50°C. $SFCO_2$ flow rate: 0.2 ml/min. The PEEK column was wound eccentrically. ■ = Methanol–water (50:50); ◆ = methanol–water (30:70).

improved and *Sf* values between 25 and 40% were obtained under two different pressures. The reason for the higher *Sf* values at lower applied pressure was probably due to the greater density difference between the stationary and mobile phases [37], since increased pressure increases the density of $SFCO_2$. To further examine the methanol content effect on *Sf*, more measurements were performed under 360 and 435 rpm rotation speeds. Higher methanol contents gave greater *Sf* values, as shown in Fig 12.5. Between 0 and 50% v/v, as the methanol content

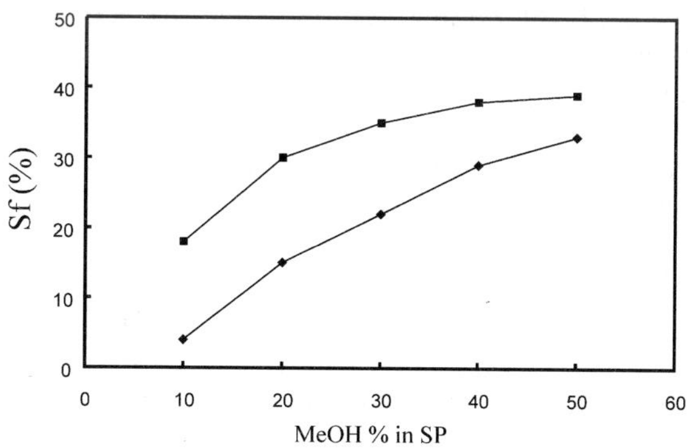

Fig. 12.5. Phase retention ratio (*Sf*) in aqueous stationary phases with various contents of methanol in the stationary phase at two different rotation speeds at 50°C and 130 bar (13.1 Mpa). $SFCO_2$ flow rate: 0.2 ml/min. The PEEK column (46 ml) was wound eccentrically. ■ = 435 rpm; ◆ = 360 rpm.

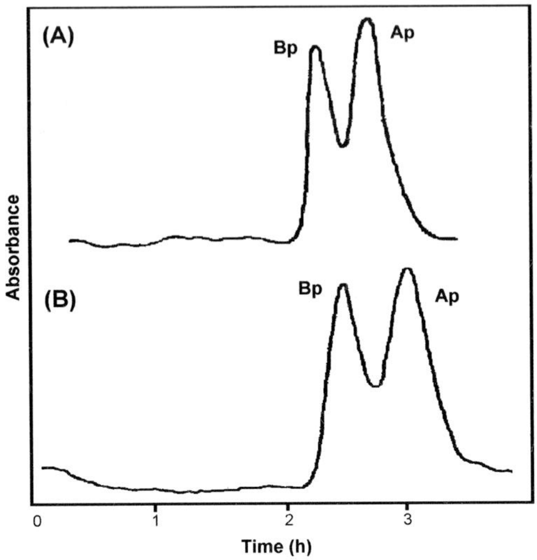

Fig. 12.6. Separation of a mixture of acetophenone (Ap) (1000 mg/l) and benzophenone (Bp) (1000 mg/l). Experimental conditions: injection volume 100 μl; rotation speed 435 rpm; applied pressure 13.1 MPa; temperature 50°C; flow rate 0.2 ml/min; monitoring wavelength 247 nm. Stationary phase: (a) methanol–water (30:70); (b) methanol–water (40:60). The PEEK column (46 ml) was wound eccentrically.

increased, the viscosity of the stationary phase decreased and resulted in higher retention and higher *Sf* values. In addition, higher rotation speeds also gave greater *Sf* values.

A mixture of acetophenone and benzophenone was injected into this system as a test. The resultant chromatograms are shown in Fig 12.6 [38]. The higher methanol content in the stationary phase should increase the solubility of both compounds, and thus their partition coefficients, and result in longer retention times. Although these two compounds were not well resolved, the chromatograms illustrated the feasibility of the original method.

12.3.2 Temperature effect on separation resolution

Temperature programming, a commonly used technique in GC, has also long been applied in optimizing SFC experiments [39–41]. Two major factors must be considered in understanding solute retention: fluid density and solute vapour pressure. Solute solubility in supercritical fluids depends on the fluid density. In general, the solvent strength of a supercritical fluid increases as a function of its density [42]. Under constant pressure, the density decreases as the temperature rises, thus resulting in lower solubility in the mobile phase and longer retention time. The second factor, i.e., vapour pressure of the solute, increases with the temperature which tends to decrease the retention time. The combination of these two opposite effects determines the solute retention in SFC. In SF/CCC, the solvent strength due to the density change

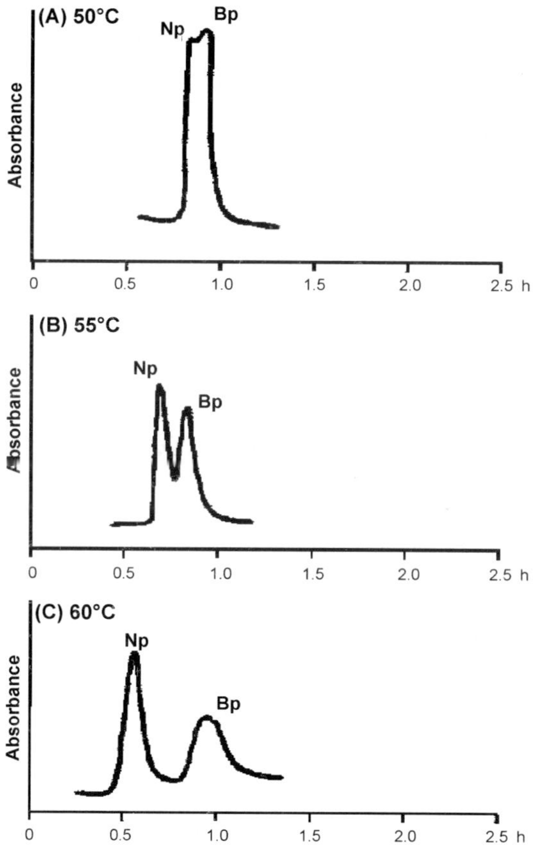

Fig. 12.7. Chromatograms of a mixture of naphthalene (Np) (1000 mg/l) and benzophenone (Bp) (1000 mg/l) under 50, 55, and 60°C. Experimental conditions: pressure 105 bar (10.5 MPa); stationary phase methanol-water (30:70); rotation speed 420 rpm; flow rate 0.5 ml/min; monitoring wavelength 247 nm. The PEEK column (46 ml) was wound eccentrically.

still plays a significant role in affecting the retention time. However, partitioning of the solute between the supercritical fluid and the stationary phase under different temperatures should be considered in SF/CCC, while only the solute vapour pressure is considered in conventional SFC.

Since our instrument was not capable of doing temperature programming operations, we studied the temperature effects by running different separations of a mixture of naphthalene and benzophenone at different temperatures [43]. As can be seen in Fig. 12.7, the peaks were well resolved at 60°C, while essentially only one peak appeared at 50°C. The naphthalene retention time decreased as the temperature rose, while that of benzophenone increased. At higher temperatures the $SFCO_2$ solvent strength declined due to the lower density. However, the elevated temperature apparently promoted the solute partitioning into the mobile phase, thus resulting in reduced retention time. Although the resolution was significantly improved, the band broadening due to greater diffusion coefficient at elevated temperatures should not be ignored.

12.3.3 Pressure effect on separation resolution

Under constant temperature, the $SFCO_2$ density increases as the applied pressure is elevated, and usually results in stronger solvation strength. The solute vapour pressure is not modified. Since the pressure has little effect on the solvent strength of the aqueous stationary phase, the pressure effect on the solute retention factor is simpler than the temperature effect. An example is shown in Fig. 12.8. The resolution between benzophenone and acetophenone was improved when the applied pressure went from 105 bar to 115 bar.

12.3.4 Stationary phase composition effect on separation resolution

One of the great advantages of CCC over packed-column LC lies in the fact that both the stationary and mobile phases can be changed easily. The operator can select various solvent systems according to the characteristics of the analytes. In this study the mobile phase, i.e., the $SFCO_2$, remained the same while the stationary phase composition was changed to improve the resolution. A series of chromatograms were obtained with different stationary phases, as shown in Fig. 12.9.

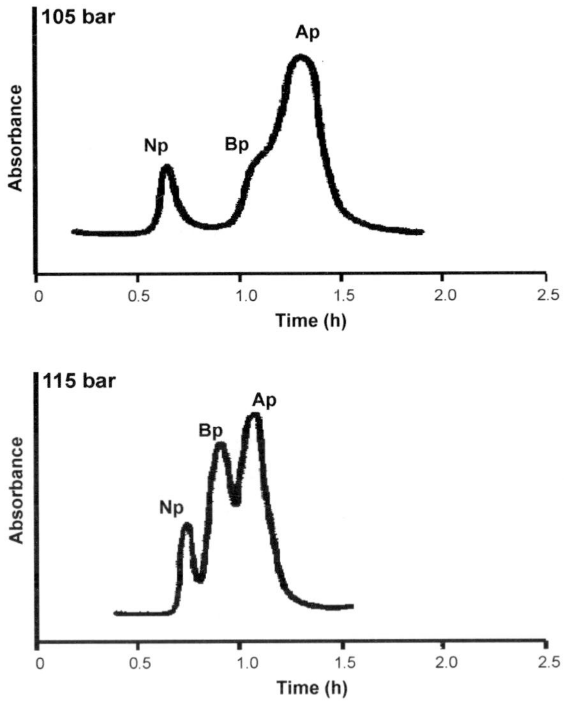

Fig. 12.8. Separation of a mixture of naphthalene (Np) (645 mg/l), benzophenone (Bp) (900 mg/l), and acetophenone (Ap) (450 mg/l) under two different pressures. Experimental conditions: temperature 60°C; rotation speed 420 rpm; flow rate 0.5 ml/min; monitoring wavelength 247 nm. The PEEK column (46 ml) was wound eccentrically.

The coupling of the temperature, pressure and solvent composition could make the separation efficiency difficult to predict; however, it may also add versatility to this technique.

12.4 SEPARATION OF STEROIDS

Three steroids, Progesterone (Pr), (+)-4-cholesten-3-one (Ch) and Δ^4-Androstene-3,17-dione (An), shown in Fig. 12.10, were separated using SF/CCC [44]. A wide-bore (5.33 mm i.d.) stainless steel tube was wound solar-coaxially instead of eccentrically on the column holders. A slower rotation speed with higher flow rate gave even higher *Sf* values

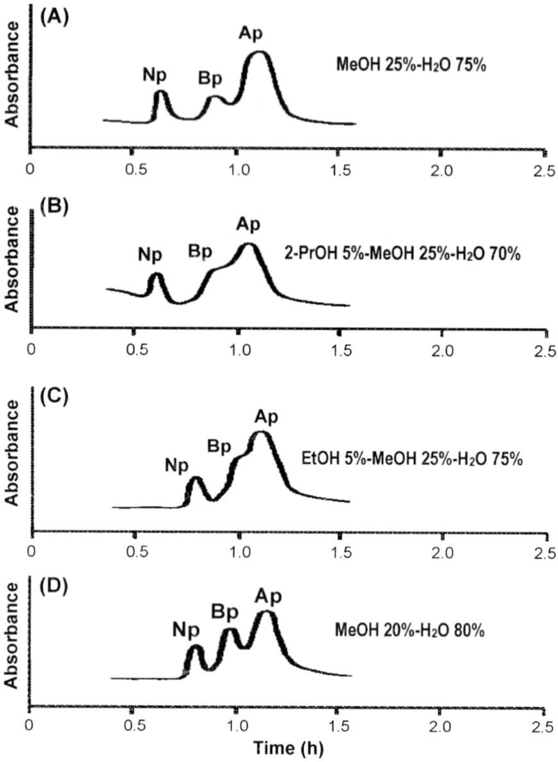

Fig. 12.9. Separation of a mixture of naphthalene (Np) (645 mg/l), benzophenone (Bp) (900 mg/l), and acetophenone (Ap) (450 mg/l) with different stationary phase modifications. Experimental conditions: pressure 105 bar; temperature 55°C; rotation speed 420 rpm; flow rate 0.5 ml/min; monitoring wavelength 247 nm. The PEEK column (46 ml) was wound eccentrically (adapted from Ref. [43]).

(*Sf* > 50%) than before. Temperature as well as pressure could be adjusted to optimize the separations. As can be seen in Fig. 12.11, the $SFCO_2$ solvent strength remained great enough at lower temperatures, i.e., 35 and 40°C, to elute these three compounds under 10 MPa. Under 40°C and 10 MPa, the three compounds could be separated to baseline resolution with *Sf* = 52%. As the temperature passed 50°C, the solvent strength became significantly lower for both An and Ch. The Pr and An peak heights were very diminished due to the large retention time and serious band broadening. These chromatograms again demonstrated the versatility of using $SFCO_2$ as the mobile phase.

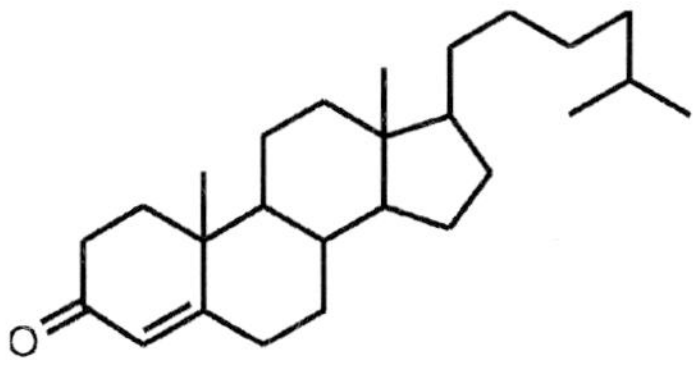

(+)-4-cholestene-3-one (Ch)

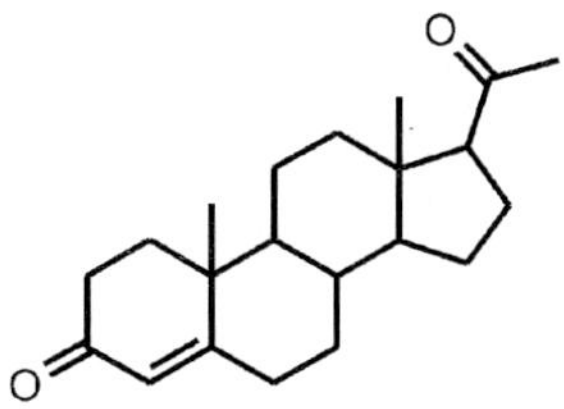

Progesterone (Pr)

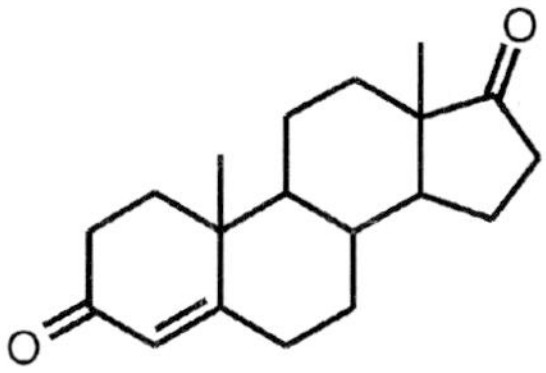

Androstene-3-17-dione (An)

Fig. 12.10. The chemical structures of the steroids separated with $SFCO_2$ mobile phases.

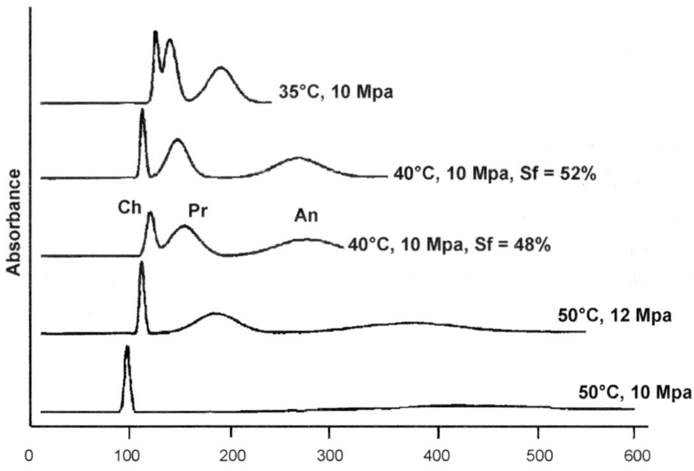

Fig. 12.11. Separation of steroids Ch (2000 mg/l), Pr (4000 mg/l) and An (4000 mg/l) at different temperatures, pressures, and phase retention ratios. Experimental conditions: stationary phase, methanol–water (30:70); mobile phase $SFCO_2$, 1.2 ml/min measured at the pump head; rotation speed 70 rpm. A stainless column (240 ml) was wound coaxially on the column holders. In all chromatograms, Ch is the first eluting compound, Pr the intermediate, and An the final compound.

12.5 SUPERCRITICAL FLUID EXTRACTION USING A CCC APPARATUS

In addition to solid samples, SFE also finds application for liquid samples. Instead of just using a CCC apparatus as an efficient cell for SFE, we try to optimize the extraction by developing a mathematical model to interpret the extraction recovery data obtained using SF/CCC. In fact, some fundamental researches [45–47] were involved in understanding the extraction process. Several articles [48–50] propose models allowing one to transpose SFC data to SFE. Although the SFE process has been considered to be much more complicated than that of SFC, both processes share the fact that analytes are extracted or separated by partitioning either between matrix and effluent (SFE) or stationary phase and mobile phase (SFC). One of the major differences, however, lies in the fact that analytes are scattered in an extraction vessel for SFE, while a sample plug is injected from the inlet of a chromatographic column in SFC. Therefore, the retention factor (the retention time) and the peak width (broadening) of an analyte obtained in SFC should be in a way related to the extraction recovery obtained in SFE. For example, McNally and Wheeler [48] predicted the SFE extraction efficiency using the analyte retention factor obtained by SFC. The SFC retention characteristics indeed reflected extraction recoveries to a certain extent. Lou et al. [50] examined the relationship between SFE and SFC using the extraction cell as a chromatographic column. Using SFC retention factors and peak broadening data, they successfully predicted the extraction profile for homogeneous samples with a model derived from convolution theory. The same model also provided helpful information for inhomogeneous samples. In our studies, we investigated the correlation of SFE and SFC in aqueous matrices. The CCC apparatus in our laboratory functioned ideally for both extraction and chromatography in aqueous matrix and aqueous stationary phase.

12.5.1 Theory

The separation mechanism of conventional CCC as well as SF/CCC involves only pure partitioning of analytes in the mobile and stationary phases. The extraction efficiency of SFE from aqueous samples depends also only on the analyte distribution between the two phases without any possible re-adsorption that often occurs for solid samples. In addition, analyte is homogeneously distributed in the aqueous sample, unlike the heterogeneous nature of solid samples. These characteristics simplify greatly the mathematical approaches to relate SFE with SFC data.

Correlation of extraction with chromatography

The model [51,52] is developed by dividing the extraction container, i.e., the separation column, into finite layers, as shown in Fig. 12.12. During the extraction, each layer with finite thickness was assumed to undergo an independent chromatographic elution. Analytes in the layers at the column outlet side travelled a shorter time in the column, therefore their band broadening was lower than that of the analytes located at the inlet side. Analyte molecules in the top layer of the inlet side should be considered just as the sample plug injected from a sample valve. Accordingly, the corresponding peak could be obtained in a true chromatographic experiment. Basically, a chromatographic peak, assuming a Gaussian shape, can be defined by its retention factor and the peak

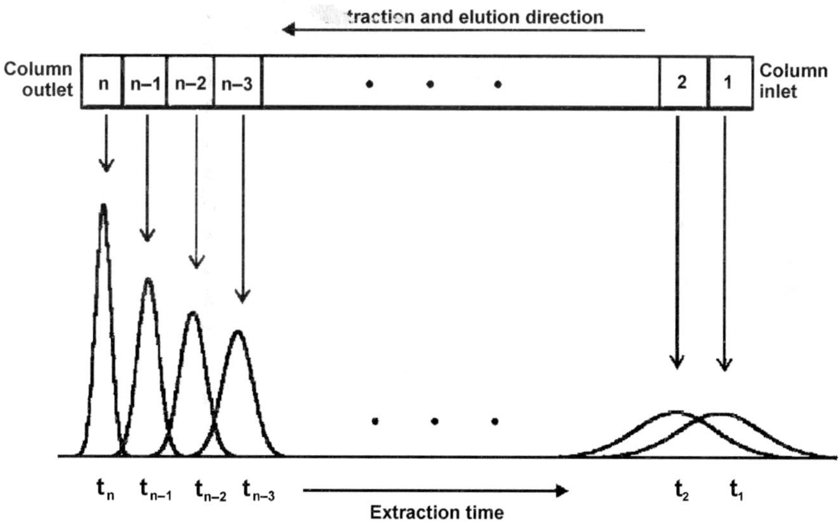

Fig. 12.12. Theoretical model correlating extraction with chromatography in liquid matrix. The extraction cell (i.e., the CCC column) was divided into *n* layers. During extraction, analyte molecules in all layers were assumed undergoing separate chromatographic process, and formed Gaussian peaks at the outlet. Layers close to the column outlet (left), travelling shorter distances, elute first giving relatively sharp peaks while those close to the column inlet gave relatively broad peaks.

standard deviation. The retention factors and standard deviations of the said peaks vary linearly according to the initial position of the analyte in the column. Therefore they can easily be computed from the values obtained from a true chromatographic run.

The Gaussian distributions for all the chromatographic elution of the assumed layers can be expressed as

$$y_i(t) = h_i \mathrm{e}^{-((t-t_{\mathrm{R}_i})^2/2\sigma_i^2)} \quad i = 1, \ldots, n \tag{12.1}$$

where $y_i(t)$ represents the analyte mass under the peaks as a function of time t; h_i, the peak heights, t_{R_i}, the peak retention times; and σ_i, the standard deviations. The band broadening per unit length due to the motion of the solute through the column, H, is defined as:

$$H = \sigma_{\mathrm{c}}^2 / L \tag{12.2}$$

where L is the column length, i.e. the elution distance for the true chromatographic solute plug; σ_{c} is the band broadening or standard deviation for the whole column. If H is constant throughout the column, σ_i for each layer should be related to σ_{c} by the following equation:

$$\sigma_i^2 = \frac{L_i}{L} \sigma_{\mathrm{c}}^2 \quad i = 1, \ldots, n \tag{12.3}$$

where L_i, the elution distance for analyte molecules in layer i, i.e. the distance from the column outlet to the head of layer i, is given by:

$$L_i = L\left(1 - \frac{i-1}{n}\right) \quad i = 1, \ldots, n \tag{12.4}$$

Of course, for the first layer, $L_{i=1}$ equals L, the column length. Combining Eqs. (12.3) and (12.4) yields

$$\sigma_i^2 = \sigma_{\mathrm{c}}^2 \left(1 - \frac{i-1}{n}\right) \quad i = 1, \ldots, n \tag{12.5}$$

In addition, the retention time for the individual peak is linearly related to the elution distance from the layer head to the vessel outlet:

$$t_{\mathrm{R}_i} = t_{\mathrm{R}_{\mathrm{c}}} \left(1 - \frac{i-1}{n}\right) \quad i = 1, \ldots, n \tag{12.6}$$

The standard deviation, σ_i, for the individual peak can be expressed as

$$\sigma_i = A_i / h_i \sqrt{2\pi} \quad i = 1, ..., n \tag{12.7}$$

where A_i is the individual peak areas for each layer i, and h_i, the peak heights. If analyte molecules are distributed uniformly in the column, the area A_i under individual peaks should be equal, and can be expressed as:

$$A_i = A_n = A_c \quad i = 1, ..., n \tag{12.8}$$

where A_c is the peak area of the true chromatographic peak obtained experimentally. Since A_c, h_c and t_{R_c} can be acquired through a real chromatographic experiment, the standard deviation, σ_c, can be calculated using Eq. (12.7), only by changing the index i's into c's in the equation. Parameters including standard deviation, retention time and peak height for each layer can then be computed using Eqs. (12.5)–(12.8). All these values are then substituted into Eq. (12.1) to calculate $y_i(t)$ for the individual layers. Extraction recovery, $r(t)$, can thus be calculated by taking the quotient of the summation of $y_i(t)$ divided by the total area as a function of time:

$$r(t) = \sum_i y_i(t) / \sum_i A_i \quad i = 1, ..., n \tag{12.9}$$

However, this simulation is made possible by assuming analyte molecules in each layer are undergoing chromatography at the same time. It may not come this way unless the mobile phase also uniformly distributes in the column in the meantime. In other words, Eq (12.9) starts to become effective only when the mobile phase is already evenly distributed in the column. The analyte molecules were homogeneously dispersed in the "stationary" phase, but the mobile phase was not. However, once the solvent front moves from the column inlet to the outlet, the mass (or concentration) in each layer no longer stays uniform as assumed before. Accordingly, mass redistribution calculations after the $SFCO_2$ moved to the column outlet must be executed before the above simulation can be carried out.

Mass redistribution by countercurrent distribution approach

The concept of countercurrent distribution (CCD) was originally initiated to fabricate the chromatographic process using the concept of stepwise stage extraction. Thermodynamic equilibrium is assumed and longitudinal diffusion of analyte is forbidden at each transfer. An equation was derived [51] using the CCD approach to calculate the analyte quantity at each stage:

$$Q'_i = Q_i \frac{k'}{1+k'} + Q_{i-1} \frac{1}{1+k'} \tag{12.10}$$

where Q'_i is the total analyte quantity remaining in both mobile and stationary phases of the ith layer after each CCD step, Q_i is the total analyte quantity in the ith layer before each CCD step, and Q_{i-1} is the total analyte quantity in the $(i-1)^{th}$ layer before each CCD step. In other words, the total analyte quantity in the ith layer after each CCD step (the solvent front advances one layer) was equal to the amount left in the stationary phase (first term on the right-hand side of the equation) plus the amount carried over from the $(i-1)^{th}$ by the mobile phase (second term on the right-hand side of the equation). Iterative calculations of Eq. (12.10) using a computer program gave the analyte mass distribution when the solvent front reached the column outlet and exited the column. Once the analyte mass redistribution was obtained, an adjustment was made by multiplying the term $y_i(t)$ in Eq. (12.9) by the analyte mass fraction to calculate the extraction recovery curve.

Mass redistribution by chromatographic approach

The CCD approach, although providing a reasonable calculation for extraction efficiency, did not really reflect the nature of the chromatographic process proposed in this mathematical model. In fact, the same concept described in the previous section can well be adopted to estimate the analyte mass distribution as the solvent front just reached the column outlet. The detailed derivation can be found elsewhere [52], and will not be described here. Again, the analyte mass redistribution thus obtained was introduced in Eq. (12.9) to calculate the extraction recovery curve.

12.5.2 Results

The extraction experiment was done first, followed by a chromatographic run using the same stationary phase, i.e., the sample solution without analyte. A typical extraction curve and chromatogram are shown in Fig. 12.13 [51]. Naphthalene is insoluble in water so it was

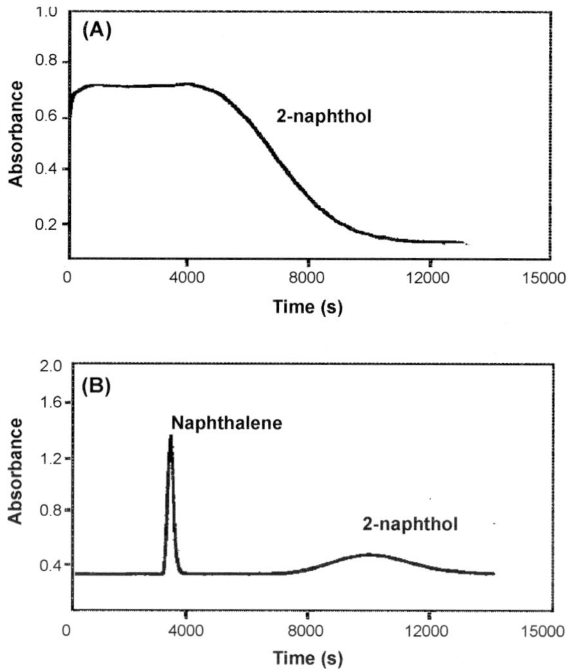

Fig. 12.13. (A) Extraction curve for 2-naphthol. Sample volume: 40 ml (10 mg/l). (B) Chromatogram for 2-naphthol. Sample volume: 100 μl (500 mg/l). Naphthalene was the solvent front tracer. Experimental conditions for both experiments: pressure 100 bar; temperature 40°C; rotation speed 80 rpm; flow rate 0.7 ml/min measured at the pump head. A nickel column (95 ml) was wound coaxially on the column holders.

used as the $SFCO_2$ solvent front marker. Since we found that the simulation deviation was dependent on the analyte retention time (the retention factor), we therefore tried to obtain experiments with different retention times. By changing the system temperature and/or pressure, the solvation strength of $SFCO_2$ was easily tuned to achieve different retention factors even using the same sample. The extraction curves obtained experimentally were compared with those calculated from the mathematical model using the two approaches. The deviations obtained using the chromatographic approach, shown as a function of the experimental recovery in Fig. 12.14 [52], were indeed lower than those using the CCD approach. However, the deviation became smaller as the retention factor, k', increased for both approaches. When the k' values were greater than 1, the deviations were minimized to less than

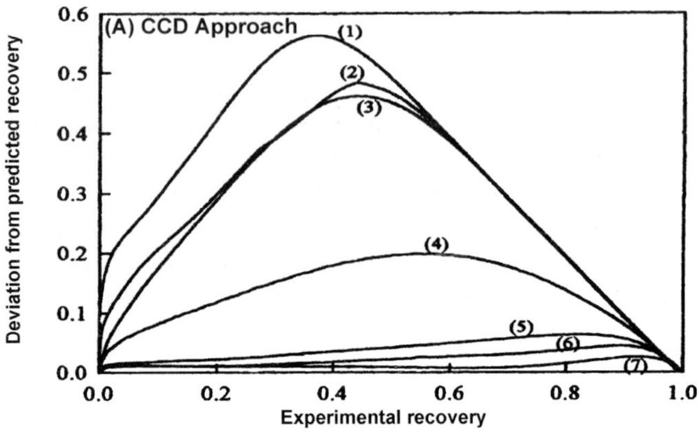

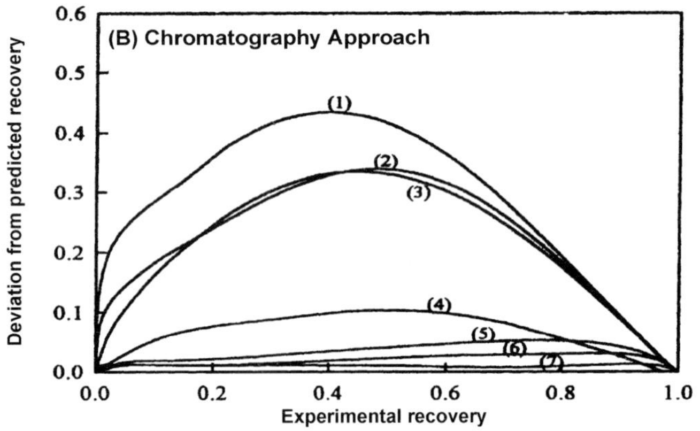

Fig 12.14. Deviation between the simulated and the experimental recovery curves as a function of the experimental recovery using CCD approach (A) and chromatography approach (B). Experimental conditions: (1) (for $k' = 0.097$) acetophenone, 130 bar, 50°C; (2) (for $k' = 0.140$) benzaldehyde, 90 bar, 43°C; (3) (for $k' = 0.222$) benzaldehyde, 100 bar, 50°C; (4) (for $k' = 0.585$) benzaldehyde, 100 bar, 55°C; (5) (for $k' = 1.451$) 2-naphthol, 130 bar, 50°C; (6) for ($k' = 1.470$) *m*-cresol, 130 bar, 50°C; (7) (for $k' = 1.847$) 2-naphthol, 100 bar, 40°C.

0.05 for the whole recovery curve. This deviation should result from our basic assumption: the divided independent layers. The diffusion of the analyte molecules in an individual layer should behave like a sample plug in a chromatographic process, and form essentially a Gaussian shape. However, the layers were not truly independent, therefore the diffusion of the analyte molecules would not follow a Gaussian trend. In the case of small k' values, the analyte molecules had a high affinity for the mobile phase and rapidly converged to the column outlet with the solvent front of $SFCO_2$. Accordingly, the diffusion behaviour departed from the Gaussian-like assumption. The solutes with larger k' interacted significantly with the stationary phase as described by the models. As for the two approaches to calculate the mass redistribution, the chromatographic approach prevailed because of its greater similarity to the true extraction process. The models were also used to correlate liquid–liquid extraction and chromatography system (data not shown). The deviations were even smaller than those of the SF/CCC system [52].

12.6 CONCLUSIONS AND PERSPECTIVES

The solvation strength of the mobile phase $SFCO_2$ can be modified by changing the methanol content, temperatures and pressures. This makes the SF/CCC technique very versatile. In addition, the mobile phase becomes completely volatile and isolates from the analytes immediately after separation. However, the SF/CCC technique needs modification for high-pressure operations, which makes the system build-up and maintenance more troublesome. In addition, band broadening due to the higher diffusivity of supercritical fluids may jeopardize the resolving power. At present, the pressure must be kept below 130 bar (13 MPa) to maintain acceptable *Sf* values under the current stationary phases (methanol–water solutions). Pressures much higher than 130 bar (13 MPa) are required to elute higher molecular weight or more polar analytes. Accordingly, the SF/CCC can only be applied to less polar compounds at the moment.

In addition to chromatography, the CCC apparatus, as described before, can act as an efficient extraction vessel. Except for low-polarity compounds, highly polar compounds, such as proteins, can also be solubilized in $SFCO_2$ with suitable surfactants [31–33] under medium pressure. Both extraction and concentration can be achieved simultaneously for analytes in dilute aqueous solutions, such as highly valued enzymes in a fermentation broth. We are currently more involved in extraction studies than chromatography with different CCC apparatuses.

ACKNOWLEDGEMENT

Former students of T. Yu, including Shu-En Li, Yi-Hsing Chen, Chiuan Shiuan Yeh, Sheng-Kang Luo, Shin Jyh Chen, and Chia Shan Liang, who were involved in SF/CCC studies are greatly appreciated. The authors also thank the National Science Council of Taiwan, the Republic of China for the financial support.

REFERENCES

1 E. Klesper, A.H. Corwin and D.A. Turner, *J. Org. Chem.*, 27 (1962) 700.
2 T.A. Berger, *Chromatographia*, 41 (1995) 133.
3 E. Pocurull, R.M. Marce, F. Borrull, J.L. Bernal, L. Toribio and M.L. Serna, *J. Chromatogr.*, 755 (1996) 67.
4 J. Thompson, B. Strode III, L.T. Taylor and T.A. van Beek, *J. Chromatogr.*, 738 (1996) 115.
5 C. Blum, K.-H. Kubeczka and K. Becker, *J. Chromatogr.*, 773 (1997) 377.
6 N. Wu, R. Yee and M.L. Lee, *Chromatographia*, 53 (2001) 197.
7 K. Hatada and T. Kitayama, *Polym. Int.*, 49 (2000) 11.
8 M.A. Saleh, A. Kamel, A. El-Demerdash and J. Jones, *Chemosphere*, 36 (1998) 1543.
9 B.Wenclawiak and A. Otterbach, *J. Biochem. Biophy. Meth.*, 43 (2000) 197.
10 D.X. Tong, K.D. Bartle, A.A. Clifford and R.E. Robinson, *Analyst*, 120 (1995) 2461.
11 K. Yaku and F. Morishita, *J. Biochem. Biophy. Methods*, 43 (2000) 59.
12 A. Salvador, B. Herbreteau and M. Dreux, *Chromatographia*, 53 (2001) 207.
13 T.L. Chester, J.D. Pinkston and D.E. Raynie, *Anal. Chem.*, 70 (1998) 301R.
14 E.G. van der Velde, M.R. Ramlal, A.C. van Beuzekom and R. Hoogerbrugge, *J. Chromatogr.*, 683 (1994) 125.
15 J.M. Snyder, J.W. King, L.D. Rowe and J.A. Woerner, *J. AOAC Int.*, 76 (1993) 888.
16 M.T. Tena, M.D. Luque de Castro and M. Valcarcel, *Chromatographia*, 40 (1995) 197.
17 D. Sterzenbach, B.W. Wenclawiak and V. Weigelt, *Anal. Chem.*, 69 (1997) 831.
18 J.J. Langenfeld, S.B. Hawthorne, D.J. Miller and J. Pawliszyn, *Anal. Chem.*, 67 (1995) 1727.
19 B. Vanbavel, M. Jaremo, L. Karlsson and G. Lindstrom, *Anal. Chem.*, 68 (1996) 1279.
20 A.T. Champagne and P.R. Bienkowski, *Sep. Sci. Technol.*, 30 (1995) 1289.
21 S.B. Hawthorne and D.J. Miller, *Anal. Chem.*, 66 (1994) 4005.
22 I.J. Barnabas, J.R. Dean, W.R. Tomlinson and S.P. Owen, *Anal. Chem.* 67 (1995) 2064.
23 S.R. Sargenti and F.M. Lancas, *J. Chromatogr.*, 667 (1994) 213.
24 M.-K. Chun, H.-W. Shin, H. Lee and J.-R. Liu, *Biotechnol. Tech.*, 8 (1994) 547.

25 J. Hedrick and L.T. Taylor, *Anal. Chem.*, 61 (1989) 1986.
26 D. Thiebaut, R.W. Vannoort, U.A. Th. Brinkman, J.P. Chervet, G.J. De Jong and R.W. Frei, *J. Chromatogr.*, 477 (1989) 151.
27 I.J. Barnabas, J.R. Dean, S.M. Hitchen and S.P. Owen, *J. Chromatogr. A*, 665 (1994) 307.
28 E.J. Roggeman, A.M. Scurto and J.F. Brennecke, *Ind. Eng. Chem. Res.*, 40 (2001) 980..
29 C.Y. Tai, G.S. You and S.L. Chen, *J. Supercrit. Fluids*, 18 (2000) 201.
30 C.M. Wai and S.F. Wang, *J. Chromatogr. A*, 785 (1997) 369.
31 K.P. Johnston, K.L. Harrison, M.L. Clarke, S.M. Howdle, M.P. Heitz, F.V. Bright, C. Carlier and T.W. Randolph, *Science* 271 (1996) 624.
32 E.G. Ghenciu and E.J. Beckmann, *Ind. Eng. Chem. Res.*, 36 (1997) 5366.
33 E.G. Ghenciu, A.J. Russell, E.J. Beckmann, L. Steele and N.T. Becker, *Biotechnol. Bioeng.*, 58 (1998) 572.
34 Y. Ito and R. Bhatnagar, *J. Liq. Chromatogr.*, 7 (1984) 257.
35 Y. Liu, V. Lopezavila, M. Alcaraz and T.L. Jones, *Anal. Chem.*, 66 (1994) 4483.
36 M.K. King, A. Mubarak, J.D. Kim and T.R. Bott, *J. Supercrit. Fluids*, 5 (1992) 296.
37 Y. Ito and W.D. Conway, *J. Chromatogr.*, 301 (1984) 405.
38 T. Yu, S.-E. Li, Y.-H. Chen and H.P. Wang, *J. Chromatogr. A*, 724 (1996) 91.
39 K.E. Markides, E.D. Lee, R. Bolick and M.L. Lee, *Anal. Chem.*, 58 (1986) 740.
40 P.A. David and M. Novotny, *J. Chromatogr.*, 461 (1989) 111.
41 S.V. Olesik, L.A. Pekay and E.A. Paliwoda, *Anal. Chem.*, 61 (1988) 58.
42 J.C. Giddings, M.N. Myers, L. McLaren and R.A. Keller, *Science* 162 (1968) 67.
43 T. Yu and Y.-H. Chen, *J. Chromatogr. A*, 790 (1997) 31.
44 C.S. Yeh, T. Yu and A. Berthod, *J. Liq. Chrom. Relat. Technol.*, 22 (1999) 345.
45 J. Pawliszyn, *J. Chromatogr. Sci.*, 31 (1993) 31.
46 J.J. Langenfeld, S.B. Hawthorne, D.J. Miller and J. Pawliszyn, *Anal. Chem.*, 67 (1995) 1727.
47 A.A. Clifford, M.D. Burford, S.B. Hawthorne, J.J. Langenfeld and D.J. Miller, *J. Chem. Soc.*, 91 (1995) 1333.
48 M.E.P. McNally and J.R. Wheeler, *J. Chromatogr.*, 447 (1988) 53.
49 K.G. Furton and Q. Lin, *J. Chromatogr. Sci.*, 31 (1993) 201.
50 X. Lou, H.-G. Janssen and C.A. Cramers, *J. High Resolut. Chromatogr.*, 18 (1995) 483.
51 T. Yu, S.-K. Luo and S.J. Chen, *Anal. Chem.*, 70 (1998) 2115.
52 T. Yu, C.S. Liang and S.-K. Luo, *Anal. Chem.*, 71 (1999) 507.

Appendix

Commercially available countercurrent chromatographs

Alain Berthod

The interest in the CCC technique is abundantly demonstrated in this book and its predecessors. Clearly, the major problem hindering the development of CCC is the lack of reliable machines. The primary cause of this problem is that there is still fundamental knowledge to be gained on the delicate liquid–liquid stationary–mobile phase equilibrium occurring in any CCC machine. British teams are working on the equilibrium in hydrodynamic machines (Chapter 6); the French school is working on the equilibrium in hydrostatic machines (Chapter 5). The knowledge is rapidly progressing. The second cause is that, to date, no major instrumentation company is involved in CCC apparatus production. It is therefore important to support the small industrial companies who build and sell CCC machines around the world. A tentatively exhaustive presentation of the commercially available CCC machines in 2002 is presented in this appendix. Although it is always difficult to separate scientific information from commercial advertising, the reader should know that the author has gathered the following information without interference by the industrial companies. The order in which the machines are listed is alphabetically by company name; it is not a reflection of their quality.

A.1 TWENTIETH CENTURY CCC TECHNOLOGY

A.1.1 Discontinued apparatuses

As Walter Conway quoted in his book, Peter Carmeci (1927–1988) was, "a pioneer in the commercial development of CCC". Using Yoichiro Ito's coil planet centrifuge scheme IV, he started the company P.C. Inc. in Potomac, Maryland (USA) which built and sold hundreds of Multi-layer Coil Separator–Extractors for more than 15 years. This CCC machine had only one spool equilibrated by a counterweight. The maximum rotation speed was around 1000 rpm. The main volume was 240 ml, but different coils and bobbins with two different coiled volumes were later marketed. Many of these machines are still in use today.

Figure 6.2 shows a 1955 Craig machine, the true countercurrent distribution machine. Figure 1.6 shows a 1989 SFCC2000, a three-spool coil-planet-centrifuge machine. These machines are no longer commercially available, but are still in use in several laboratories.

A.1.2 Hydrostatic machines

Eyela Tokyo Rikakikai (3-3-4 Honcho Nihonbashi, Chuo-ku, Tokyo 103-0023, Japan, tel: 81 3 5201 6421, fax: 81 3 3241 0177) still sells droplet CCC apparatuses. In particular, their model DCC-3000 is a 3-l DCCC machine that can be ordered with glass or Teflon® tubes. This company is the only one still selling a Rotation Locular CCC apparatus, the RLCCC-1000 with a volume of 1 litre (Fig A.1). The US dealer is Dychrom (265 Meridian Avenue, #8, San Jose, California 95126, USA, tel: 1 408 279 1611, fax: 1 408 279 6236). Web site: www.eyela.co.jp

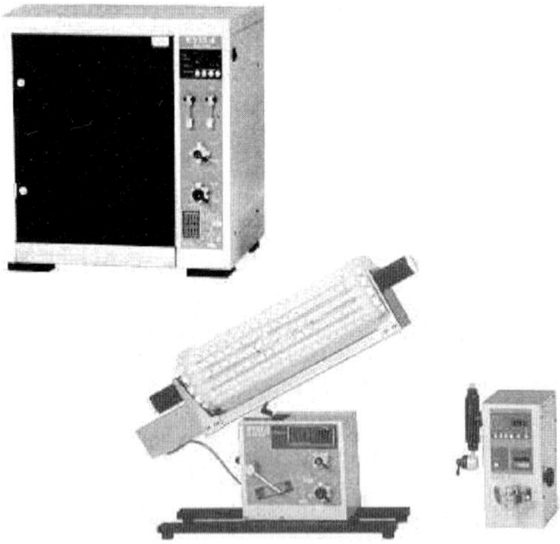

Fig. A.1. Top: The Eyela droplet CCC machine with connected glass or Teflon tubes. Bottom: The Eyela rotating locular CCC machine.

A.2 TWENTY-FIRST CENTURY CCC MACHINES

A.2.1 Hydrodynamic machines

Beijing Institute of New Technology Application

The Beijing Institute of New Technology Application, whose Chief Expert is TianYou Zhang, author of Chapter 8 of this book, is located in Beijing (N. 16, South Street, Xizhimen, Beijing 100035, P.R. China, tel: 661 78485, fax: 661 63035). It sells, mainly to the Asian market, two coil planet centrifuge CCC machines: the Model GS10A2, with a maximum rotation speed of 1000 rpm, 240 ml volume with one spool and a counterweight; and the Model GS20, a 30 ml mini two-spool CCC machine (Fig. A.2). See Chapter 8, Section 8.2, for information about these machines. Web site: www.binta.com.cn

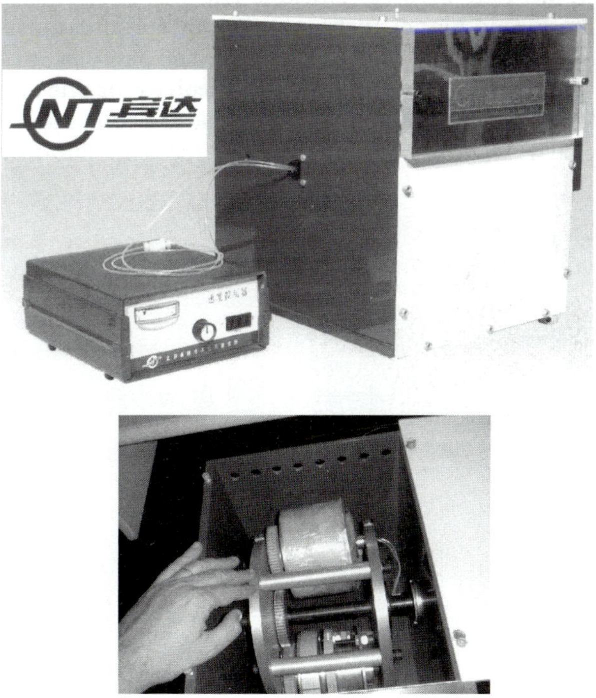

Fig. A.2. Top: The coil planet centrifuge Model GS10A2 of the Beijing Institute of New Technology Application. Bottom: View of the rotor of the 30 ml Model GS20 CCC machine.

Brunel Institute of Bioengineering
The Brunel Institute of Bioengineering (BIB), whose director is Ian Sutherland, author of Chapter 6 of this book, is located in the London suburbs (Brunel University, Kingston Lane, Uxbridge, Middlesex UB8 3PH, United Kingdom, tel: 44 1895 271 206, fax: 44 1895 274 608). It will commercialize the improved coil planet centrifuge machines after the results of the British BBSRC/DTI LINK program are fully exploited. Industrial requirements will be taken in account. BIB already markets the Brunel CCC machine whose volume is custom-tailored, but it is a dedicated preparative machine with two self-equilibrating spools of significant volume (Fig. A.3). Web site: www.brunel.ac.uk

Conway Centrichrom Inc.
The Conway Centrichrom Inc. (52 MacArthur Drive, Buffalo, New York, 14221, tel: 1 716 635 0883, fax: 1 716 635 0886) is directed by Walter Conway, a pioneer in CCC. It markets the DP-1000 CCC machine, a self-balancing double planet instrument with an adjustable volume. It can be used with a variety of coils of either the same or different type. Coil choices include two Unicoils® or two Bicoils®, or one of each, or one of either plus a Tripple® coil. Unicoils consist of a continuous multilayer winding of either 1.6 mm i.d. PTFE tubing (approx. 350-ml volume) or

Fig. A.3. The Brunel CCC machine is a coil planet centrifuge machine with two large self-equilibrating spools rotating vertically.

Fig. A.4. The DP-100 coil planet centrifuge machine by Conway Centrichrom Inc. Two coils rotate horizontally in a well insulated box.

2.6 mm i.d. tubing (approx. 0.5-l volume). Unicoils® can be operated independently, or can be connected in series to provide a system volume up to 1 litre. In Bicoils®, the coil winding is divided to provide two process streams per spool. The individual windings can be either equal or different in volume. The user can operate these windings in parallel or can connect them in series to obtain a single coil of larger volume. Two Bicoils, in double planet configuration, will provide four simultaneously operable process streams. Similarly, the Tripple® coil contains three windings of about 12, 80 and 320 ml. Low operating noise is achieved through the use of advanced gearing and an enclosure, which provides both thermal and acoustical insulation (Fig. A.4). Web site: www.centrichrom.com

Pharma Tech Research

The Pharma Tech Research Corp. (6807 York Road, Baltimore, Maryland 21212-4715, USA, tel: 1 410 377 7018, fax: 1 410 377 7224) is directed by Edward Chou who always worked in close collaboration with Yoichiro Ito. PharmaTech CCC instruments have been marketed for more than 15 years. The TCC-1000 machine is a toroidal coil CCC, with extremely low volume (3–5 ml) for fast experiments at 1000 rpm. The CCC-3000 and the CCC-1000 are coil planet centrifuge self-equilibrating

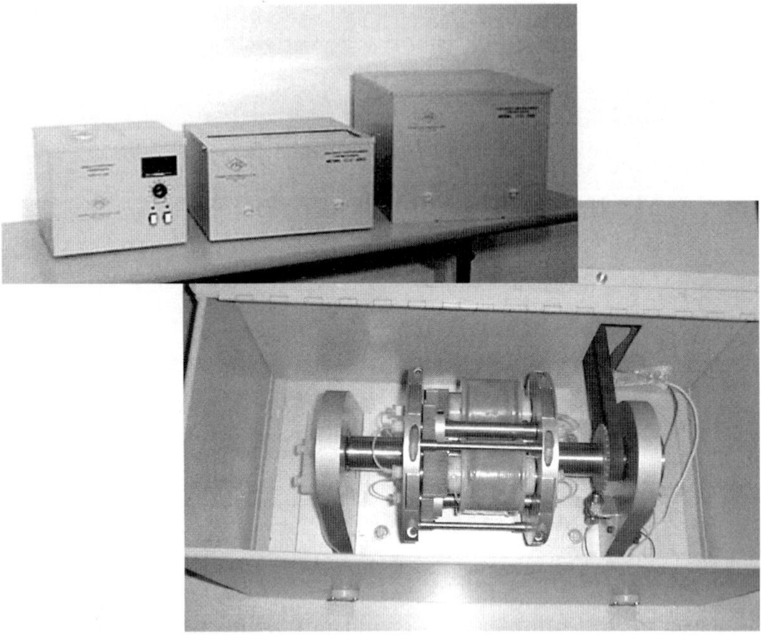

Fig. A.5. Top: The three CCC machines from Pharma Tech Corp. From left to right, the micro toroidal TCC-1000 machine (3–5 ml), the CCC 3000 (30 ml) and the CCC 1000 (120–850 ml) coil planet centrifuge machines. Bottom: View of the three coil rotor of the CCC 3000 machine.

machines with 50 ml or 120 to 850 ml volumes (Fig. A.5). The volumes are changed by installing different spools. Web site: www.pharma-tech.com

Quattro AECS Ltd.

The AECS company (P.O. Box 80, Bridgend, Mid-Glamorgan, South Wales CF31 4XZ, United Kingdom, or c/o MicroSolv, 101 Brighton Avenue, Long Branch, New Jersey 07740, USA, tel: 1 732 229 3400, fax: 1 732 229 2403) is directed by Les Brown, coauthor of Chapter 2 in this book. AECS markets a line of tailor-made coil planet centrifuge machines. The size and volume of the two-spool self equilibrating machines range from analytical (in the 150 ml range) to preparative (up to 3 litres). The bobbins can be easily interchanged to obtain the desired volume (Fig. A.6). Web site: www.ccc4labprep.com

Fig. A.6. The Quattro Labprep apparatus loaded with two different bobbins coiled with different diameter tubing. They can be used in parallel (two separations at the same time), or serially connected (larger volume).

Tauto Biotech Ltd.

The Tauto Biotech Ltd. is an industrial company marketing the products developed at the Research Institute of the Tsinghua University (High-Tech National Park, Tsinghua University, Shenzhen 518057, P.R. China, tel: 86 755 6551407, fax: 86 755 6551496-22). These products are mainly pure natural compounds and standards that were obtained using CCC as the purification tool. In collaboration with Amersham Biosciences (ex Pharmacia), they started to sell a 100-ml and a 1-l coil planet centrifuge machine (Fig. A.7). Web sites: www.tautobiotech.com and www.amershambiosciences.com

Fig. A.7. Top: The Tauto analytical coil planet centrifuge machine with a cooling (or heating) unit. Bottom: The 1-l preparative Tauto machine (left) and the Amersham Bioscience LC system (right, with a fraction collector, pump, UV detector and Ätka Explorer® GLP software).

A.2.2 Hydrostatic machines

In Chapter 4 of this book, Alain Foucault extensively describes the two commercially available hydrostatic machines called centrifugal partition chromatographs (CPC).

Kromaton Technologies
Kromaton Technologies (9 rue Alexander Fleming, 49066 Angers, France, tel: 33 241 774 148, fax: 33 241 739 623) markets the Fast CPC hydrostatic machine which is pictured in Figs. 1.8 and 4.19. It was first presented in London at the 1st International Symposium on CCC (Fig.

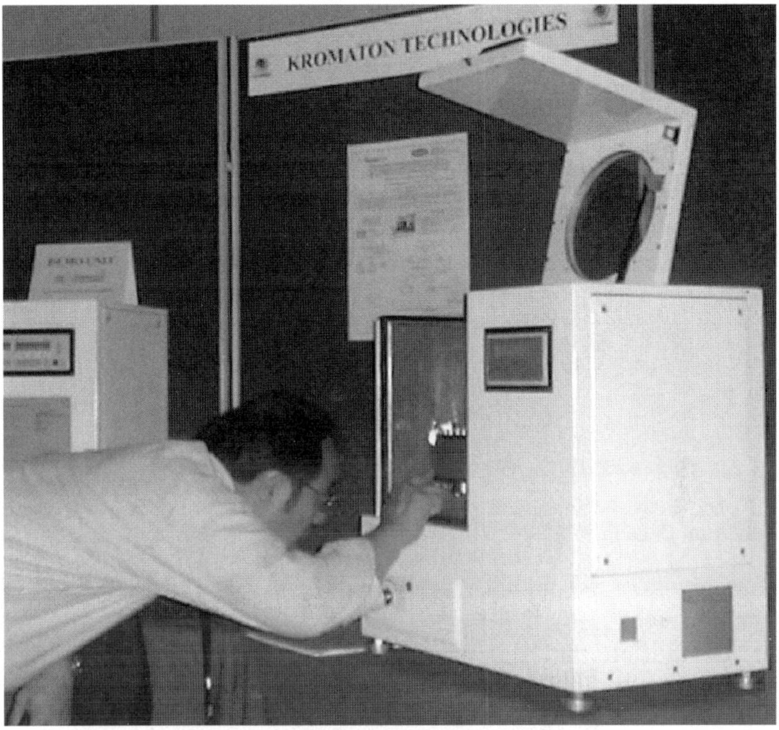

Fig. A.8. Dr. Chou, president of the Pharma Tech Research company, is caught closely examining the Kromaton FCPC shown at the CCC 2000 International Conference held in London (photo by A. Berthod).

A.8). This machine has two rotary seals. Its channel design is shown in Fig. 5.15 (French patent 02791578A1). The US Kromaton representative is Richard Scientific (366-A Bel Marin Keys Blvd, Novato, California 94949, USA, tel: 1 415 883 2888, fax: 1 415 382 1922). Web site: ccc.kromaton.com

Sanki Engineering Ltd.

The story of the Sanki company who, for 15 years, was the only one selling CPC machines, is described in detail in Chapter 4 of this book. Their products, shown in Figs. 4.7, 4.8, 4.10, 4.14, 4.16 and 5.1 and listed in Table 4.1, are marketed by the EverSeiko Corporation (4F, Tagagi Building, 4-36-17, Ikebukuro Toshima-ku, Tokyo 171, Japan, tel: 813 3988 3771, fax: 813 3988 3773). Web site: www.dhsi.com.cn

Fig. A.9. Dr. Udaya Wanasundara of the POS Pilot Plant Corporation (Saskatoon, Saskatchewan, Canada S7N 2R4, web site: www.pos.ca) adjusts the rotor of a 1.3-l LLI-5 Sanki CPC unit (courtesy of POS).

More information and the prices of the machines can be obtained directly contacting the companies through their web sites.

Glossary

SYMBOLS USED

C_D	concentration of displacer in the mobile phase in pH-zone refining
C_m	solute concentration in the mobile phase (M)
C_S	solute concentration in the stationary phase (M)
C_R	concentration of retainer in the mobile phase in pH-zone refining
D_{ax}	axial solute diffusion coefficient (cm^2/s)
D_m	diffusion coefficient in the mobile phase (cm^2/s)
D_S	diffusion coefficient in the stationary phase (cm^2/s)
F	mobile phase flow rate (ml/min)
g	constant acceleration in a CPC machine (m/s^2)
k or k'	solute retention factor
k'_f	kinetic reaction constant (s^{-1})
k_m	mobile phase mass transfer coefficient (s^{-1})
k_o	overall mass transfer coefficient (s^{-1})
k_r	kinetic reaction constant (s^{-1})
k_S	stationary phase mass transfer coefficient (s^{-1})
K	equilibrium constant
K^+	equilibrium constant of the + enantiomer with the chiral selector
K^-	equilibrium constant of the – enantiomer with the chiral selector
K_1	distribution constant of Solute 1
K_2	distribution constant of Solute 2
K_i	distribution constant of Solute i
K_A	acid dissociation constant
K_D	solute distribution constant, also called partition coefficient
K'_D	$= 1/K_D$ after inverting the role of the liquid phases in dual-mode

K_{DM}	theoretical dual-mode distribution constant
$K_{oct/w}$	octanol/water partition coefficient
K_R	distribution constant of the retainer (acid or base) in pH-zone
L	column length (m)
N	plate number or chromatographic efficiency (plate)
P_c	critical pressure for a pure compound (Pa, bar, kg/cm^2 or p.s.i.)
P_d	pressure drop in a CCC column (Pa, bar, kg/cm^2 or p.s.i.)
P_e	Peclet number
P_{exp}	apparent partition coefficient (not a constant)
pK_A	$= -\log K_A$, the acid dissociation constant
Q	analyte concentration (g/l)
Q'	analyte quantity in SFC fluid (g/l)
R	rotor radius of a CPC machine (m)
Rs	resolution factor
Sf	stationary phase volume retention factor ($= V_S/V_M$)
T_c	critical temperature (K or °C)
t_o	dead time (min)
t_R	solute retention time (min)
u	linear mobile phase velocity (cm/s)
$\bar{u}$	average mobile phase velocity (cm/s)
V_C	column or CCC machine volume (ml)
V_d	dead volume (ml)
V_{DM}	dual-mode volume after which the phase role is inverted (ml)
V_M	mobile phase volume inside the CCC machine (ml)
V_M^{min}	minimum volume of V_M in a CPC machine (ml)
V_r, V_R	solute retention volume (ml)
V_S	stationary phase volume inside the CCC machine (ml)
W, W_b	peak width at baseline (ml or s or min)
$W_{0.6h}$	peak width at 60% of the peak height (ml or s or min)

Greek Letters

α	selectivity factor
Γ_{cor}	Coriolis acceleration (m/s^2)
Δd	density difference between the two liquid phases (g/cm^3)
$\Delta \rho$	density difference between the two liquid phases (g/cm^3)
η	viscosity (cP or s·Pa/100)
μ_n	nth statistical moment
σ^2	variance of a Gaussian peak (ml^2 or s^2)
Φ	liquid phase ratio

ω	rotor spin rate (rd/s or rpm)
Ω	rotation vector (rd/s)

ABBREVIATIONS

ACN	acetonitrile
AES	atomic emission spectroscopy
APCI	atmospheric pressure chemical ionization
ATPS	aqueous two-phase system
AZ	Arizona (heptane–methanol–EtOAc–water) liquid system
BAW	butanol–acetic acid–water biphasic liquid system
BSA	bovine serum albumin
n-BuOH	*n*-butanol
t-BuOH	*tert*-butanol
C18	octadecyl bonded silica HPLC column (reversed phase)
CCC	countercurrent chromatography
CCD	countercurrent distribution, Craig machine
CEC	capillary electrochromatography
CETP	channel equivalent to a theoretical plate
CNRS	Centre National de la Recherche Scientifique (French National Center for Scientific Research)
CPC	centrifugal partition chromatography (or coil planet centrifuge)
CS	chiral selector
DAB	deacetyl-baccatine
DEHPA	diethylhexyl phosphoric acid
DMF	dimethylformamide
DMSO	dimethylsulfoxide
DNB	3,5-dinitrobenzoyl
EGCG	epigallocatechin gallate
ESI	electrospray ionization
EtOAc	ethyl acetate
EtOH	ethanol
FCPC	fast centrifugal partition chromatography
FID	flame ionization detector
Glyme	ethylene glycol dimethyl ether
HBAW	heptane–butanol–acetonitrile–water liquid system
HETP	height equivalent to a theoretical plate (m)
HM	hexane–methanol liquid system

HMW	hexane–methanol–water liquid system
HSCCC	high speed countercurrent chromatography
HTU	height of a transfer unit
ICP	inductively coupled plasma
LC/MS	liquid chromatography with a mass spectrometer detector
NPEO	nonylphenol ethoxylate
MeOH	methanol
MIBK	methyl isobutyl ketone
MP	mobile phase
MTBE	methyl *tert*-butyl ether
NTU	number of transfer unit
PCTFE	polychlorotrifluoroethylene
PDA	photodiode array detector
PEEK	polyether ether ketone
PEG	polyethylene glycol
PPS	polyphenylene sulfide
QN	quinine
QD	quinidine
REE	rare earth elements (lanthanides)
RTD	residence time distribution
SFE	supercritical fluid extraction
SFC	supercritical fluid chromatography
SP	stationary phase
TFA	trifluoroacetic acid
THF	tetrahydrofuran
TOPO	trioctylphosphine oxide

See also Table 9.4, page 293, for ligand abbreviations.

Index